# MOLECULES
# AND
# MEDICINE

BICENTENNIAL
**1807**
**⊛WILEY**
**2007**
BICENTENNIAL

## THE WILEY BICENTENNIAL–KNOWLEDGE FOR GENERATIONS

Each generation has its unique needs and aspirations. When Charles Wiley first opened his small printing shop in lower Manhattan in 1807, it was a generation of boundless potential searching for an identity. And we were there, helping to define a new American literary tradition. Over half a century later, in the midst of the Second Industrial Revolution, it was a generation focused on building the future. Once again, we were there, supplying the critical scientific, technical, and engineering knowledge that helped frame the world. Throughout the 20th Century, and into the new millennium, nations began to reach out beyond their own borders and a new international community was born. Wiley was there, expanding its operations around the world to enable a global exchange of ideas, opinions, and know-how.

For 200 years, Wiley has been an integral part of each generation's journey, enabling the flow of information and understanding necessary to meet their needs and fulfill their aspirations. Today, bold new technologies are changing the way we live and learn. Wiley will be there, providing you the must-have knowledge you need to imagine new worlds, new possibilities, and new opportunities.

Generations come and go, but you can always count on Wiley to provide you the knowledge you need, when and where you need it!

**WILLIAM J. PESCE**
PRESIDENT AND CHIEF EXECUTIVE OFFICER

**PETER BOOTH WILEY**
CHAIRMAN OF THE BOARD

# MOLECULES
# AND
# MEDICINE

## E.J. Corey, B. Czakó and L. Kürti

Department of Chemistry
and
Chemical Biology
Harvard University

**WILEY**

Published by John Wiley & Sons, Inc., Hoboken, New Jersey.
Published simultaneously in Canada.

For general information on our other products and services or for technical support, please contact our Customer Care Department within the United States at (800) 762-2974, outside the United States at (317) 572-3993 or fax (317) 572-4002.

Wiley also publishes its books in a variety of electronic formats. Some content that appears in print may not be available in electronic format. For information about Wiley products, visit our web site at www.wiley.com.

Wiley Bicentennial Logo: Richard J. Pacifico

*Library of Congress Cataloging-in-Publication Data is available.*

ISBN 978-0-470-22749-7

Printed in the United States of America.

10 9 8 7 6 5 4 3 2 1

# PREFACE

This book is intended for a broad readership, starting with curious and thoughtful college undergraduates and, reaching beyond, to professionals and researchers in the life sciences and medicine. It is hoped that it will also be useful to the educated lay person with an interest in health and medicine.

An effort has been made to integrate chemistry, biology, drug discovery and medicine in a way that is clear and self-explanatory. Heavy use has been made of chemical structures, since they provide a fundamental key to the language of life and the human activities that flow from it. Our age has seen the rapid evolution of molecular medicine as a critical part of the broader fields of health care and the biochemical basis of human disease. The understanding of human illness at the molecular level has brought, and will bring, great benefit to mankind.

There is a price to be paid in any attempt to understand molecular medicine, because that comprehension requires an ability to decipher chemical structures, which many have regarded as too onerous. One purpose of this book is to demonstrate that an adequate understanding of chemical structures is easily within the reach of most educated people, and well worth the effort. Pages 4-31 of this book aim to provide the insights and background required to appreciate the architecture of therapeutic molecules and their target proteins, as they parade though the subsequent pages of this book.

"*Molecules and Medicine*" delves into the discovery, application and mode of action of well over one hundred of the most significant molecules now in use in modern medicine. It is limited to *centamolecules*, i.e. molecules with molecular weights in the hundreds (several hundred times more than a hydrogen atom). The important and rapidly developing area of macromolecular therapy, which involves *much larger molecules* (macro-molecules), such as biologically active proteins and monoclonal antibodies, is a different story, and another book.

We have tried to minimize the amount of prior knowledge required of the reader by providing much background information, both chemical and biomedical. An effort has also been made to be concise as well as clear. A large amount of material has been compressed into a small space for each therapeutic agent. Generally, each medicine is allotted just one page. One advantage of this modular arrangement is that it facilitates the reading of the book in small installments and also its use as a reference work. The therapeutic agents in this book are arranged in sections according to the type of medical condition they treat.

There are also numerous sections in the book that provide biomedical background. For instance, there are two-page to eight-page summaries of topics such as inflammation, metabolic syndrome, immunology, drug resistance, cancer and neurotransmission. These are placed at strategic locations throughout the book.

The structural representations of proteins in this book are in the public domain and may be downloaded from http://www.pdb.org.

In the process of writing this book we have come to appreciate ever more keenly the enormous amount of human talent and effort that has enabled the extraordinary advances in molecular medicine since its advent several decades ago. To the countless chemists, physicians, biologists, educators and other professionals who have participated in this venture we extend our gratitude and thanks. This book is a tribute to them all.

The writing of this work has also been motivated by the realization that the advance of molecular medicine can be even more remarkable during the coming decades of this century, if a steady flow of dedicated and able young people into all the key areas of research in the life sciences can be maintained. If this book plays just a minor part in enhancing progress in molecular medicine and human well-being, our small effort will have been amply rewarded.

# A NOTE ON THE USE OF THIS BOOK

### Structures of Molecules and Proteins

Those who are not familiar with the notation used to describe the structures of organic compounds will profit from a close reading of pages 4-22 of this book. Others can read through this part quickly and proceed to the next section (pages 26-31) that reviews the notation used in the book for the structures of proteins.

The coordinates for each of the protein structures shown in the book are in the public domain and can be accessed electronically.

### Accessing X-Ray Crystal Structure Data Files Online – The Protein Data Bank

All the crystal structure files can be downloaded from http://www.pdb.org by entering the four-character PDB ID that is indicated in red at the bottom of the page where the protein is displayed or in the reference section. For example, on page 63 in the entry for sitagliptin (Januvia™), the X-ray crystal structure of sitagliptin bound to the target protein DPP-4 is displayed. The access code (PDB ID) for this crystal structure appears below the picture as **1X70** along with the corresponding reference.

### Graphic Rendering of X-Ray Crystal Structure Data Using PyMol

The renderings of the X-ray crystal structure data were developed by the authors using the software PyMol v0.99 (DeLanoScientific LLC, http://www.delanoscientific.com). For a description of the Protein Data Bank, *see*:

H.M. Berman, J. Westbrook, Z. Feng, G. Gilliland, T.N. Bhat, H. Weissig, I.N. Shindyalov, P.E. Bourne: The Protein Data Bank. *Nucleic Acids Research, 28,* 235-242 (**2000**).

### Organization and References

The discussion of each of the therapeutic agents in the book is limited to one page, and, consequently, much in the way of detail has been omitted. For this reason, a number of up-to-date references on other aspects of

each agent are given both at the bottom of the appropriate page and at the end of each section. The page numbers on which detailed references are listed for each therapeutic agent are highlighted in green at the bottom of the pages. Additional information on the pharmacology and properties of each medicinal agent can be found in:

*Goodman & Gilman's The Pharmacological Basis of Therapeutics.* Laurence L. Brunton, John S. Lazo and Keith L. Parker (Editors); (McGrawHill, 11th Edition, **2006**).

Other recent texts that provide much useful background information are:

(1) L. Stryer et al. *Biochemistry,* (W.H. Freeman, 6th Edition, **2007**)

(2) B. Alberts et al. *Molecular Biology of the Cell* (Garland Science, 4th Edition, **2002**)

(3) Weinberg, R.A. The Biology of Cancer (Garland Science, **2007**)

(4) *The Merck Manual of Diagnosis and Therapy.* M.H. Beers and R.S. Porter (Editors). (Merck & Co., 18th Edition, **2006**)

The references that appear in this book can serve as a portal to a great deal of information on the chemistry, biology and medicine relevant to the therapeutic agent and disease area. Additional material can be located by appropriate database- and internet searching (e.g., using SciFinder Scholar™, MEDLINE™ or Google™).

A **glossary** and an **index** appear at the end of the book.

### Online Reader Feedback for the Authors

There is a website for this book maintained by the authors, the address for which is:

http://moleculesandmedicine.info/

Readers can use this website to provide comments or feedback on "*Molecules and Medicine*". The website will also contain certain useful updates. The authors invite the views of non-scientists who have read pages 4-31.

# CONTENTS

## PART I.

## PART II.

## INFLAMMATORY, CARDIOVASCULAR AND METABOLIC DISEASES

## GLAUCOMA AND ANTIULCER AGENTS

# PART IV.

## AUTOIMMUNE DISEASE AND ORGAN TRANSPLANT

## INFECTIOUS DISEASES

# PART V.

## MALIGNANT DISEASE

# PART VI.

## DRUGS ACTING ON THE NERVOUS SYSTEM

### PAIN AND ANALGESIA

### HYPNOTICS (INSOMNIA) AND ANTISMOKING

### NEURODEGENERATIVE AND PSYCHIATRIC DISEASES

# PART I.

# INTRODUCTION

We live in a troubled, but wonderful time. It is our good fortune to witness and benefit from scientific advances that would have been literally unimaginable to our grandparents. However, there are dark clouds on the horizon. The rate of growth of scientific knowledge has been so great as to outstrip the ability of our society to assimilate it, the capacity of the educational system to teach it properly, and the wisdom of government adequately to sustain and apply it. There is growing indifference to science among the young. Even medical science, which touches the lives of us all, is generally left to the practitioners. Whatever the reason for this disparity between the importance of science and the lack of general public understanding, it is important to address it.

In this book we try to take a few steps in this direction. Specifically, the pages to follow tell the tales of many molecules that can qualify as miracles of modernity. These relatively small, highly-structured clusters of atoms, the principal therapeutic agents of modern medicine, can perform in a way that would have been considered miraculous to our ancestors. Such "miracle molecules" can save countless human lives, prolong human life, alleviate pain and suffering, control cells, tissues and organs millions of times their size, and bring enormous material gains through commercial sales of billions of dollars per year. Such molecules also can serve as tools to probe the molecular nature of life processes and disease states and pave the way for the discovery of other effective medicines.

The molecules at the core of this book have been carefully selected from several thousand therapeutic agents that have been used in medicine at one time or another. The development of each of them, arduous and costly though it might have been, represents an enormously valuable investment with very large and ongoing benefits. In the course of discovering all these wonderfully useful molecules, we have learned more about the discovery process itself and have developed an ever expanding set of new discovery tools. The invention of these new platforms for innovation is being powered by dramatic advances in technology, computing and the underlying chemical and biomedical sciences.

The very next section of this book provides a step-by-step introduction to the understanding of the architecture of organic molecules and the general principles that govern structures of molecules. In addition, we explore the fundamental forces that hold molecules together and that allow them to recognize and bind to one another. The affinity of molecules for one another is central to the biological activity of therapeutic agents and to life itself. The section on how to read the chemical diagrams of small molecules is followed by another tutorial on understanding much larger structures, the proteins of life.

It seems quite possible that, in the next century or so, effective treatments for most illnesses will emerge. Disease, premature death, suffering and pain may no longer be a part of the human condition. Humans will as a matter of course live out a full and healthy lifespan, and then depart with grace and dignity. The famous poem of Lady Gio in "The Tale of the Heike" describes the life process and its end in an eloquent and happy way:

> Grasses of the plain,
> Springing up and withering,
> They all fare alike.
> Indeed the lot of all things
> Is but to wait for autumn.

An impossible dream? Perhaps, but the immense effort required will be well worthwhile, because the gain will be incalculable. The achievements of modern science and technology provide both encouragement and inspiration.

For instance, we now can trace our universe back some 14 billion years to an unbelievably hot object, with a temperature of about $10^{32}$ Kelvin (10 followed by 32 zeros), and more than a million times smaller than the period at the end of this sentence. From this inferno of exceedingly small and simple objects, the first elements, hydrogen and helium, formed about a million years later, to be followed by all the other objects of the universe – the chemical elements, stars and galaxies, and an unknown collection of other forms of matter and energy, and finally the earth and life upon it. Surely, a time will come when our knowledge of life, intelligence, disease and health will dwarf that of the present.

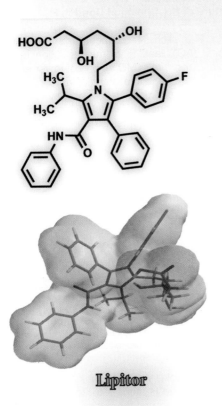

Penicillin V

Viagra

MIRACLE
MOLECULES

Lipitor

Cortisol

# UNDERSTANDING STRUCTURAL DIAGRAMS OF ORGANIC MOLECULES

## Introduction. Some Sample Molecules

### The Simplest Molecule, $H_2$

Hydrogen, the simplest element, has atomic number *one* because it contains just *one* proton in the nucleus and *one* electron that surrounds it. A hydrogen atom is so reactive that it will *totally* combine with another hydrogen atom to form the simplest molecule, $H_2$. This process, the simplest of all chemical reactions, takes place extremely rapidly and with the release of much energy because the energy content of $H_2$ is much less than two isolated hydrogen atoms. The chemical equation for the reaction is:

$$H\cdot \ + \ H\cdot \ \longrightarrow \ \begin{matrix} H:H \\ \text{or} \\ H-H \end{matrix} \ + \ \text{Energy}$$

**Equation 1.** Two hydrogen atoms combine to form one hydrogen molecule (H-H) with the evolution of energy.

In equation 1, the dots represent electrons. The chemical bond that holds $H_2$ together is designated by a double dot, or simply, by a line between the hydrogens. Because the electrons are shared equally, the molecule is nonpolar. This type of two-electron bond is called a covalent single bond. Although an electron has a mass, it is very small ($9\times10^{-28}$ g, at rest), and behaves like a wave. In addition, because of the Heisenberg uncertainty principle, its position and momentum cannot both be known precisely. Its location is best described in a probabilistic way as a cloud-like representation (Figure 1).

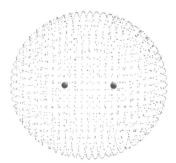

**Figure 1.** Approximate representation of the spatial distribution of electrons in an $H_2$ molecule about the nuclei (red spheres). The outer layer of dots encloses about 90% of the total electron cloud.

## Helium, a Chemically Inert Element

Helium (He), atomic number *two*, has *two* protons in the nucleus and *two* surrounding electrons. The atom is stable chemically because it neither combines with itself nor reacts with other elements. The reason for this inertness is that two is the maximum number of electrons that can be accommodated in the available orbital. This orbital allows the electrons to distribute themselves like a cloud with spherical symmetry about the nucleus. This orbital is called the 1s orbital. The other orbitals of He are so high in energy that they are not accessible for chemical bonding with any atom or chemical fragment. Helium is the simplest of the inert elements. These elements share the feature of having the full complement of electrons in the available orbitals.

## Nature of the Chemical Bond in $H_2$

In $H_2$, each H contributes a 1s *atomic orbital* (AO), leading to the formation of two *molecular orbitals* (MO's), one of lower energy than the AO and the other of higher energy. The two electrons of $H_2$ can occupy the lower energy MO, and so $H_2$ is stabilized relative to two H atoms, as the diagram in Figure 2 illustrates.

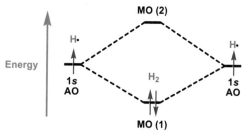

**Figure 2.** Energy diagram for the combination of two H atom 1s AO's to form two MO's of diatomic $H_2$. The electrons occupy the lower, bonding MO(1) and are shown as blue arrows. MO(2) is an orbital that is not occupied by electrons because of its high energy. MOs can hold only two electrons. These electrons must have opposite spins.

One simple way of thinking about the electron in the H atom is to consider it like a cloud of gas around the nucleus that is kept in place by the electrostatic attraction between the negative electron and the positively charged proton. Once $H_2$ is formed, the two electrons of $H_2$ become delocalized over a greater volume around the two-proton axis of the diatomic molecule, which leads to great stability. For a simple analogy, recall that the rapid expansion of a gas lowers its temperature (i.e., energy content). Conversely, compression of a gas raises its temperature/energy content.

The energy of the $H_2$ molecule is at a minimum when the nuclei are separated by $0.75 \times 10^{-8}$ cm or 0.75 Angströms (Å). That bond length is determined by a balance between two factors:

(1) there is electrostatic repulsion between the two positively charged nuclear protons which tends to keep them apart and

(2) the electronic stabilization that results from formation of a molecular orbital, increases with the interpenetration (or overlap) of the individual atomic orbitals, an effect that draws the nuclei closer.

In general, chemical bonds between two atoms have characteristic bond lengths ($d_0$), which result from the balancing of these two effects. A typical graph of the stabilizing energy of a bond as a function of bond length is shown in Figure 3.

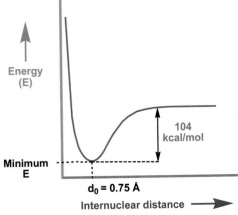

Figure 3. Energy of two hydrogens as a function of internuclear distance.

A large amount of energy is required to break the H–H bond of $H_2$ – 104 kcal/mol – (a mol is defined as $6.02 \times 10^{23}$ atoms, Avogadro's number). In the pure state the $H_2$ molecule can be heated to 500 °C without any decomposition.

The bonding diagram for $H_2$ allows us to explain why $H_2$ does not react with another H atom to form $H_3$ as a stable molecule. The bonding orbital MO(1) of $H_2$ is filled since no more than two electrons can occupy a MO. No significant bonding can be attained by combining MO(2) of $H_2$ with the AO of the H atom.

Most of the molecules in this book contain hydrogen. In every case each hydrogen forms one *and only one* electron-pair bond with its partner leading us to the first rule:

**Rule #1:** *H forms just one electron-pair bond — called a covalent single bond.*

This rule is useful in deriving the actual structures of molecules. Thus, the formula of water, $H_2O$, and Rule #1 lead to the structure H–O–H, rather than O–H–H, or something else. Similarly, since methane has the composition $CH_4$, we can draw the structure:

$$
\begin{array}{c}
\text{H} \\
| \\
\text{H}-\text{C}-\text{H} \\
| \\
\text{H}
\end{array}
$$

### Three Depictions of $H_2$

There are a number of alternative ways of representing $H_2$. In addition to a line drawing H–H, we can use a ball-and-stick drawing or a space-filling diagram, as follows:

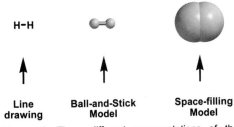

Figure 4. Three different representations of the hydrogen molecule.

The space-filling representation contains additional information because it tells us about the shape and size of a molecule. These features are exceedingly important in determining how two molecules can fit together or interact and also how close they can get. With molecules, as with macroscopic objects, two things cannot occupy the same space at the same time.

## Bonding in Carbon Compounds

Carbon, the element with atomic number *six* has *six* protons in the nucleus and *six* electrons surrounding it. *Two* of these fill the low energy 1s orbital and play no role in chemical bonding. The remaining *four* electrons can form bonds utilizing a spherical 2s orbital and three dumbbell-shaped, mutually perpendicular 2p orbitals, the shapes of which are illustrated in Figure 5. The boundary of each dumbbell encloses about 90% of the total electron cloud. The three p orbitals are usually designated as $2p_x$, $2p_y$ and $2p_z$ because they can be placed along the axes of a rectilinear coordinate system. The sum of the electron densities for all three 2p orbitals is a *spherical* cloud of larger diameter than the 2s orbital.

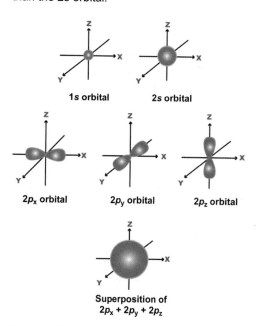

**Figure 5.** Representation of s and p atomic orbitals. The boundary of each sphere and dumbbell encloses about 90% of the total electron cloud.

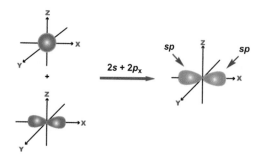

**Figure 6.** Combination of one 2s and one $2p_x$ atomic orbital (AO) to form two sp hybrid orbitals.

The 2s orbital can mix with three 2p orbitals to form hybrid atomic orbitals in *three* different ways. A sequence for generating the three different 2s/2p hybrids is shown in Figures 6-8. The number of hybrid orbitals always equals the number of AOs from which they derive.

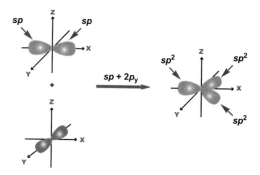

**Figure 7.** Combination of two sp hybrid orbitals and a $2p_y$ atomic orbital to form three $sp^2$ hybrid orbitals.

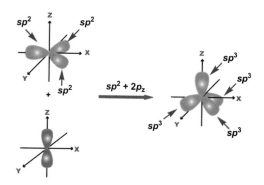

**Figure 8.** Combination of three $sp^2$ hybrid orbitals and a $2p_y$ orbital to form four $sp^3$ hybrid orbitals.

## Hybrid Orbitals for Tetracoordinate Carbon

If all three of the 2p orbitals and the 2s orbital are hybridized, a set of 4 equivalent orbitals results, the axes of which are directed at the vertices of a tetrahedron (Figure 9a). This arrangement is used in forming methane and other compounds having four atoms attached to carbon (tetracoordinate carbon). The attachment of four hydrogen atoms (4 H) to a carbon atom results in the formation of four electron-pair bonds and the complete filling of the 4 bonding MOs by eight electrons (4 from C and 4 from H).

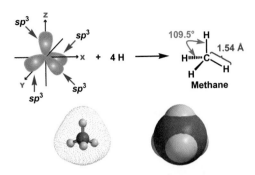

**Ball-and-Stick Model** **Space-filling Model**

**Figure 9a.** Formation of methane from four $sp^3$ hybrid orbitals of carbon and four $1s$ orbitals of hydrogen.

The angle between any two of the C–H bonds in methane is 109.5°, the angle between the center of a tetrahedron and any two of the vertices (Figure 9b). The internuclear distance of each C–H bond is 1.54 x 10⁻⁸ cm or 1.54 angstroms (Å). A second useful bonding rule emerges from these facts.

**Rule #2:** *Carbon (C) can bond to a maximum of four atoms (tetracoordination). The preferred angle between any two of the bonds is 109.5°. Tetracoordinate carbon utilizes sp³ hybrid orbitals.*

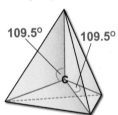

**Figure 9b.** The angle between the center of a tetrahedron and any two of its vertices is 109.5°.

Deviation of the bond angles at an $sp^3$-hybridized carbon atom from the preferred value of 109.5° leads to a higher energy (i.e., less stable) structure. The destabilization increases to a value of about 6 kcal/mol for an angle of 90°. This destabilization, called *angle strain*, influences the chemistry and properties of a compound.

Carbon forms strong bonds to most atoms, including H, oxygen (O), nitrogen (N), chlorine (Cl), and by no means least, to itself. Thus, methane is just the first in a large family of compounds of carbon and hydrogen (hydrocarbons). That family includes the straight-chain saturated hydrocarbons, the first five members of which are shown in Figure 10.

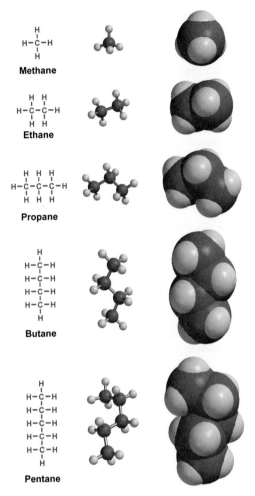

**Methane**

**Ethane**

**Propane**

**Butane**

**Pentane**

**Figure 10.** The simplest straight-chain hydrocarbons.

## Branching of Carbon Chains

Carbon chains can also be branched.

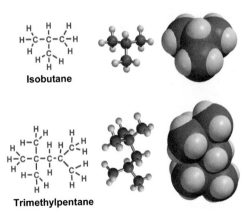

**Isobutane**

**Trimethylpentane**

**Figure 11a.** Two simple branched hydrocarbons, isobutane and trimethylpentane.

7

There is a simpler notation for depicting the structures of carbon compounds in which the hydrogens are omitted. This shorthand notation leaves it to the reader to add the number of hydrogens corresponding to tetracoordination (Figure 11b).

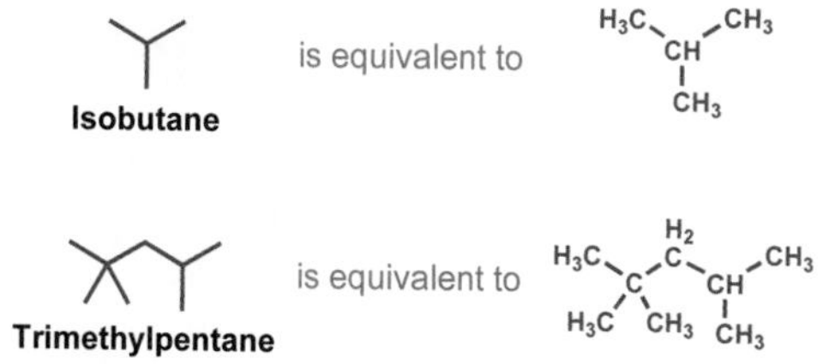

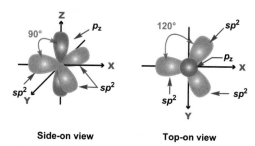

**Isobutane**       is equivalent to       $H_3C$ $CH_3$ $CH$ $CH_3$

**Trimethylpentane**   is equivalent to   $H_3C$ $C$ $H_2$ $CH_3$ $H_3C$ $CH_3$ $CH$ $CH_3$

**Figure 11b.** Simplified notation of carbon compounds in which the hydrogens are omitted.

## Cyclic Structures

Carbon can form rings by bonding with itself in a cycle as well as chains. The simplest members of the family of cyclic hydrocarbons are shown in Figure 12.

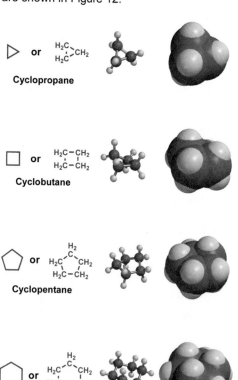

**Cyclopropane**  or  $H_2C$ $CH_2$ $H_2C$

**Cyclobutane**  or  $H_2C-CH_2$ $H_2C-CH_2$

**Cyclopentane**  or  $H_2$ $C$ $H_2C$ $CH_2$ $H_2C-CH_2$

**Cyclohexane**  or  $H_2$ $C$ $H_2C$ $CH_2$ $H_2C$ $CH_2$ $C$ $H_2$

**Figure 12.** The first four members of cyclic hydrocarbons: cyclopropane, cyclobutane, cyclopentane and cyclohexane.

## Tricoordinate Carbon Compounds. The Double Bond

Carbon can form compounds in which *three atoms* are linked to it using hybrid orbitals generated from the combination of the 2s atomic orbital with two of the 2p atomic orbitals. The orbitals of trigonally hybridized carbon are shown in Figure 13.

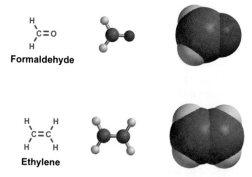

**Side-on view**       **Top-on view**

**Figure 13.** Side-on and Top-on views of the orbitals of a trigonally hybridized (sp²-hybridized) carbon atom.

Two of the simplest carbon compounds that involve tricoordinate (trigonally hybridized) carbon atoms are formaldehyde ($H_2C=O$) and ethylene ($H_2C=CH_2$). These are planar molecules that contain a double bond to carbon as well as single (electron-pair) bonds to the hydrogens (Figure 14).

**Formaldehyde**

**Ethylene**

**Figure 14.** Formaldehyde and ethylene contain trigonally hybridized carbon.

Bonding to tricoordinate carbon utilizes three sp² hybrid orbitals and the remaining 2p AO. These orbitals allow the derivation of the correct geometry of molecules with tricoordinate carbon. For example, the planar structure of ethylene results because overlap of the p-orbitals is maximum when they are parallel, as shown in Figure 15a.

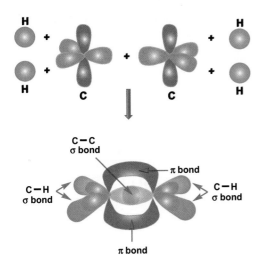

**Figure 15a.** Formation of ethylene by the combination of two carbon atoms and four hydrogen atoms.

The linkage between the two carbons of ethylene is called a double bond because it involves four electrons. The double bond can be represented by two lines, as in the drawing in Figure 14, or by σ and π bonds as shown in Figure 15a. The π bond, formed by the side-by-side combination of two parallel $p$ atomic orbitals (shown in blue in Figure 15a), has two lobes, one above and one below the molecular plane. The σ bond, formed by the combination of two colinear $sp^2$ orbitals, is symmetric about the C-C axis (axial symmetry).

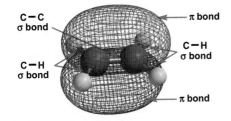

**Figure 15b.** π-Bond of ethylene.

**Rule #3:** *Tricoordinate carbon is connected to each of the three attached atoms in a planar arrangement. The bonding involves three in-plane hybrid $sp^2$ orbitals and an orthogonal p atomic orbital.*

## Dicoordinate Carbon Compounds. The Triple Bond.

Dicoordinate carbon compounds utilize two $sp$ orbitals formed from the hybridization of the $2s$ orbital with one $2p$ orbital, and also the remaining two $2p$ orbitals. The orbitals for this

type of carbon are shown in Figure 16; note that the angle between the two $sp$ orbitals is 180°. The $p_y$ and $p_z$ orbitals that are not involved in hybridization remain unchanged.

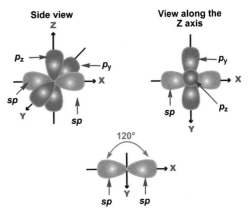

**Figure 16.** Two views of the orbitals of a dicoordinate ($sp$-hybridized) carbon atom.

Two simple examples of dicoordinate carbon compounds are hydrogen cyanide and acetylene, both of which possess triple bonds and linear geometry because of the 180° angle between the $sp$ orbitals of carbon (Figure 17).

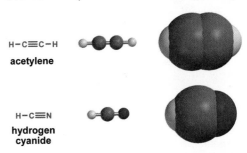

**Figure 17.** The two simplest dicoordinate carbon compounds, acetylene and hydrogen cyanide.

The triple bond consists of six electrons; two of these are in an axially symmetric MO formed from the combination of two $sp$ AOs. The remaining 4 electrons are in two bonding π-MOs formed from overlap of the four $2p$ AOs.

The linear structure of acetylene follows from the use of the $sp$ hybrid orbitals and $p$ orbitals of each carbon and two H $1s$ orbitals to assemble the molecule, as shown in Figure 18a.

An electron cloud representation of the two π bonds of acetylene is shown in Figure 18b. These two π-bonds and the $sp$-$sp$ σ bond of acetylene hold six electrons and constitute a C-C triple bond.

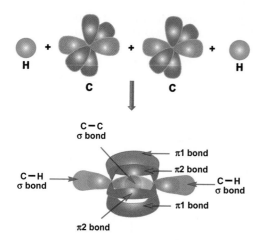

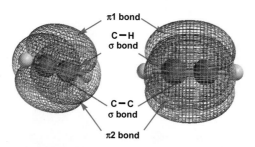

**Figure 18a.** Formation of acetylene by the combination of two *sp*-hybridized carbon atoms and two hydrogen atoms.

**Figure 18b.** The two π-bonds of acetylene.

**Rule #4:** *Dicoordinate carbon forms bonds to the two attached atoms in a colinear arrangement. The bonding involves two colinear sp orbitals and two p atomic orbitals at carbon.*

Carbon dioxide ($CO_2$) is another linear molecule in which carbon is *sp*-hybridized (Figure 19).

O=C=O
carbon
dioxide

**Figure 19.** Structure and shape of carbon dioxide.

**The Common Chemical Elements in Living Systems**

Most of the common elements that make life possible fall within the first three rows of the Periodic Table of Elements. These are shown in Figure 20 along with the

corresponding atomic numbers. The atomic number of an atom is identical to the number of protons in the nucleus or the number of orbiting electrons.

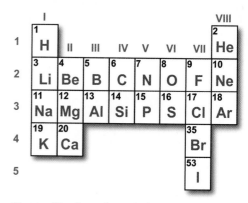

**Figure 20.** A portion of the Periodic Table of Elements.

**Some Simple Compounds of Hydrogen and Non-Carbon Elements**

All the elements shown in Figure 20 combine with hydrogen, with the exception of the inert gases He, Ne and Ar. Some examples of the simplest of these are the following (Figure 21).

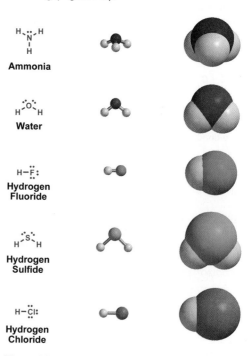

Ammonia

Water

Hydrogen
Fluoride

Hydrogen
Sulfide

Hydrogen
Chloride

**Figure 21.** Simple compounds of hydrogen and non-carbon elements.

The dot pairs in the above structures represent electrons in the outer valence shell that are not needed in bonding. The structure of water, for example, involves an $sp^3$ hybridized oxygen atom connected to two hydrogen atoms. The single bonds to the hydrogens use up two out of the six available electrons of an O atom. The remaining four oxygen electrons are located in the remaining (nonbonding) $sp^3$ orbitals. Since the four $sp^3$ orbitals are filled by eight electrons, no further electrons, for instance from a hydrogen atom (H·), can be added. (*Reminder:* each orbital can hold only two electrons.)

## Carbon Bonding to Elements Other than Hydrogen

Carbon can also bond to most of the elements, for instance replacing hydrogen in the compounds shown in Figure 21.

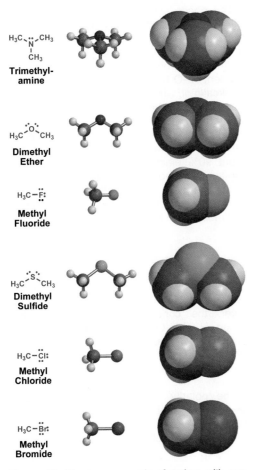

**Trimethyl-amine**

**Dimethyl Ether**

**Methyl Fluoride**

**Dimethyl Sulfide**

**Methyl Chloride**

**Methyl Bromide**

**Figure 22.** Simple compounds of carbon with non-carbon elements.

**Rule #5:** *Carbon can bond to itself to form either straight or branched chains or rings. Carbon can also bond to many other atoms.*

## Ionic Bonds

The metallic elements at the left of the Periodic Table lose an electron very readily and tend to form positively-charged ions (cations), rather than covalently bonded compounds. The elements F, Cl, Br and I at the right of the Periodic Table, in contrast, have high electron affinity and readily accept an electron to form negative ions (anions). In the case of fluoride ion (F⁻) the available orbitals are filled, as with the inert gas neon (Ne) with which it is isoelectronic (i.e., both F⁻ and Ne have a total of 8 outer shell electrons). The bonding between sodium and chlorine, for example, is essentially electrostatic, and the bond is described as ionic in character. It is the extreme of the covalent bond of $H_2$ which involves no charge separation. Sodium chloride (NaCl, salt) in solid form is a crystalline structure containing Na⁺ and Cl⁻ ions in an indefinitely repeating lattice in which each Na⁺ is surrounded by 6 Cl⁻, and vice versa. It is so stable that the melting point of salt is about 800 °C. The energy that holds Na⁺Cl⁻ in the crystal lattice is 187 kcal/mol, much greater than the H–H covalent bond energy (104 kcal/mol).

## Bonds of Intermediate Polarity

Hydrogen chloride (HCl) is a gas at room temperature, in contrast to the ionic solid sodium chloride. The bonding in H-Cl is best described as a covalent bond with appreciable (but far from full) ionic character or charge separation. The electron pair between H and Cl is not shared equally. It is a polarized molecule with more electron density at the Cl end and less at the H end (Figure 23).

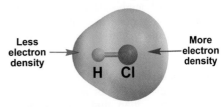

Less electron density

More electron density

H   Cl

**Figure 23.** The electron density around HCl. In this computer-generated electron-density map, the blue color represents the lowest electron density whereas the red color represents the highest electron density.

In aqueous solution HCl ionizes to form a hydrated proton and a hydrated chloride ion. Thus, it is a strong acid.

$$HCl + H_2O \longrightarrow HOH_2^+(H_2O)_n + Cl^-(H_2O)_n$$

**Equation 2.** Dissociation of HCl in water.

The polarization of the bond in gaseous hydrogen chloride, often indicated using the notation $H^{\delta+}-Cl^{\delta-}$, is a consequence of greater electron affinity of a chlorine atom as compared to a hydrogen atom. Expressed in another way, chlorine is more strongly electron-attracting than hydrogen.

Polarization of covalent bonds is very common. Four examples are shown in Figure 24.

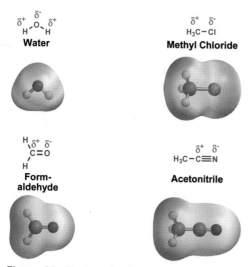

**Water**

**Methyl Chloride**

**Form-aldehyde**

**Acetonitrile**

**Figure 24.** Electron density maps of four simple compounds. The highest electron density is shown in red whereas the lowest electron density is shown in blue.

## Molecular Polarity and Hydrogen Bonding

Oil and water do not mix because neither can dissolve the other. The former is essentially hydrocarbon-like and nonpolar, whereas water is polarized with the oxygen relatively negative and the two hydrogens positive.

The polarity of the O–H bonds in water causes the boiling point of water (100 °C) to be much higher than that, for instance, of methane ($CH_4$, -161 °C) which has about the same size, or ammonia ($NH_3$, -33 °C). The O–H bond polarity of water causes molecules of $H_2O$ to associate with one another,

primarily because of electrostatic forces, forming an extended three-dimensional network, a small part of which is shown in Figure 25.

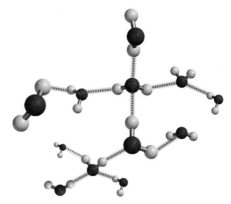

**Figure 25.** An extended three-dimensional network of molecules in liquid water involves "hydrogen bonds" (blue dashes) as shown.

The bonds between molecules of $H_2O$ in the liquid – called "hydrogen bonds" – are much weaker than the H–H bond (104 kcal/mol), or the C–H bond in methane (105 kcal/mol). The bond dissociation energy of the covalent H–O bond in gaseous $H_2O$ is about 117 kcal/mol, whereas the energy of attraction of an H in water with an O atom of a neighboring water molecule is about 6 kcal/mol. Intermolecular hydrogen bonds between water molecules in the liquid are about 90% electrostatic and 10% covalent. The hydrogen bonds in liquid water stabilize it by an energy of cohesion that is responsible for the unusually high heat of vaporization (9.7 kcal/mol, or 538 kcal/liter).

## Aqueous Solvation of Ions

Another unique property of water is the existence of ionized species. In pure, neutral water, the concentration of the hydrated proton $H_3O^+(H_2O)_n$ and hydrated hydroxide ion $HO^-(H_2O)_n$ are each ca. $10^{-7}$ mol/liter. These species are essentially hydrogen-bonded clusters.

The polarity of water makes it a good solvent for polar ionic molecules because water can form electrostatic or hydrogen bonds to the dissolved species. For example:

$$NaCl + H_2O \longrightarrow Na^+(H_2O)_n + Cl^-(H_2O)_n$$

$$HCl + H_2O \longrightarrow H_3O^+(H_2O)_n + Cl^-(H_2O)_n$$

**Equation 3.** Formation of hydrated ions in water.

One of the simplest indications that solutions of NaCl or HCl in water contain ions is their high electrical conductivity. Pure water is only a weak conductor of electricity because the concentrations of the ions $H_3O^+$ and $HO^-$ are only $10^{-7}$ mol/liter. Seawater is a much better conductor because it contains 0.1 mol/liter of $Na^+$ and $Cl^-$ ions.

### Solvation Energies in Water

Water is unique as a solvent.

Despite the fact that solid NaCl is bound in the crystal lattice with an energy of 187 kcal/mol, it dissolves in water to give solutions of hydrated $Na^+$ and $Cl^-$. The reason for this is that the energy of solvation of these ions in water (191 kcal/mol) overcomes the high lattice energy by 4 kcal/mol. The sum of energies of solvation by water of a proton (269 kcal/mol) and of a chloride ion (89 kcal/mol) are greater than the dissociation energy of gaseous hydrogen chloride (358 kcal/mol vs 103 kcal/mol), and so it is clear why HCl is both soluble in water and fully ionized. The take-home lesson is that solvation by water strongly stabilizes both positive and negative ions.

The high solvation energy of sodium chloride in water is largely electrostatic. Solvated $Na^+$ is surrounded by a cluster of at least six $H_2O$ molecules with oxygen in proximity to $Na^+$. Solvation of $Cl^-$ similarly involves a cluster of $H_2O$ molecules with hydrogen in proximity to $Cl^-$.

### Interactions Between Nonpolar Molecules

There are also attractive forces that operate between nonpolar molecules such as straight-chain hydrocarbons. These intermolecular attractions, sometimes called van der Waals forces, are very much *weaker* than covalent or ionic bonds, or even hydrogen bonds. An instructive example is the attraction between two atoms of the inert gas argon (Ar) which has been measured as ca. 0.28 kcal/mol. This attraction arises not from covalent bonding (because the atomic orbitals of Ar are filled), but from fluctuations in electron density around Ar that create transient imbalance, with one side of the atom being more negative than the other. This polarity induces an opposite distribution of charge on a nearby Ar atom, and the result is a transient attraction (Figure 26).

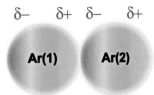

$\delta-$    $\delta+$  $\delta-$    $\delta+$

**Figure 26.** Electron density fluctuation in Ar(1) induces opposite electron density in Ar(2), leading to net attraction between them.

The attraction between two adjacent non-polar molecules increases in proportion to the area of contact and is usually on the order of ca. 1 kcal per square Å of close contact. One manifestation of van der Waals attraction is the steady increase in boiling point temperatures (in °C) with increasing molecular size for the series of straight-chain hydrocarbons (Figure 27).

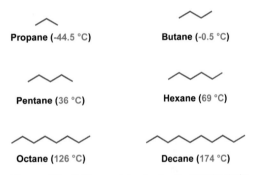

Propane (-44.5 °C)          Butane (-0.5 °C)

Pentane (36 °C)          Hexane (69 °C)

Octane (126 °C)          Decane (174 °C)

**Figure 27.** Boiling points (red) of straight-chain hydrocarbons (blue).

Although van der Waals attractions between molecules are weaker than hydrogen bonding or electrostatic interactions, they can become significant when two nonpolar molecular surfaces are complimentary in shape of sizeable area. These forces play a major role in determining three-dimensional protein geometry and specificity of drug action. In addition, this type of interaction is what makes it possible for gecko lizards to walk across smooth ceilings or vertical walls.

### Functional Groups, Subunits Within Structures that Confer Characteristic Properties and Reactivity.

Compounds containing a carbon-carbon *double bond* within the structure show characteristic chemical behavior. For instance, ethylene, 1-pentene and cyclohexene all react with $H_2$ in the presence of finely divided nickel (Ni) as a catalyst to form products in which hydrogen has been added to the $sp^2$ carbons (Figure 28). Such addition reactions to C=C are so common that these compounds are called unsaturated.

There are many other characteristic reactions of the C=C subunit, often called an *olefinic functional group* (or alkene). In addition, there are many other types of functional groups in organic compounds. A tabulation of some of the most common ones are shown in Figures 29a and 29b

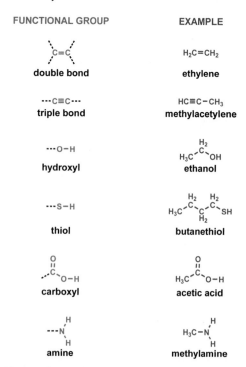

**Figure 28.** Reaction of compounds containing a C=C double bond with hydrogen gas ($H_2$) in the presence of Ni catalyst.

| FUNCTIONAL GROUP | EXAMPLE |
|---|---|
| double bond | ethylene |
| triple bond | methylacetylene |
| hydroxyl | ethanol |
| thiol | butanethiol |
| carboxyl | acetic acid |
| amine | methylamine |

**Figure 29a.** A small sample of functional groups in organic compounds.

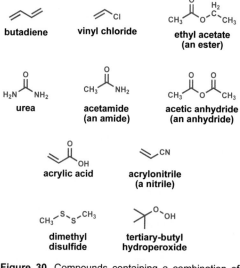

**Figure 29b.** A small sample of functional groups in organic compounds.

Many of the compounds shown in Figures 29a and 29b are familiar to most people, especially ethanol (alcohol), butanethiol (butylmercaptan, the essence of the odor of skunks), and acetic acid (vinegar). They are also widely useful articles of commerce. For instance ethylene is the building block from which giant molecules of the plastic polyethylene are made.

Often, combinations of directly connected functional groups occur in molecules. Some examples of such combinations are shown in Figure 30.

**butadiene**    **vinyl chloride**    **ethyl acetate (an ester)**

**urea**    **acetamide (an amide)**    **acetic anhydride (an anhydride)**

**acrylic acid**    **acrylonitrile (a nitrile)**

**dimethyl disulfide**    **tertiary-butyl hydroperoxide**

**Figure 30.** Compounds containing a combination of functional groups shown in Figures 29a and 29b.

Functional groups containing sulfur and phosphorus can also exist in states of higher oxygenation since these elements have five 3*d*-orbitals available for bond formation in addition to the 3*s* and *three* 3*p* orbitals. Some examples of these more highly oxidized functional groups are shown in Figure 31.

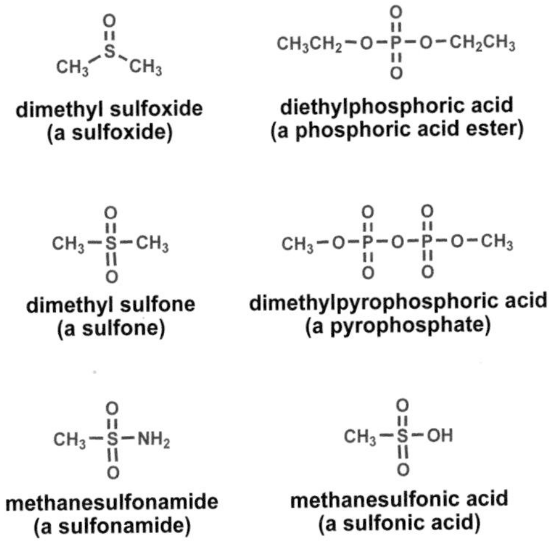

**dimethyl sulfoxide**
**(a sulfoxide)**

**diethylphosphoric acid**
**(a phosphoric acid ester)**

CH₃CH₂−O−P−O−CH₂CH₃

**dimethyl sulfone**
**(a sulfone)**

**dimethylpyrophosphoric acid**
**(a pyrophosphate)**

CH₃−O−P−O−P−O−CH₃

**methanesulfonamide**
**(a sulfonamide)**

CH₃−S−NH₂

**methanesulfonic acid**
**(a sulfonic acid)**

CH₃−S−OH

**Figure 31.** Compounds containing highly oxidized functional groups derived from sulfur and phosphorus.

## Carboxylic Acids. Part I. Acidity

The classification of structural subunits as functional groups is a very useful technique for organizing the enormous amount of information that is encompassed by organic chemistry, the chemistry of the vast family of carbon compounds (carbogens). In this section the carboxylate functional group is examined in some detail using a few of the smaller molecules of the class (carboxylic acids) as representative.

Formic acid, the simplest carboxylic acid, is a proton donor. In aqueous solution it ionizes partially to hydrated negative formate ion and hydrated $H_3O^+$, as Equation 4 indicates.

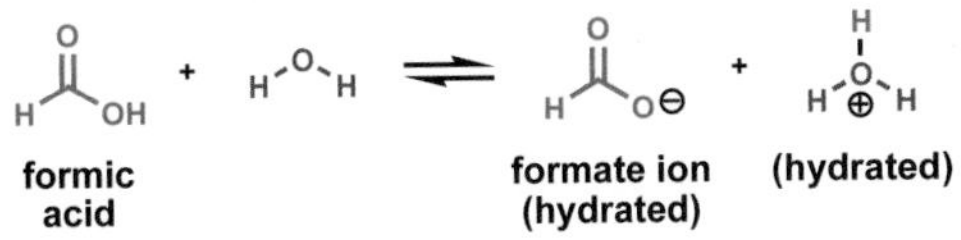

**formic acid**    **formate ion (hydrated)**    **(hydrated)**

**Equation 4.** Dissociation of formic acid in water.

The reaction is rapid (submillisecond time scale) and reversible, i.e., the components are in very fast equilibrium with one another. One reason for the stability of formate ion is that the negative charge is spread equally between the two oxygens. The carbon is $sp^2$ hybridized and there are two delocalized π-bonding MOs as expressed in formulas **A**, **B** and **C** in Figure 32.

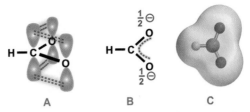

**Figure 32.** Three different representations of the delocalization of electrons in formate ion. In formula **A**, atomic orbitals combine to form two 3-center bonding MOs. In formula **B**, the delocalization over O-C-O is illustrated with dotted lines and the oxygen atoms share the negative charge equally. Formula **C** is the computer-generated electron density map of formate ion in which red indicates high electron density.

Delocalization of electrons or charge in organic molecules is, in general, strongly stabilizing because it leads to a lower energy structure than the hypothetical electron-localized version(s).

Formic acid is the acidic ingredient that causes the immediate sting in the bite of a bee or hornet. It can be neutralized by a base such as ammonia ($NH_3$) or sodium hydroxide (NaOH) to form formate salts. (*See* Figure 33.)

**ammonium formate**         **sodium formate**

**Figure 33.** Two simple salts of formic acid, ammonium and sodium formate.

## Carboxylic Acids. Part II. General Reactions and Derivatives

The conversion of a carboxylic acid to a carboxylate salt by treatment with a base is a general property of this functional group class. As might be expected, the salts are generally much more soluble in water than the corresponding carboxylic acids. Solutions of carboxylic acids in water are acidic, just as for formic acid, although the degree of acidity varies from one compound to another. Acetic acid, $CH_3COOH$, is less acidic than formic acid. Trifluoroacetic acid, $CF_3COOH$, is a much stronger acid than either formic or acetic acid. The reason for this is that fluorine is powerfully electron attracting, and considerable negative charge is delocalized to the three fluorines in $CF_3COO^-$, conferring extra stabilization. Figure 34 shows the electron density maps for acetate, formate and trifluoroacetate.

15

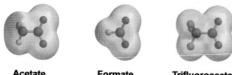

**Acetate**     **Formate**     **Trifluoroacetate**

**Figure 34.** Electron density maps for acetate, formate and trifluoroacetate anions, displayed in order of increasing stability.

The subfragment RCO of carboxylic acids is called an *acyl group* (the R of RCO can be any carbon group, Figure 35).

**carboxylic acid**     **carboxylic acid derivative**

**Figure 35.** Formation of carboxylic acid derivatives. The acyl group is highlighted with the yellow box.

There are many reactions of carboxylic acids that form compounds of structure RCOX which are called carboxylic acid derivatives. Some examples are shown in Figure 36.

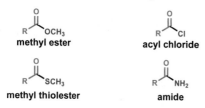

**methyl ester**     **acyl chloride**

**methyl thiolester**     **amide**

**Figure 36.** Examples of simple carboxylic acid derivatives.

These compounds can all be made by standard reactions from carboxylic acids (RCOOH).

## Polyfunctional Carboxylic Acids

The essentially infinite diversity of organic compounds is made possible in part because the various functional groups can occur together in all possible numbers and combinations. Several important specific examples are shown in Figures 37a and 37b.

| COMPOUND | DESCRIPTION |
|---|---|
| 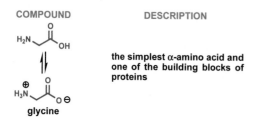 | the simplest α-amino acid and one of the building blocks of proteins |

**glycine**

**Figure 37a.** Polyfunctional carboxylic acids.

| COMPOUND | DESCRIPTION |
|---|---|
| 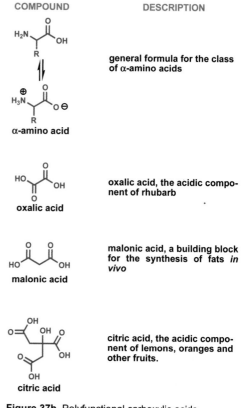 | |

general formula for the class of α-amino acids

**α-amino acid**

oxalic acid, the acidic component of rhubarb

**oxalic acid**

malonic acid, a building block for the synthesis of fats *in vivo*

**malonic acid**

citric acid, the acidic component of lemons, oranges and other fruits.

**citric acid**

**Figure 37b.** Polyfunctional carboxylic acids.

## Sulfur and Phosphorus Acids and their Derivatives

Sulfuric acid is a strong acid that dissolves in water to form the hydrated negative ions (anions) bisulfate and sulfate.

**sulfuric acid**     **bisulfate ion**     **sulfate ion**

**Equation 5.** Dissociation of sulfuric acid in water.

Similarly, sulfonic acids, such as methanesulfonic acid, ionize completely in water and form salts with bases.

**methanesulfonic acid**     **sodium methanesulfonate**

**Equation 6.** Neutralization of methanesulfonic acid with aqueous sodium hydroxide to form sodium methanesulfonate.

Long chain sulfonate salts such as $C_{18}H_{37}SO_3^-Na^+$ (sodium octadecane sulfonate, *see* Figure 38) are important detergents (e.g., Tide) that interact by van der Waals forces to attract greasy deposits and carry them into a water wash because of the strong aqueous solvation of the sulfonate ion.

**Figure 38.** Line drawing and space-filling representation of sodium octadecane sulfonate ($C_{18}H_{37}SO_3Na$).

The negative charge in bisulfate, sulfate and sulfonate ions is spread out over oxygens in delocalized and highly stabilized molecular orbitals.

**Bisulfate ion**  **Sulfate ion**  **Methanesulfonate ion**

**Figure 39.** Computer-generated electron density maps of bisulfate, sulfate and methanesulfonate ions.

The behavior of phosphoric acid is similar. Phosphoric acid reacts with sodium hydroxide in water to form soluble mono-, di- or trisodium salts.

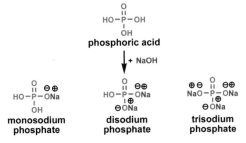

**Equation 7.** Neutralization of phosphoric acid with sodium hydroxide to form phosphate salts.

Phosphate esters can be formed from the attachment of phosphate to a hydroxyl group in an organic molecule (e.g., alcohols), as shown in the examples that follow.

**diethylphosphate**       **phosphoglyceric acid**

**Figure 40.** Simple esters of phosphoric acid.

Phosphorylation of hydroxyl groups in proteins is important in living organisms as a key reaction in chemical signaling.

**Benzene, Structure and Stabilization by $\pi$-Delocalization**

Benzene is a remarkably stable hydrocarbon of formula $C_6H_6$. The six carbons are held together in a planar hexagonal arrangement as shown in Figure 41. That geometry corresponds to $sp^2$ hybridization of the carbons in the ring.

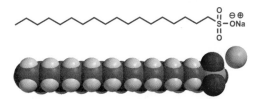

| A | B | C |
|---|---|---|
| $\pi$-bonds localized | $\pi$-bonds delocalized | space-filling model |

**Figure 41.** Structure of benzene.

Although analogy with ethylene (*see* page 8) might suggest structure **A** for benzene, it has been shown experimentally that the six C-C bonds in benzene are equal in length and that it is a delocalized $\pi$-electron structure. Quantum chemical calculations show that the six electrons not used for C-H and C-C $\sigma$-bonding lie in three bonding orbitals formed by the overlap of the $p$ AOs on each carbon. A MO energy diagram is shown in Figure 42.

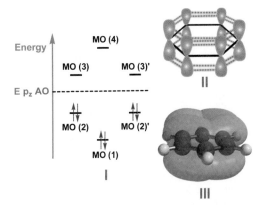

**Figure 42.** I: Energy levels of the six $\pi$-MOs of benzene that result from the combination of the six $2p_z$ atomic orbitals of the six carbons in the ring. II: An MO picture of the $\pi$-electron clouds above and below the benzene ring. III: A picture of the $\pi$-electron clouds above and below the benzene ring.

Each carbon contributes a p-AO to form six π-MOs. Three of these will be of lower energy than the original $p_z$ atomic orbitals and three will be higher (*see* diagram **I.** in Figure 42). There are six electrons available for π-bonding. Since these will be accommodated in the lower energy orbitals MO (1) and MO (2) and MO (2'), the resulting π-delocalized structure will be stabilized. In fact, simple calculations show that the delocalized structure **B** is more stable than the localized three C=C structure **A** (*see* Figure 41) by about 35 kcal/mol.

Despite the fact that structure **A** is less realistic than structure **B** in Figure 41, it is commonplace to draw the structure of benzene and related compounds with alternating double bonds. The structures of benzenoid compounds in the later sections of this book are drawn with the traditional double/single bond notation.

### Chemical Consequences of the Stability of the Benzene Ring

The delocalization of the six π-electrons of benzene over the whole ring in low-energy orbitals bestows special properties and huge importance to this structural unit. Typically the C=C unit tends to undergo reactions in which groups add to the two carbons of the double bond. However, benzene generally reacts with these same reagents differently. For instance, 1-pentene reacts with chlorine gas ($Cl_2$) to form 1,2-dichloropentane, but benzene reacts to form chlorobenzene and HCl as shown in Figure 43.

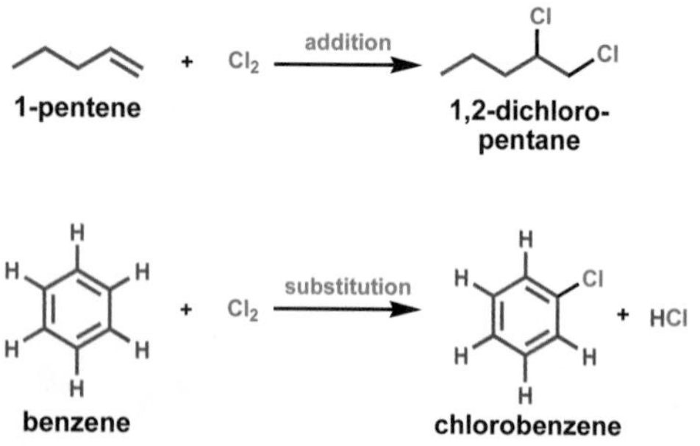

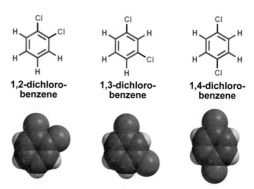

**Figure 43.** Reaction of 1-pentene and benzene with chlorine gas ($Cl_2$). In the case of benzene one of the hydrogen atoms (H) is replaced with a chlorine (Cl) atom – a process which is known as substitution.

Most reactions of benzene are of the substitution type (replacement of atoms) with retention of the very stable π-system, whereas

a double bond in a typical non-benzenoid hydrocarbon undergoes addition reactions.

There are innumerable chemical reactions that replace the hydrogens of benzene by some other groups, and countless thousands of substituted benzene compounds can be made (synthesized). The six carbons of the benzene ring allow different positioning of groups. For instance, there are three distinctly different dichlorobenzenes, as indicated in Figure 44.

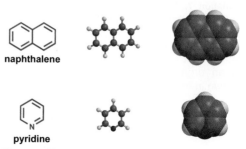

1,2-dichloro-benzene    1,3-dichloro-benzene    1,4-dichloro-benzene

**Figure 44.** Three distinctly different position isomers of dichlorobenzene.

These compounds, called *position isomers*, have the same molecular formula ($C_6H_4Cl_2$) but different physical and chemical properties.

### More Complex Benzenoid Compounds. Heterobenzenoid Ring Systems.

A closed circuit of six π-electrons confers stabilization of the ring systems of many compounds other than benzene. The rings can be a different size or contain non-carbon atoms, especially O, N and S. One or more such rings can be connected to one another. Some examples of important ring systems are shown in Figures 45a and 45b.

naphthalene

pyridine

**Figure 45a.** Important benzenoid and hetero-benzenoid ring systems.

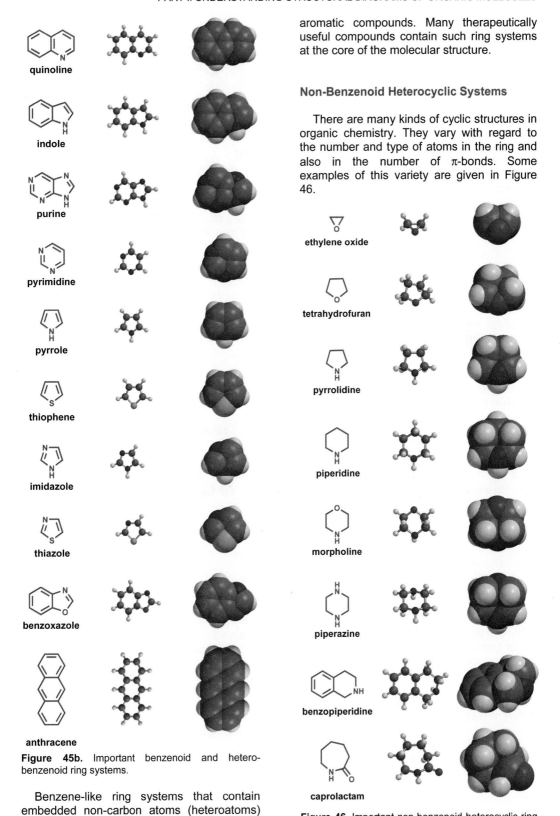

quinoline

indole

purine

pyrimidine

pyrrole

thiophene

imidazole

thiazole

benzoxazole

anthracene

**Figure 45b.** Important benzenoid and hetero-benzenoid ring systems.

Benzene-like ring systems that contain embedded non-carbon atoms (heteroatoms) are termed heterobenzenoid or hetero-aromatic compounds. Many therapeutically useful compounds contain such ring systems at the core of the molecular structure.

## Non-Benzenoid Heterocyclic Systems

There are many kinds of cyclic structures in organic chemistry. They vary with regard to the number and type of atoms in the ring and also in the number of π-bonds. Some examples of this variety are given in Figure 46.

ethylene oxide

tetrahydrofuran

pyrrolidine

piperidine

morpholine

piperazine

benzopiperidine

caprolactam

**Figure 46.** Important non-benzenoid heterocyclic ring systems.

## A Useful Notation for the Representation of Structures with Delocalized π-Electrons

π-Electron delocalization is commonplace in organic structures and generally results in greater stability (i.e., lower energy content) than in the corresponding localized forms. The stabilization of carboxylate ions (*see* page 15) is an example. Another way of representing a delocalized carboxylate ion is to draw two line structures which depict the hypothetical localized forms with a double-headed arrow connecting them to indicate that the *true* structure is a hybrid of the extremes. This method can be illustrated with carboxylate ion (Figure 47a).

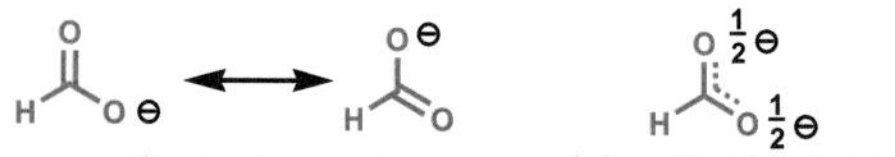

**two extremes**          **delocalized hybrid structure**

**Figure 47a.** Hypothetical electron-localized structures of formate ion shown along with the delocalized hybrid structure.

Similarly, we can represent the delocalized forms of sulfate ion, benzene or formamide as shown in Figure 47b.

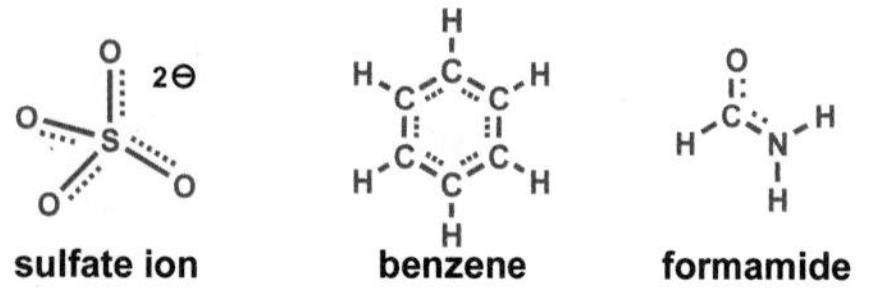

**sulfate ion**     **benzene**     **formamide**

**Figure 47b.** Representation of electron-delocalized structures by line drawings.

Delocalized formulas provide useful information with regard to charge distribution, bond lengths or molecular shape. For instance, formamide can be expected to be a planar molecule because of the partial π-bond between C and N. Also, rotation about the C-N bond is greatly retarded, since such rotation would break the C-N π-bond and require an energy input of about 17 kcal/mol.

## Structures with the Same Connectivity but Different Geometry – Geometrical Isomers

Compounds with the same atom connectivity but different geometry are important in chemistry and medicine. Such compounds are called geometrical isomers. A simple example, the *cis*- and *trans*-1,3-dimethylcyclobutanes, is shown in Figure 48.

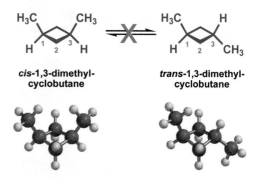

*cis*-**1,3-dimethyl-cyclobutane**          *trans*-**1,3-dimethyl-cyclobutane**

**Figure 48.** *Cis*- and *trans*-isomeric cyclic compounds. Representation of two geometrical isomers of 1,3-dimethylcyclobutane.

This kind of geometrical isomerism is very common in compounds containing double bonds. Thus, there are two different compounds with the 2-butene structure. These isomers, called *cis*- and *trans*-2-butene, are depicted in Figure 49.

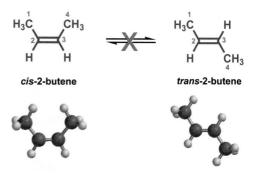

*cis*-**2-butene**          *trans*-**2-butene**

**Figure 49.** *Cis*- and *trans*-isomeric acyclic compounds. Representation of two geometrical 2-butene isomers.

Each of these isomers is stable, even upon heating, and their properties are distinctly different. Their existence as separate compounds depends on the fact that no rotation is possible about the 2,3-double bond, because that necessitates destroying the π-bonding which requires that the p-orbitals be parallel to one another. It would cost about 60 kcal/mol to effect such a rotation about a C=C double bond, a prohibitively high value.

In contrast, there is a low barrier to rotation about C-C single bonds because these correspond to an axially symmetrical σ-MO. That barrier is just 3 kcal/mol for ethane, and so ethane exists in just one form, even though different three-dimensional geometries are possible, as shown in Figure 50.

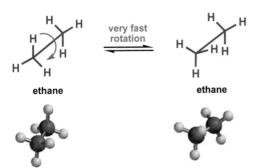

**Figure 50.** Since the rotation about the C-C axially symmetric σ-bond of ethane is fast, it is just a single compound.

### Another Kind of Isomerism – Chirality Isomerism (Stereoisomerism)

The fact that the geometry of organic molecules is rigidly controlled by the preferred bond lengths and angles has numerous and profound consequences. One of these is that life, as we know it, would not be possible without such constraints. There is another, and much more subtle reality, that comes from the tetrahedral bond angle preference for tetracoordinate carbon. If such a carbon has *four* different groups attached to it, there can be *two* different arrangements of the groups in space, corresponding to *two* different compounds. The relationship between these is analogous to that between a person's right and left-handed gloves which are not superimposable, but mirror images of one another (Figure 51).

**Figure 51.** Non-superimposable objects. Mirror images of gloves and corkscrews. Mirror planes are indicated as gray lines.

The same would hold for two corkscrews that are identical except for having right or left-handed threads, or two cars that have the steering wheel on the left-side looking forward or the right-side. You would be very surprised to enter your car and find the steering wheel on the side opposite to the usual one, and you would know that the two cars are different.

A hypothetical molecule having *four* different groups on carbon is shown in both possible forms in Figure 52.

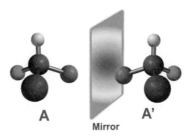

**Figure 52.** Two molecules having four different groups attached to the central carbon atom which differ only in the spatial arrangement of those groups. The two molecules, **A** and **A'**, are mirror images of each other.

The two possible mirror image forms of the α-amino acid alanine, are shown in Figure 53.

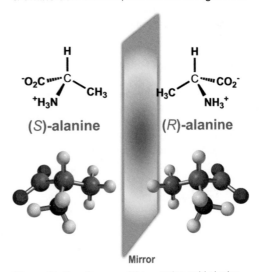

**Figure 53.** Enantiomers of the α-amino acid alanine.

The two structures **A** and **A'** in Figure 52 are symmetrical about a mirror plane, but they cannot be superimposed and, therefore, must be different (*see* Figure 54). The same is true for the α-amino acids shown in Figure 53. These amino acids are called (*S*)- and (*R*)-alanine to distinguish them; they are isomers that have different biological and biochemical properties. One, the (*S*)-isomer is naturally occurring, and is one of the twenty amino acids coded for by DNA and used as a building block for proteins. The other isomer (*R*) is a different compound that cannot be substituted for natural (*S*)-alanine in living systems.

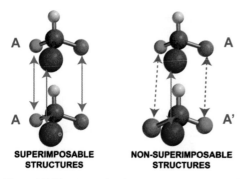

**SUPERIMPOSABLE STRUCTURES**　　**NON-SUPERIMPOSABLE STRUCTURES**

**Figure 54.** The two molecular structures on the left (A and A) are superimposable, whereas the two structures on the right (A and A') are not.

Any two isomers of different "handedness" or *chirality* are generally called "enantiomers". Such molecules do not have a plane or point of symmetry. Many important therapeutic agents and naturally occurring compounds possess chirality. The chirality can involve one or more $sp^3$ centers of chirality.

## Conformations of Molecules

The geometry of the ring in cyclohexane is unlike that of planar benzene. Planar cyclohexane would involve the vertex angles of the regular hexagon of 120° rather than the preferred tetrahedral angle of 109.5°.

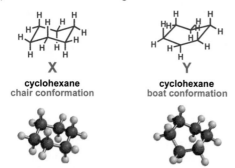

**X**　　　　　　　　**Y**
**cyclohexane**　　　**cyclohexane**
chair conformation　boat conformation

**Figure 55.** The chair (X) and boat (Y) conformations of cyclohexane. The chair conformation is significantly more stable than the boat conformation at room temperature.

The distortion from 109.5° to 120° is so destabilizing that the ring distorts to a non-planar, puckered form with vertex angles of 109.5° (structures X and Y, Figure 55).

Structure X is called a *chair form* (*chair conformation*) to distinguish it from another nonplanar structure (Y) that also has 109.5° vertex angles, but is less stable (by about 6 kcal/mol). This energy difference is so large

that more than 99.99% of cyclohexane molecules have a chair form at room temperature. The structures X and Y are termed *conformers* to indicate that the molecules have different *shapes*.

## The Importance of Molecular Shape

The shape and size of a molecule are very important in determining its affinity for a biomolecular target. In general, the binding between the two increases with increasing area of van der Waals contact attraction and also with the degree of hydrogen bonding. The better the molecular fit, the stronger is the affinity. It is also true that a therapeutic agent in aqueous solution will be stabilized by hydrogen bonding to water and dipolar solvation. Thus, there is a trade-off for therapeutic agents: they must be sufficiently well solvated to be soluble in water, but not so strongly solvated that they cannot be pulled from solution by the target biomolecule.

Noncyclic organic molecules tend to be quite flexible because of the low energy barrier to rotation about single bonds. For this reason, most therapeutically useful structures possess cyclic subunits with a modest number of preferred conformations, or even just one. The conformation of prednisone, a very important anti-inflammatory and immuno-suppressive drug is shown in Figure 56, which also shows the preferred conformation of glucose.

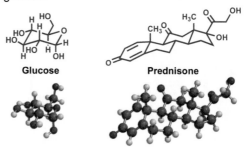

**Glucose**　　　　　　　　**Prednisone**

**Figure 56.** The preferred conformations of glucose and prednisone.

The polycyclic framework of prednisone is quite rigid and gives the molecule a characteristic shape. At the same time, a number of polar functional groups are positioned at specific sites in space so that they can bind optimally to the target biomolecule.

1. Smith, J. G. Organic Chemistry (McGraw-Hill, 2nd Edition, **2007**); 2. Atkins, P. Atkins' Molecules (Cambridge University Press, 2nd Edition, **2003**).

# SOME COMMON MOLECULES

There are many small non-medicinal compounds encountered in everyday life that are required by the body. A sampling of these is shown on this and the following page in the space-filling format.

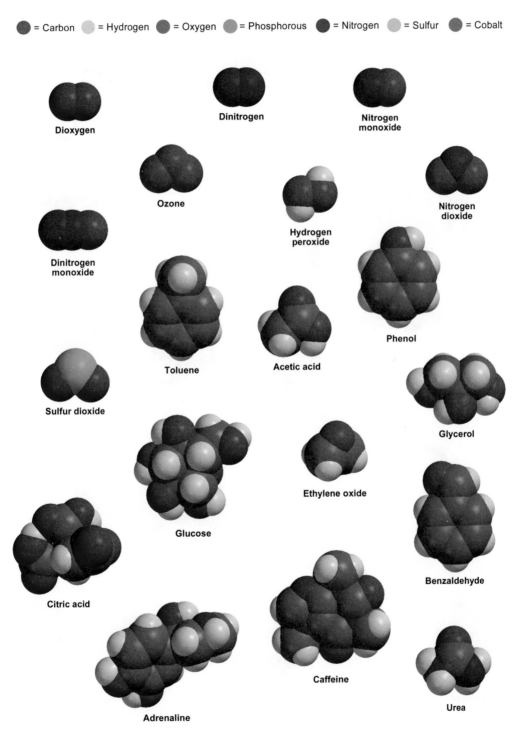

● = Carbon  ○ = Hydrogen  ● = Oxygen  ● = Phosphorous  ● = Nitrogen  ○ = Sulfur  ● = Cobalt

Dioxygen

Dinitrogen

Nitrogen monoxide

Ozone

Nitrogen dioxide

Dinitrogen monoxide

Hydrogen peroxide

Phenol

Toluene

Acetic acid

Sulfur dioxide

Glycerol

Glucose

Ethylene oxide

Benzaldehyde

Citric acid

Caffeine

Adrenaline

Urea

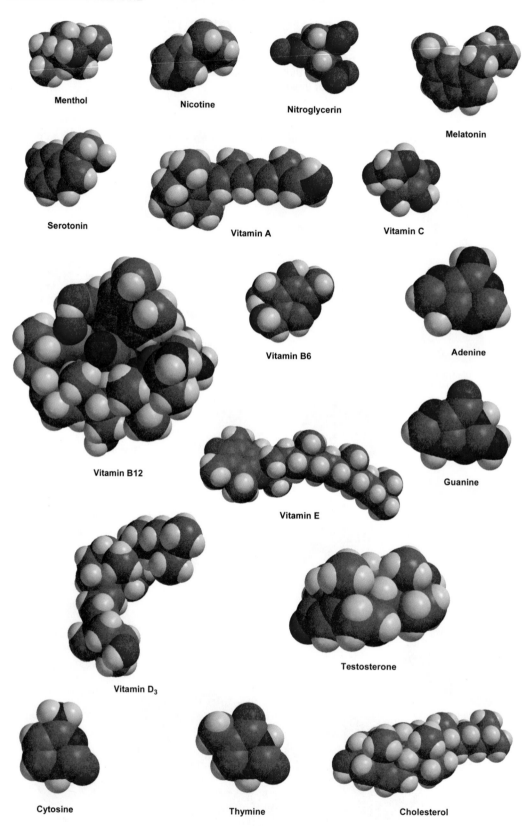

Menthol

Nicotine

Nitroglycerin

Melatonin

Serotonin

Vitamin A

Vitamin C

Vitamin B12

Vitamin B6

Adenine

Vitamin E

Guanine

Vitamin D$_3$

Testosterone

Cytosine

Thymine

Cholesterol

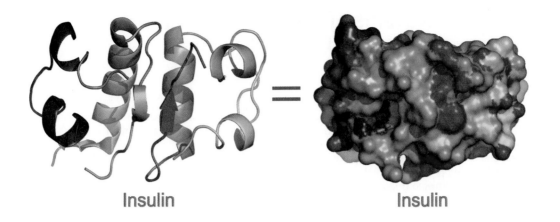

Insulin = Insulin

# THREE-DIMENSIONAL PROTEIN STRUCTURES

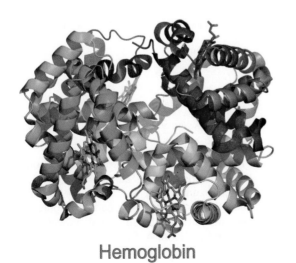

Hemoglobin

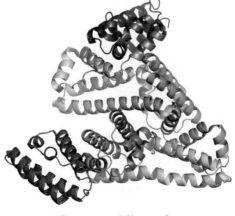

Serum Albumin

# PROTEINS AND THREE-DIMENSIONAL PROTEIN STRUCTURES

## Introduction

Proteins are very large molecules formed by the joining of many α-amino acid subunits through amide linkages. A subsection of a protein chain can be represented in a general- and highly simplified way by a two-dimensional line drawing (Figure 1).

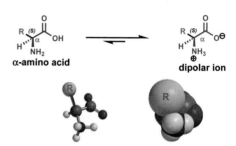

**Figure 1.** Portion of a protein/polypeptide chain.

Proteins are essential to life and perform innumerable critical functions. For instance, they serve as catalysts for the formation of the molecules of life (biosynthesis), for the disassembly and recycling of such molecules and for energy production. They act as signaling molecules to transfer information and as regulators of cell function or division. They are key structural components in all living organisms.

Proteins are generally large molecules of sizes ranging from a hundred to several thousand α-amino acid subunits. Smaller collections of linked α-amino acids (10-100) are called peptides. Proteins are biosynthesized from twenty different α-amino acid building blocks by an elaborate and rigorously controlled process that is guided and directed by the genetic material deoxyribonucleic acid (DNA), its offspring ribonucleic acid (RNA) and other proteins. The exact sequence of the various α-amino acids along the protein chain is determined by the sequences of purine and pyrimidine bases in DNA and the genetic code. An almost infinite number of different protein molecules are possible from the 20 different amino acids by variations of α-amino acid sequence and protein size.

## The Proteinogenic Amino Acids

The twenty α-amino acids that are the building blocks for proteins differ in the carbon group attached to the α-carbon (Figure 2). They all have the configuration at the α-car-

bon atom which is shown in Figure 2, and *not* the mirror image arrangement.

One group of α-amino acids consists of those (beyond glycine) in which **R** contains tetracoordinate carbons and *no* functional groups. These α-amino acids are shown in Figure 3.

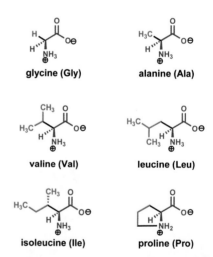

**Figure 2.** General structure of α-amino acids. These compounds exist predominantly in the form of dipolar ions.

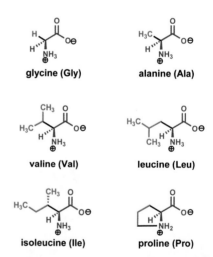

glycine (Gly)    alanine (Ala)

valine (Val)    leucine (Leu)

isoleucine (Ile)    proline (Pro)

**Figure 3.** α-Amino acids with hydrophobic (nonpolar) side chains.

Four of the 20 fundamental proteinogenic α-amino acids possess either a benzenoid or heterobenzenoid ring as part of the variable group **R** attached to the α-carbon (Figure 2). The structures of these α-amino acids are displayed in Figure 4. Each of these α-amino acids plays a unique role in living systems. For example, the hydroxyl group on the benzenoid ring of tyrosine can be transformed into a phosphate ester, and if this happens, the three-dimensional shape (conformation) of a protein may change drastically. This phenomenon is one of the biochemical ways in which cells in living systems receive information from one another and from their environment. In addition, the five-membered imidazole ring of histidine has basic properties because of the nitrogen atoms in it and is extremely important in the catalysis of biochemical reactions by enzymes, a key family of proteins.

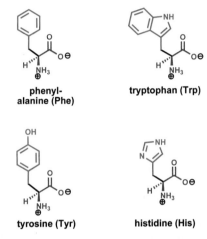

**Figure 4.** Amino acids with aromatic side chains.

The rings of phenylalanine and tryptophan are nonpolar and oily (hydrophobic). They serve several critical functions. Because of their nonpolar nature, size and shape they can help stabilize a particular protein conformation by van der Waals attraction with themselves, each other or other nonpolar side chains. Such an interaction is shown in Figure 5.

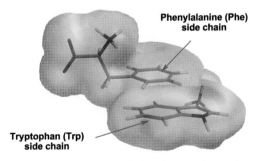

**Figure 5.** An example of attractive π-π interaction between the aromatic rings of phenylalanine and tryptophan.

Hydrophobic groups such as those in Phe, Trp, Leu and Ile tend to avoid the interface of proteins with water, i.e., the outside of a folded protein structure, and to be buried in the interior, where there is little or no water and where they pack close to one another.

Four of the proteinogenic amino acids contain hydroxyl or dicoordinate sulfur functional groups, as displayed in Figure 6. The hydroxyl groups of serine and threonine can be phosphorylated, as described above for tyrosine, resulting in a change of protein shape, properties and function. The remaining six α-amino acids are displayed in Figure 7.

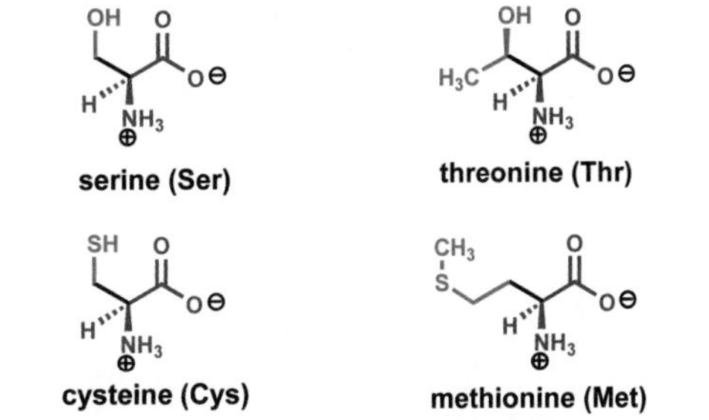

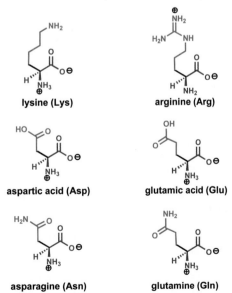

**Figure 6.** Amino acids containing hydroxyl or dicoordinate sulfur functional groups.

The α-amino acids in Figure 7 all contain polar side chains and play a very critical role in enzymes and biochemical catalysis, and also in determining the three-dimensional structure of proteins. Since the side chain of arginine is generally positively charged, it is capable of donating a proton in hydrogen bonding, or catalysis. The amino group of lysine is also uniquely reactive in a chemical sense and crucial to protein function.

**Figure 7.** Hydrophilic amino acids with polar side chains.

## The Simplest Peptide

The dipeptide glycylglycine consists of two glycine units connected through an amide (peptide) bond. The amide functional group involves electron delocalization over three atoms as shown in Figure 8 (*see* also page 15). The bonds to the amide subunit lie in one plane, as imposed by the π-bond between C and N, and the C-N bond distance is shortened (from 1.46 Å to 1.32 Å).

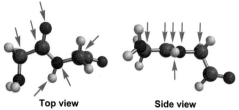

X            glycylglycine            Y

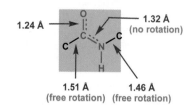

Top view            Side view

1.24 Å → 
1.32 Å (no rotation)

1.51 Å (free rotation)    1.46 Å (free rotation)

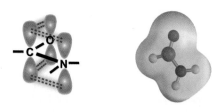

**Figure 8.** Glycylglycine and the structure of the peptide bond. The atoms that are part of the planar amide subunit are indicated with red arrows in the ball-and-stick models.

The planar amide linkage is a critical organizing element of protein structure because of its rigidity. In contrast, rotation about the bonds attaching carbon to it is rapid and lends flexibility in protein structures.

Figure 9 displays a general formula for a tripeptide and the formula for the specific tripeptide, Ala-Gly-Phe.

N-terminus            C-terminus

**generic tripeptide**

**Ala-Gly-Phe**
**(alanyl-glycinyl-phenylalanine)**

**Figure 9.** Structures of a generic tripeptide and of the specific tripeptide Ala-Gly-Phe.

## Amino Acid Sequence (Primary Structure) of Proteins

The sequence of the individual amino acids in a polypeptide or protein, the *primary structure*, is critical to its preferred three-dimensional shape and properties. It is the very large number of possible sequences that enables the existence of an unimaginably large number of possible proteins ($100^{20}$ for a 100 amino acid protein). There are six (3x2x1, 3!) possible primary structures for the tripeptides containing a single Ala, Gly and Phe, as shown in Figure 10.

**Ala-Gly-Phe**            **Ala-Phe-Gly**

**Gly-Phe-Ala**            **Gly-Ala-Phe**

**Phe-Ala-Gly**            **Phe-Gly-Ala**

**Figure 10.** Six tripeptides may be generated using only three amino acids, Ala, Gly and Phe.

## Preferred Conformations (Secondary Structure) of Proteins: The α-Helix, β-Turn and β-Strand as Motifs

The most common structural motif found in proteins is the α-helix. Figure 11 shows an α-helical polypeptide with 30 amino acid residues in various representations. In structure **A** the elements in an α-helix are shown, with carbon as gray, oxygen as red, nitrogen as blue and sulfur as yellow. Structure **B** displays the backbone of the polypeptide without any of the side chains (for clarity). Structure **C** illustrates the polypeptide chain with stabilization by hydrogen bonds between the carbonyl groups (C=O) and amino groups (N-H).

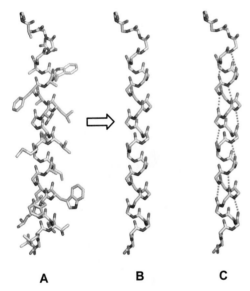

**A**          **B**          **C**

**Figure 11.** Three different representations of an α-helical polypeptide with 30 amino acid residues. Structure **A** displays the entire polypeptide with the side chains. Structure **B** shows only the backbone of the polypeptide. Structure **C** illustrates that the α-helix is held together by a network of hydrogen bonds.

For clarity, peptides and proteins are often represented by a ribbon diagram in which side chains are omitted and the backbone is replaced by a ribbon. Figure 12 shows how these two representations are related to the structure of the polypeptide.

The α-helices in proteins are all right-handed, with the backbone of the polypeptide chain receding from view in a clockwise fashion (Figure 13). The α-helical structure is compact and highly stabilized by hydrogen bonding between the amide C=O and N-H groups.

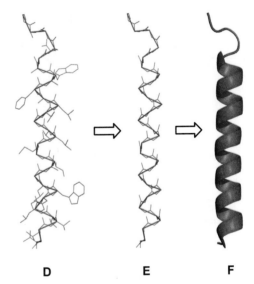

**D**          **E**          **F**

**Figure 12.** Generation of a ribbon diagram of a polypeptide. Structure **D** shows the entire polypeptide with the backbone. Structure **E** shows the backbone and its tracing. Structure **F** shows only the ribbon diagram of the original polypeptide.

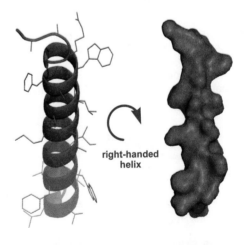

right-handed helix

**Side-view cartoon diagram**          **Side-view space-filling diagram**

**Figure 13.** An α-helix in ribbon and space-filling representations.

α-Helices can form bundles when the aggregate is stabilized by attractive interactions between neighboring side chains. An important example of such aggregation is the G protein-coupled receptor, rhodopsin, which has seven α-helical domains (Figure

14). Each helix is shown in a different color (*see* page 78 for more on G proteins).

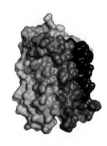

**Figure 14.** Ribbon and space-filling diagrams of the G protein-coupled receptor rhodopsin. The seven α-helical domains (different colors) occur embedded in cellular membranes.

The other structural motif that is abundant in proteins is the β-sheet which is formed by the stabilizing hydrogen bonds between the backbones of two β-strands (Figure 15).

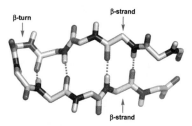

**Figure 15.** β-Sheet stabilized by hydrogen bonding between two β-strands connected by a β-turn.

β-Sheets occur in proteins both in an antiparallel arrangement, as in Figures 15 and 16, and in a parallel sense.

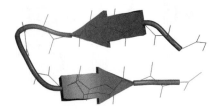

**Figure 16.** Simplified representation of a β-sheet. The arrows point into the direction of the C-terminus of the polypeptide chain.

### The Structure of Insulin

One of the simplest and most important protein-type structures is the hormone insulin. Produced in the islet cells of the pancreas, it is essential for life and health and has many different actions in the body. It is a fundamental regulator of energy production, metabolism and muscle function. Insulin is composed of two chains of 21 and 30 amino acids, held together by two disulfide bonds (S-S) formed from the pairing of S-H groups of four of its six cysteines. The other two, which are located in the smaller chain, form another disulfide bond, giving a tricyclic structure. The X-ray crystal structure of insulin, determined by Dorothy Hodgkin and her team in 1969, revealed a molecule with three short α-helices, which are indicated in the shorthand ribbon diagram of Figure 17. Many of the nonpolar (hydrophobic) side chains of insulin are buried in the interior region. There are also hydrophobic groups outside the core in a non-polar surface which are involved in the binding of insulin to its receptor. It is the monomeric (single molecule) form of insulin that circulates in the body and activates the insulin receptor to produce the vital biological responses.

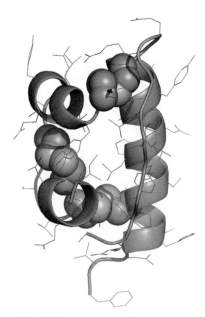

**Figure 17.** Ribbon representation of monomeric insulin. The α-helices are shown in green and the three disulfide (S-S) bridges are represented as yellow spheres.

An example of a larger and more complicated protein, TolC, is shown in Figure 18. TolC is a bacterial protein that plays a role in bacterial resistance to antibiotics. The TolC protein combines with two other proteins (MexA and AcrB) to form a giant assembly (or protein machine) to pump antibiotics or other molecules harmful to a bacterium out of the organism (*see* page 143 for more on drug resistance).

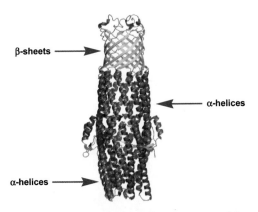

β-sheets

α-helices

α-helices

**Figure 18.** The protein TolC is a component of the bacterial drug efflux pump and has several α-helical domains (red) as well as a number of β-sheets (yellow).

## Some Other Aspects of Protein Structure

There are an *enormous number* of possible conformations for a protein with a particular amino acid sequence because there can be fast rotation about the N-C$_\alpha$ and C$_\alpha$-CO single bonds of each amino acid subunit. However, most naturally occurring proteins exist in one, or a small number of preferred conformations. The reason for this is that there are many factors that operate to favor a particular three-dimensional geometry. Among the most important determinants of protein shape are the following:

(1) Sulfur-sulfur linkages may form between the SH groups of two cysteine subunits in the protein. These disulfide (S-S) bridges generate a ring that greatly limits molecular shape.

(2) The interior of proteins is generally rich in hydrophobic groups. Polar or charged side chains tend to be exposed to water or on the surface of the protein.

(3) Organized substructural motifs, such as the α-helix, β-sheet (both with parallel and antiparallel β-strands) organize and favor a particular conformation for large subsections of the protein.

(4) These organized substructural motifs can adhere to one another through attractive interactions (H-bonding, electrostatic, van der Waals) leading to the favoring of a particular giant domain structure. For instance, attractions between α-helical subunits can result in the formation of α-helical bundles, such as the 7-helix bundle of rhodopsin.

(5) Repulsions between groups set in when they are closer than the optimal van der Waals contact distance. These repulsions, known as steric repulsions, occur basically because atoms behave as very hard objects once they are within a certain critical distance of one another. Although close packing is essential for intramolecular attraction of protein subsections, it is also limited by steric repulsion.

(6) Substructural motifs such as α-helices or β-sheets can be *connected* by short flexible loops or β-turns.

(7) α-Helices have polarity because the C=O groups all point in one direction along the helical axis, and they usually pack together in an antiparallel way.

## Determination of the Structures of Proteins

The three-dimensional structures of proteins are most generally elucidated by the use of X-ray diffraction analysis using high-speed computers to decode the data on X-ray beam scattering by a single crystal of the protein. Since the first determination of a protein structure (hemoglobin by John Kendrew and Max Perutz in 1962), thousands of protein structures have been determined. The number of new known protein structures has been growing sharply year by year thanks to the convergence of several key circumstances.

(1) Modern molecular biology and protein science provide access to adequate quantities of pure protein.

(2) Improved techniques for the crystallization of proteins are available.

(3) The use of high-intensity, narrow-beam X-ray sources coupled to advances in analytical software and the ever increasing speed of computers has greatly simplified the determination of protein structure at atomic resolution.

The structure of a crystalline protein can now be determined in just a few weeks. The availability of a three-dimensional protein structure greatly assists in the discovery of molecules that can bind to the protein and affect its structure. Structure-guided design of therapeutic molecules is now a crucial tool for molecular medicine.

1. Whitford, D. Proteins: *Structure and Function*, Hoboken: John Wiley & Sons; **2005**.

# SOME OF THE PROTEIN STRUCTURES
# THAT APPEAR IN THIS BOOK

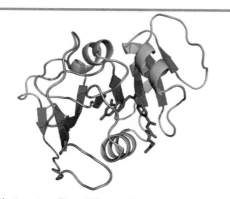

**Methotrexate** (Trexall™), used to treat rheumatoid arthritis, is shown in red bound in the active site of its target, the enzyme dihydrofolate reductase (*see* page 46; PDB ID: **1RG7**). The image shows the full protein. The α-helical domains are colored cyan, the β-sheets are magenta and the loops are orange.

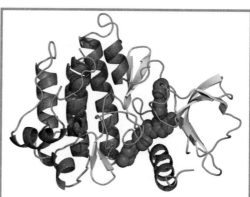

**Imatinib** (Gleevec™) used to treat leukemia, is shown in magenta bound in the active site of its target, the enzyme tyrosine kinase (*see* page 195; PDB ID: **1IEP**). The image shows the full protein. The α-helical domains are colored red, the β-sheets are yellow and the loops are green.

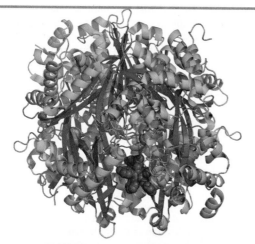

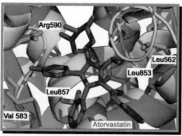

**Atorvastatin** (Lipitor™), used for the reduction of LDL cholesterol levels, is shown in red bound in the active site of its target, the enzyme HMG-CoA reductase (*see* page 64; PDB ID: **1HWK**). The top image shows the whole enzyme, whereas the bottom image is a close-up view.

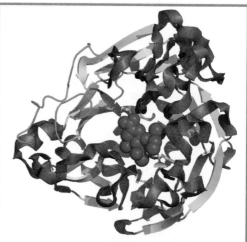

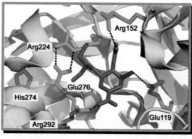

**Oseltamivir** (Tamiflu™), used to prevent influenza A and B viral infections, is shown in magenta bound in the active site of its target, the viral enzyme neuramidinase (*see* page 150; PDB ID: **2HT8**). The top image shows the whole enzyme whereas the bottom image is a close-up view in which oseltamivir is colored red.

# PART II.

# INFLAMMATORY, CARDIOVASCULAR AND METABOLIC DISEASES

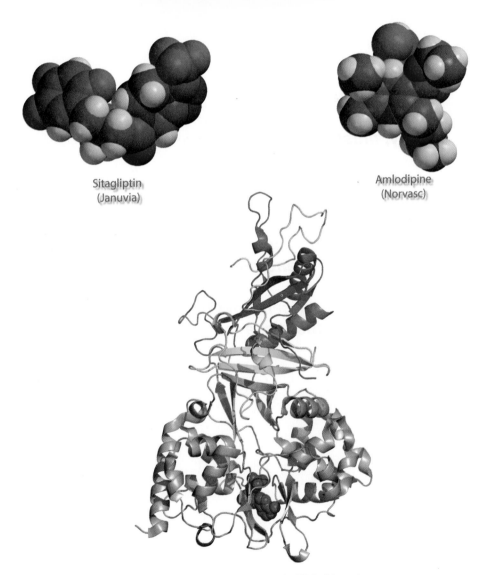

Sitagliptin
(Januvia)

Amlodipine
(Norvasc)

AMP-Activated Protein Kinase with AMP (red) bound
in the catalytic subunit

*Inflammation, a biological response needed in the protection or repair of the body, must be exquisitely regulated.*

*Disregulation of the process, either chronic or acute, plays a role in a wide variety of disease states from cardiovascular disease to arthritis to psoriasis to degenerative illnesses.*

*Inflammation can now be alleviated by various therapeutic agents, but none of these is ideal.*

# ANTI-INFLAMMATORY AGENTS

# ACETYLSALICYLIC ACID (ASPIRIN™)

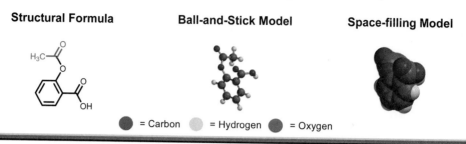

| Structural Formula | Ball-and-Stick Model | Space-filling Model |

● = Carbon  ○ = Hydrogen  ● = Oxygen

*Year of discovery:* 1897; *Year of introduction:* 1899 (Bayer); *Drug category:* Non-steroidal anti-inflammatory drug (NSAID); *Main uses:* Treatment of pain, inflammation and fever; also as an anticlotting agent for the prevention of heart attack. *Approximate number of people using the drug regularly:* Over 100 million; *Related drugs:* Ibuprofen (Advil), Naproxen (Aleve), Celecoxib (Celebrex), Clopidogrel (Plavix).

The acetyl ($CH_3CO$) derivative of salicylic acid has been widely used as a general purpose pain reliever for over a hundred years. It is potent, relatively safe and inexpensive. Annual production of aspirin is in excess of 40,000 tons worldwide.

Written records from 500 B.C. indicate that the Greek doctor Hippocrates used the bark of the willow tree as a pain reliever for individuals suffering from rheumatism and various forms of inflammation.[1] Research showed later that the active ingredient was salicylic acid (*see* formula below).

**Salicylic acid**

By the end of the 19th century, doctors regularly prescribed salicylic acid for the treatment of arthritic pain. However, salicylic acid is no longer used as an oral medicine, since it is very irritating to the stomach and can cause serious gastrointestinal bleeding. Its main use is in topical medications to remove warts and callouses.

Acetylsalicylic acid was discovered by the German chemist Felix Hoffmann, who tried to make a less irritating medicine for his arthritic father. In 1897 he prepared aspirin, a more potent and less irritating anti-inflammatory agent, and just two years later Bayer & Co. began marketing it as Aspirin. Aspirin acts by inhibiting the enzyme cyclooxygenase (COX) that directs the synthesis of a family of cell regulators called prostaglandins (PGs). In the stomach, a particular PG ($PGE_2$) is beneficial because it inhibits the excessive production of hydrochloric acid (HCl) and enhances the formation of a protective layer of mucus. [For more detail regarding the function of COX, *see* page 40.]

Aspirin was widely used during the flu epidemic in Europe in 1917-1918 because it effectively lowers dangerously high fevers. Such fevers are caused by elevated levels of $PGE_2$ in the brain which are decreased by aspirin. By the 1950s aspirin became by far the most widely used painkiller globally. That massive usage allowed the detection of aspirin's anticlotting properties and the realization that it could be used to lower the risk of heart attack due to the clotting of blood in disease-narrowed arteries. Taken soon after a heart attack, aspirin may also limit the size of the infarcted area.[2]

Subsequent research indicated that aspirin inhibits blood clotting at low dosages (80-90 mg per day). During the late 1980s studies also showed that aspirin can limit brain damage due to occlusive stroke caused by a blood clot, if taken early. The use of aspirin is contraindicated in hemorrhagic stroke, because it may increase bleeding.[3]

The anticlotting action of aspirin at low doses is due to the irreversible inhibition of cyclooxygenase in blood platelets by transfer of the acetyl group from aspirin to a critical serine hydroxyl group at the catalytic site of the enzyme (*see* page 40). Since mature platelets have a lifetime of only about 2 weeks and are not able to synthesize new protein, the clotting ability of aspirin-treated platelets is permanently blocked.

1. *Curr. Opin. Invest. Drugs (Thomson Curr. Drugs)* **2003**, *4*, 517-518; 2. *Drugs of Today* **2006**, *42*, 467-479; 3. *Mini-Rev. Med. Chem.* **2006**, *6*, 1351-1355; **Refs. p. 80**

# NAPROXEN (ALEVE™)

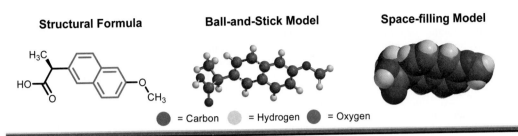

**Structural Formula**   **Ball-and-Stick Model**   **Space-filling Model**

● = Carbon    ○ = Hydrogen    ● = Oxygen

*Year of discovery:* Early 1970s; *Year of introduction:* 1976 (as Naprosyn by Syntex); *Drug category:* Non-steroidal anti-inflammatory drug (NSAID); *Main uses:* For many different types of pain, mild fever or minor inflammation; *Other brand names:* Anaprox, Naprelan and Naprogesic; *Related drugs:* Aspirin, Ibuprofen (Advil) and Celecoxib (Celebrex).

The remarkable success of aspirin as a medicine for the relief of inflammation, pain and fever stimulated the search for even more effective agents starting in the mid-twentieth century. The overall objective was the discovery of safer and more potent anti-inflammatory compounds, and especially of compounds that are devoid of the erosive gastric side effects of aspirin. Thousands of compounds were synthesized and tested in mice over more than two decades. Many active compounds were discovered, including phenylbutazolidin, indomethacin and piroxi-cam, but all exhibited side effects, especially gastric irritation.

**Indomethacin (Indocin™)**

**Piroxicam (Feldene™)**

In one approach, the structure of acetylsalicylic acid (aspirin) was methodically modified in the hope that these changes would result in better properties. The replacement of the hydroxyl group of salicylic acid by nitrogen-containing groups led to a class of compounds known as anthranilic acid derivatives that retained most of the desirable properties of aspirin (*see* below).

**Salicylic acid**

**Anthranilic acid derivatives**

Acetaminophen (Tylenol, *see* page 210), ibufenac and ibuprofen (Motrin or Advil) also

emerged. Ibufenac (isobutylphenylacetic acid) is comprised of three subunits: (1) acetic acid (blue), (2) a benzene ring (red) and (3) a branched chain attached to the benzene ring (green, *see* structures below). Although this drug was several times more potent than aspirin, it showed occasional hepatotoxicity in humans. When a methyl group (orange) was added to the acetic acid subunit (forming a propionic acid subunit), a much safer drug (ibuprofen) resulted with diminished gastro-intestinal irritation and no hepatotoxicity, even when administered in large doses (over 1 gram/day).[1] Although ibuprofen was a great success, continuing research led to more potent molecules. During the early 1970s, the Syntex Co. prepared naproxen, a propionic acid derivative with a naphthalene nucleus (two benzene rings fused together, shown in red).[2] It had twice the potency of ibuprofen and, in addition, a longer half-life (ca. ~12 hours), allowing a once-daily dosing. Not long after its introduction as Naprosyn in 1976 (and later as Aleve), the sales of naproxen exceeded $1 billion annually. Because of their significantly better potency and safety profile, ibuprofen and naproxen are today the most widely used non-steroidal anti-inflammatory agents.

**Acetic acid**

**Ibufenac**

**Propionic acid**

**Ibuprofen (Advil™)**

Extra carbon atom

1. *Int. J. Clin. Pract., Suppl.* **2003**, *135*, 3-8; 2. *J. Am. Pharm. Assoc. (Wash).* **1996**, *NS36*, 663-667; Refs. p. 80

# HOW DO ANTI-INFLAMMATORY DRUGS WORK?

Non-steroidal anti-inflammatory drugs (NSAIDs) function by inhibiting the enzyme cyclooxygenase (COX) which directs the formation of several key mediators of inflammation called prostaglandins, for example $PGE_2$.[1] Prostaglandins are synthesized locally in cell membranes by a remarkable sequence that starts with arachidonic acid (AA), a twenty-carbon fatty acid, and leads to the oxygenated products prostaglandin $G_2$ ($PGG_2$) and prostaglandin $H_2$ ($PGH_2$). This transformation proceeds *via* hydrogen atom abstraction from AA by an activated tyrosine oxygen (radical) and subsequent rapid combination with two molecules of oxygen to give $PGG_2$. Subsequently, $PGH_2$ and the other prostaglandins are formed (*see* scheme below). Using X-ray crystallography, researchers were able to capture AA (shown in red, Panel **A**) bound in the active site of COX with tyrosine

385 (green) in close proximity to the #13 carbon atom.[2] This process is completely blocked by ibuprofen (shown in red, Panel **B**), which binds tightly to the active site of COX and denies access to the normal substrate AA.[3] Naproxen inhibits COX in the same way, but aspirin blocks the enzyme by an entirely different mechanism. After aspirin binds to the enzyme, it transfers its acetyl ($CH_3CO$) group to a critical hydroxyl (OH) group, leading to a catalytically inactive form of COX that is no longer able to convert AA to prostaglandins. John Vane who discovered that aspirin works by blocking prostaglandin synthesis shared the Nobel Prize in Medicine in 1982 with Sune Bergström and his student Bengt Samuelsson who clarified the structures of the prostaglandins.

1. *Prostaglandins & Other Lipid Mediators* **2007**, *82*, 85-94; 2. *Science* **2000**, *289*, 1933-1938 (**1DIY**); 3. *Biochemistry* **2001**, *40*, 5172-5180 (**1EQG**); Refs. p. 80

**Arachidonic acid**

Cyclooxygenase-1 Enzyme (COX-1)

Carbon radical

$2 O_2$

Oxygen radical

Tyr 385 in COX-1

reduction

Endoperoxide unit

$PGG_2$

$PGH_2$

PGE$_2$ and other prostaglandins

**A.** Arachidonic Acid in the Active Site of Cyclooxygenase-1 Enzyme[2]

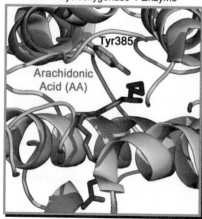

**B.** Ibuprofen in the Active Site of Cyclooxygenase-1 Enzyme[3]

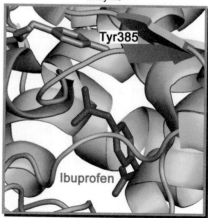

# OTHER EICOSANOIDS IN INFLAMMATION

During inflammation many local chemical mediators are produced. These substances are often referred to as local hormones since they are potent and have a specific effect on target cells close to their site of formation. They are also short-lived since they are degraded rapidly and not transported to other sites in the body. In the early stages of inflammation, various highly potent eicosanoids (compounds derived from 20-carbon fatty acids) are generated, including the proinflammatory prostaglandins (such as $PGE_2$) and various leukotrienes (such as $LTB_4$). Leukotrienes are involved in the regulation of blood flow by control of vessel tone and in the immune response. For example, $LBT_4$ attracts white blood cells to the infected cells or tissue and also increases the permeability of blood vessels. (Leukotrienes are also involved in the development of asthmatic conditions.) The formation of prostaglandins from an $\omega$-6 fatty acid, arachidonic acid (AA), *via* cyclooxygenase (COX) enzyme was discussed earlier (*see* page 40). The biosynthesis of leukotrienes also occurs from AA but is initiated by a different enzyme, 5-lipoxygenase (5-LOX, Figure 2).[1]

In addition to these proinflammatory mediators, anti-inflammatory protective lipid mediators (e.g., lipoxins, resolvins and protectins) are released in later stages of the inflammation process (*see* Figure 1) to restore normality.

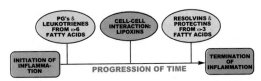

**Figure 1.** Mediators of Inflammation

Lipoxins are made from AA *via* COX and play a role in slowing down the inflammation process. Resolvins and protectins, however, are made from $\omega$-3 fatty acids such as alpha linoleic acid (ALA), eicosapentaenoic acid (EPA) and docosahexaenoic acid (DHA). $\omega$-3 Fatty acids are essential to human health, however, can not be biosynthesized in the body and must be supplied in the diet. The concentration of $\omega$-3 fatty acids is particularly high in the brain where they play a role in cognitive and behavioral function. When the body has ample supply of $\omega$-3 fatty acids, the concentration of the protectins and resolvins is increased, resulting in a faster resolution of inflammation. The formation of neuroprotectin D1 is shown on Figure 3.

Seafood is a rich source of dietary $\omega$-3 fatty acids. It is probably no coincidence that Eskimos, who consume large quantities of fatty fish and live in a harsh environment, appear to be relatively free of arthritis and heart disease.[2]

1. *Curr. Top. Med. Chem.* **2007**, 7, 297-309; 2. *Curr. Vasc. Pharmacol.* **2007**, 5, 163-172; Refs. p. 80

**Figure 2.** Formation of Various Leukotrienes from Arachidonic Acid *via* 5-Lipoxygenase

Leukotriene $B_4$ ($LTB_4$)
Arachidonic Acid (AA) [An $\omega$-6 fatty acid]
Leukotriene $E_4$ ($LTE_4$)
Leukotriene $D_4$ ($LTD_4$)
Leukotriene $C_4$ ($LTC_4$)

**Figure 3.** Formation of Neuroprotectin D1 from Docosahexaenoic Acid *via* 5-Lipoxygenase

Docosahexaenoic acid (DHA) [An $\omega$-3 fatty acid]
Neuroprotectin D1 (NPD1)

# AN OVERVIEW OF INFLAMMATION

Inflammation results from the process by which the immune system responds to foreign materials, irritation, tissue, bone or nerve damage, or infection by microorganisms. The symptoms of localized inflammation include redness, warming, swelling, pain or loss of mobility in the affected area. More widespread inflammation may bring fever, chills, fatigue, headache or loss of appetite. Inflammation may also result when the immune system malfunctions by producing antibodies against itself with pathological consequences.[1]

Inflammation is a significant factor in a number of diseases (e.g., cardiovascular, respiratory and degenerative). The sequence of biochemical events accompanying inflammation is complex and depends on the stimulus. For instance, pathogenic bacteria produce substances that activate cell surface receptors and initiate a cascade of biochemical events (i.e., a molecular signaling sequence) within the cell that produces certain potent proinflammatory proteins called cytokines such as tumor necrosis factor-$\alpha$ (TNF-$\alpha$) and interleukin-1. These then trigger the synthesis and release of other small molecular mediators including prostaglandins (PGs), histamine, bradykinin (a nine amino acid peptide) and leukotrienes. Eventually even gene expression is affected.[2]

The prostaglandins, normally present in very low concentrations, act as short-lived but powerful regulators which are produced from arachidonic acid (AA) by the cyclooxygenase enzyme (COX) *via* cyclic endoperoxides (*see below*). The biosynthesis of PGs depends on the enzymatic cleavage of membrane lipids to generate free AA and is tightly regulated. The anti-inflammatory steroid cortisol blocks the release of AA from membranes (*see* section on steroids, page 45).

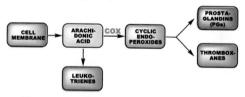

The release of histamine causes local vasodilation and increased permeability of blood vessels. It also stimulates the production of prostaglandins and bradykinin and amplifies their effects. The net result is increased blood flow to the area of the damaged tissue, warming, redness and pain, the signature of inflammation.

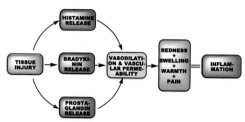

Prostaglandins and bradykinin combine to produce pain. Prostaglandins lower the sensing threshold of the nerve endings, making them more sensitive to the presence of bradykinin, thus amplifying the pain signal. This signaling process also involves changes in initial ion concentrations at ion-sensitive neuronal cells which transmit information to the brain.

Certain immune mediators attract white blood cells to a site of inflammation where they move from the blood vessels, destroy pathogens and produce collateral damage and more inflammation (page 112). In a healthy individual the immune and inflammatory response are self correcting by a sequence of molecular events which effectively downregulate the proinflammatory process. In certain disease states such as rheumatoid arthritis and osteoarthritis, or as a result of aging, this downregulation is dysfunctional. An important current treatment of rheumatoid arthritis involves the administration of a monoclonal antibody (Enbrel™, Amgen and Wyeth) against the TNF-$\alpha$ receptor protein which blocks TNF-$\alpha$ binding and the subsequent inflammation cascade.

Another macromolecule that plays an important role in inflammation is NF-$\kappa$B which, when produced and released intracellularly, migrates into the cell nucleus where it activates various proinflammatory genes, including COX. The signaling pathway that leads to free NF-$\kappa$B in the cell is complex and its suppression by synthetic molecules, which has not yet been achieved, may actually be harmful, because the accompanying immunosuppressant effect may allow infection. The immune response, in common with inflammation, is dependent on NF-$\kappa$B.

1. Inflammation: Basic Principles and Clinical Correlates (Lippincott Williams & Wilkins, **1999**); 2. Cytokines and Joint Injury (Birkhauser Basel, **2004**); Refs. p. 80

# CELECOXIB (CELEBREX™)

| Structural Formula | Ball-and-Stick Model | Space-filling Model |
|---|---|---|

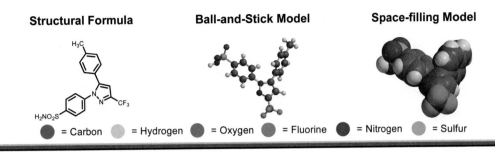

● = Carbon ○ = Hydrogen ● = Oxygen ● = Fluorine ● = Nitrogen ○ = Sulfur

*Year of discovery:* **1993**; *Year of introduction:* **1999 (Pfizer);** *Drug category:* **Non-steroidal anti-inflammatory drug (NSAID);** *Main uses:* **Treatment of osteoarthritis, rheumatoid arthritis and acute pain (e.g., painful menstruation);** *Approximate number of people taking the drug annually:* **Over 23 million;** *Older drugs:* **Aspirin, Ibuprofen (Advil) and Naproxen (Aleve).**

Non-steroidal anti-inflammatory agents such as aspirin, ibuprofen and naproxen (*see* page 39) act by inhibiting the biosynthesis of prostaglandins (PGs) from arachidonic acid (AA). There are two human enzymes that catalyze the first step in the biosynthesis of PGs, cyclooxygenase 1- and 2 (COX-1 and COX-2; *see* page 40). Although COX-1 and COX-2 catalyze the same biochemical reaction, they are distinctly different in terms of amino acid content (about 60% identity), tissue distribution, and physiological function. COX-1 has been described as a "constitutive" enzyme because it appears to have a steady presence in tissues and organs, for example in the stomach where it affects gastric acidity and mucous secretion. In contrast, COX-2 levels are normally low, but become elevated at sites of inflammation, in certain tumor cells or in response to stimuli such as growth or wound-response factors. Glucocorticoid steroids and anti-inflammatory cytokines (*see* page 45) downregulate the expression of COX-2.

COX-1 and COX-2 are both membrane-associated proteins to which the membrane-bound substrate AA is transferred directly. The two enzymes have roughly similar substrate binding sites, but that of COX-2 is slightly larger and differently shaped. Since COX-2 is not expressed in stomach and since gastric ulcers and serious gastric bleeding develop in about 1% of chronic users of COX-1 inhibitors, research to find inhibitors that are selective for COX-2 over COX-1 was initiated. The selective COX-2 inhibitor celecoxib (Celebrex) emerged from this effort in 1999, and became an important medicine for treatment of osteoarthritis in people who cannot tolerate aspirin or non-selective alternative COX-1/COX-2 inhibitors.

A competing drug, rofecoxib (Vioxx), was developed at Merck and Co. and was marketed about the same time as celecoxib. However, it subsequently had to be withdrawn when it was linked to a 1% increase in risk of heart attack. Celecoxib appears to be considerably safer than rofecoxib.[1]

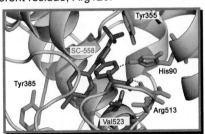

**Rofecoxib (Vioxx)** **SC-558**

Although COX-1 and COX-2 bind AA in the same three-dimensional geometry, the ligand-binding pockets are different for inhibitors of COX-1 and COX-2. The picture below shows a close structural relative of celecoxib, SC-558, bound in the active site of COX-2.[2] The selectivity results because the phenylsulfonyl group (shown in green above) binds in a pocket (formed from His90, Arg513 and Val523) that is not available in COX-1 since it is occupied by a bulky isoleucine side chain rather than the smaller isopropyl group of valine (Val523). The carboxyl group of rofecoxib interacts not with Arg513 but with a different residue, Arg120.

1. *Future Cardiology* **2005,** *1,* 709-722; 2. *Nature (London)* **1996,** *384,* 644-648. (**1CX2**); Refs. p. 80

# PREDNISONE (DELTASONE™)

| Structural Formula | Ball-and-Stick Model | Space-filling Model |

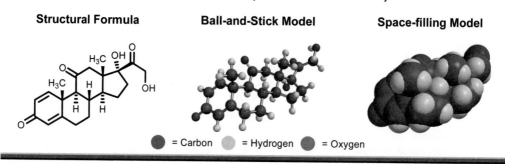

● = Carbon ○ = Hydrogen ● = Oxygen

*Year of discovery:* Early 1950s; *Year of introduction:* 1955 (as Meticorten by Schering and as Deltasone by Pharmacia and Upjohn); *Drug category:* Anti-inflammatory agent/immuno-suppressant/adrenocortical steroid; *Main uses:* For treatment of inflammatory diseases (e.g., asthma, Crohn's disease) and prevention of organ transplant rejection; *Related drugs:* Cortisol (Hydrocortisone), Fluticasone (Flonase), (and others shown on next page).

Human adrenal glands, though weighing only a few grams, are essential for life; adrenalectomized animals survive only for a matter of days. Research in the 1930s by Oskar Wintersteiner at Columbia, Edward C. Kendall at the Mayo Foundation, and Thaddeus Reichstein at the Federal Technical Institute (ETH) in Zürich resulted in the identification of more than 25 members of the adrenocortical family of steroids including cortisol and the corresponding 11-ketone, cortisone.

**Cortisol** **Cortisone**

Kendall and Philip S. Hench's demonstration in the 1940s that cortisol and cortisone exerted profound anti-inflammatory effects in humans, had a major impact on the medical sciences.[1] Reichstein, Kendall and Hench shared the Nobel Prize in Medicine in 1950 for their discoveries.

Cortisol has a wide range of activities in the body and its levels are tightly controlled. Its biosynthesis (from the early precursor cholesterol) in the adrenals is stimulated by corticotrophin, a 39 amino acid peptide produced in the brain and carried to the adrenals by blood. Cortisol levels are partly regulated by two enzymes, one that oxidizes the 11-CHOH unit to C=O and the other that catalyzes the reverse reaction. The intrinsic bioactivity of cortisol is much greater than of

cortisone, which essentially serves mainly as a storage depot. Cortisone is inactive at corticosteroid receptors.

The receptors that mediate the many biological activities of corticosteroid hormones are of two general types: (1) glucocorticoid (GR) and (2) mineralocorticoid receptors (MR).

Cortisol has both GR and MR activity, but the former dominates. Another adrenal steroid, aldosterone shows much more potent MR activity than GR activity. MR activation causes sodium retention in the body, and was of great importance during evolution because NaCl was frequently in short supply. Since an excess of NaCl in the body causes hypertension, an antagonist of aldosterone (Eplerenone, Inspra™) is used medically when the cause of elevated blood pressure is excessive body production of aldosterone.

**Aldosterone** **Eplerenone (Inspra™)**

Cortisol is associated with a remarkable variety of biological activities and plays many functional roles throughout the body, for example in the brain, peripheral muscle and the various organs. It is a principal effector in the hypothalamic-pituitary-adrenal (HPA) axis of physiological control (*see* next page). In the brain cortisol plays a crucial role in cognition, maintenance of neurons and the regulation of stress and mood.

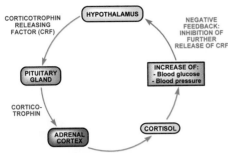

The Hypothalamic-Pituitary-Adrenal Axis.

The discovery of the anti-inflammatory properties of cortisol/cortisone in the 1940s spurred the development of several synthetic analogs of cortisone during the 1950s. Many of these compounds are several times more potent than cortisol, longer lasting and cause less sodium retention. One of the most widely used synthetic analogs, prednisone, was introduced by Schering in 1955 as Meticorten. Prednisone and other steroids of this class have many therapeutic applications. They are useful in rheumatic/inflammatory disorders, allergies, malignancies such as leukemia and multiple myeloma, and skin diseases. Prednisone can be used orally or intramuscularly to treat cases of acute inflammation. It is a powerful immunosuppressant because it reduces B- and T-cell-mediated immunity. For this reason it is administered to patients after organ transplant to prevent rejection. Fortunately, prednisone and other members of this class do not cross the blood brain barrier, and thus have minimal effects on mental function. Unfortunately, long term use of these corticosteroid hormones in systemic therapy is contraindicated because of inevitable and serious side effects (e.g., osteoporosis).

Other members of the prednisone class with longer half-lives and enhanced potency than prednisone are in common use; the most widely used of which is dexamethasone (see stereostructure at bottom left).

The anti-inflammatory activities of glucocorticoids are not fully understood because they affect the expression of numerous genes.[2] However, one way in which they exert anti-inflammatory effects is by promoting the synthesis of lipocortin-1, a calcium-binding protein that binds to cell membranes.[3] In the membrane, lipocortin-1 inhibits the enzyme phospholipase A2 (PLA$_2$) that is responsible for the selective removal of acyl groups from the C-2 position of phospholipids to release free arachidonic acid (AA). The net result is a diminished production of AA and its biooxidation products including proinflammatory eicosanoids, prostaglandins and leukotrienes. In addition, the expression of the cyclooxygenase enzyme (COX) is downregulated.

**Dexamethasone**

$OCO-nC_{17}H_{35}$

$nC_5H_{11}$

$NH_3^{\oplus}$

**Phospholipid**
(Arachidonic Acid bound at C-2)

PLA$_2$

$nC_5H_{11}$

**Arachidonic Acid**

5-Lipoxygenase (LOX)

Cyclooxygenase (COX)

Leukotrienes

Prostaglandins

(Proinflammatory agents)

Another way in which prednisone and other corticosteroids reduce inflammation is through their effect on the body's production of TNF-α and other cytokines. For certain inflammatory diseases, e.g., Crohn's disease (inflammatory bowel disease), combination therapy with prednisone and an anti-TNF monoclonal antibody allows the use of lower doses of each.

1. N. Engl. J. Med. **2005**, 353, 1711-1723; 2. Br. J. Pharmacol. **2006**, 148, 245-254; 3. Ann. N. Y. Acad. Sci. **2006**, 1088, 396-409; Refs. p. 81

# METHOTREXATE (TREXALL™)

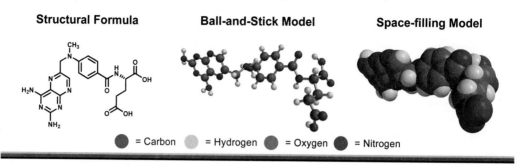

**Structural Formula**     **Ball-and-Stick Model**     **Space-filling Model**

● = Carbon     ○ = Hydrogen     ● = Oxygen     ● = Nitrogen

*Year of discovery:* 1948 (Lederle); *Year of introduction:* 1953; *Drug category:* Antimetabolite/disease-modifying anti-inflammatory agent/immunosuppressant; *Main uses:* For treatment of inflammation associated with autoimmunity (rheumatoid arthritis, Crohn's disease and psoriasis); *Other brand names:* Rheumatrex; *Related drugs:* Trimetrexate (Neutrexin), Pemetrexed (Alimta).

Methotrexate is an inhibitor of folic acid biosynthesis which slows the proliferation of cells. It has been known since the 1930s that folic acid is essential for the development of new cells. Adequate dietary intake of folic acid is absolutely necessary for human health. A deficiency of folic acid is especially serious in pregnant women, since it leads to maldevelopment of the fetus and results in defects of the spine, brain and skull.

Folic acid undergoes reduction in the body by the enzyme dihydrofolate reductase (DHFR) in two stages, giving first dihydrofolate and then the essential metabolite tetrahydrofolate. Parts of the tetrahydrofolate molecule are required as building blocks for the synthesis of the nucleoside thymidine, an ingredient for DNA formation. The figure below shows in red the segment of the six-membered ring of thymidine that originates from tetrahydrofolate. Methotrexate blocks DHFR and interferes with DNA synthesis.

**Folic Acid (Folate)**     **Tetrahydrofolate**

**Thymidine**

Because tetrahydrofolate is essential for normal cell division, it is especially critical in tissues that divide rapidly (e.g., bone marrow, blood cells, skin, gastrointestinal tissues and

the immune system). Since cancer cells divide even more rapidly, blockade of the synthesis of tetrahydrofolate by inhibition of the enzyme DHFR appeared to be promising for effective cancer therapy. Methotrexate (MTX) was successfully developed in 1948 by the Lederle Company and marketed under the trade name Trexall. Although MTX was originally used in cancer chemotherapy, its main use is now in the treatment of autoimmune and inflammatory diseases, especially psoriasis and rheumatoid arthritis.[1] The combination of the anti-TNF-α monoclonal antibody Enbrel and MTX is currently a standard treatment of rheumatoid arthritis.

Methotrexate is a disease-modifying anti-rheumatic agent that not only reduces the symptoms, such as pain and swelling, but also slows the progression of the disease by preventing further damage to joints. The effective dose of MTX is several orders of magnitude lower for inflammation than for cancer.[2] Even at low doses, MTX appears to decrease T-cell proliferation and the levels of proinflammatory factors such as TNF-α and interleukin-10. The figure below shows an X-ray picture of MTX bound to the enzyme DHFR from *E. Coli*.[3]

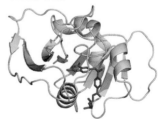

1. *Pharm. Rep.* **2006**, *58*, 473-492; 2. *Biomed. & Pharmacother.* **2006**, *60*, 678-687; 3. *Biochemistry* **1997**, *36*, 586-603. (1RG7); Refs. p. 81

# ALLOPURINOL (ZYLOPRIM™)

| Structural Formula | Ball-and-Stick Model | Space-filling Model |
|---|---|---|

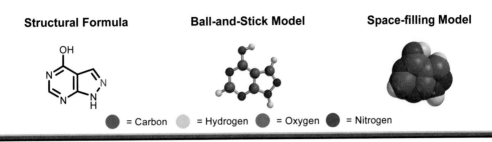

● = Carbon  ○ = Hydrogen  ● = Oxygen  ● = Nitrogen

*Year of discovery:* Early 1950s; *Year of introduction:* 1964 (Burroughs Wellcome, now GlaxoSmithKline); *Drug category:* Xanthine oxidase inhibitor for treatment of inflammatory gout; *Main uses:* For the prevention and treatment of gout attacks and certain types of kidney stones; *Related drugs:* Colchicine, Probenecid (Benemid).

Allopurinol is used for the treatment of arthritic gout. Arthritis encompasses of over one hundred different rheumatic diseases and conditions affecting joints, muscle and bone. These diseases include osteoarthritis, rheumatoid arthritis and gout. There are many symptoms associated with arthritis, but the most common are pain, aching, stiffness, and swelling of the joints. Currently over 50 million people in the US have arthritis and about 3 million of them suffer from gout, a very painful form. Unlike other forms of arthritis, the cause of gout has been pinpointed as being the deposition of needle-like crystals of uric acid in the connective tissue, joint spaces or both (*see* below). Such uric acid deposits cause inflammation of the affected joint and result in swelling, stiffness and intense pain. The first sign of gout is usually pain in the joints of the big toes.[1, 2]

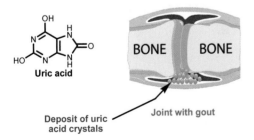

Uric acid

BONE    BONE

Deposit of uric acid crystals — Joint with gout

In healthy humans uric acid is produced by the breakdown of purines. Normally it is excreted in the urine. In individuals with gout, uric acid is overproduced and accumulates because the kidneys do not efficiently eliminate it. As a result, blood levels of uric acid increase and uric acid crystallizes in the joints and even in the kidneys, causing kidney

stones. Uric acid is formed in the last step in the metabolic pathway of purines.

The conversion of hypoxanthine to xanthine and of xanthine to uric acid is catalyzed by the enzyme xanthine oxidase (*see* scheme below).

Hypoxanthine → Xanthine Oxidase → Xanthine

Xanthine Oxidase → Uric acid

Treatment of gout can be achieved with a combination of therapies. The pain can be reduced with NSAIDs (e.g., naproxen), colchicine and, in acute cases, injection of corticosteroids into the affected joints. These medicines are not safe for long-term use and do not prevent future gout attacks.

Allopurinol, a purine derivative, was first prepared in the 1950s and shown to inhibit the enzyme xanthase oxidase (XO) and inhibit the biosynthesis of uric acid. Allopurinol is an effective inhibitor of XO because it binds to the enzyme effectively competing with xanthine. Since allopurinol has a nitrogen atom at the 2-position (shown in blue) instead of the C-H group of xanthine, it can not be converted to uric acid.[3]

1. *J. Clin. Invest.* **2006**, *116*, 2073-2075; 2. *Crystal-Induced Arthropathies* **2006**, 189-212; 3. *Pharmacol. Rev.* **2006**, *58*, 87-114; Refs. p. 81

*Chronic obstructive pulmonary disease (COPD) and its congener emphysema are the fourth leading cause of death in the US.*

*These smoker's diseases are tragic, not only because they are largely preventable, but because there is no cure.*

*The time between appearance of the first symptoms (in smokers, generally around age 50) and death is about 15-20 years.*

*Asthma is another life-threatening disease, but it is manageable. It may also be preventable, but the measures for prevention remain obscure.*

# ANTIASTHMATIC
# AND
# ANTIALLERGIC
# AGENTS

# SALMETEROL (SEREVENT™)

| Structural Formula | Ball-and-Stick Model | Space-filling Model |
|---|---|---|

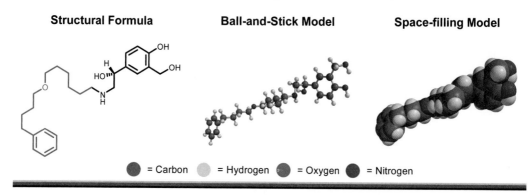

● = Carbon   ● = Hydrogen   ● = Oxygen   ● = Nitrogen

*Year of discovery:* 1980; *Year of introduction:* 1990 (as Serevent by GSK). It is marketed by GSK as Advair in combination with the corticoid fluticasone; *Drug category:* $\beta_2$-Adrenergic receptor agonist; *Main uses:* To treat the worsening of asthma and chronic obstructive pulmonary disease (COPD); *Approximate number of people treated annually:* Over 15 million; *Related drugs:* Formoterol (Foradil, Oxis), Bambuterol (Bambec), Salbutamol (Ventolin). These drugs are also used in combination with inhaled anti-inflammatory steroids of the cortisone family.

Asthma is a chronic inflammatory illness of the respiratory system that causes coughing, wheezing or shortness of breath. These symptoms result from narrowing of airways, inflammation, and accumulation of mucus. In the US alone this disease affects about 5% of the population and usually worsens with age. Asthma attacks may be triggered by allergens, dust, cigarette smoke, cold air, exercise or emotional stress. Children living in urban areas are especially at risk to develop asthma as a consequence of viral infection and urban pollution.

Treatment of asthmatic patients with bronchodilators (substances that dilate the airways and improve bronchial airflow) started in the early 1900s but it was not until the 1960s that anti-inflammatory drugs were also included in the course of therapy.[1,2]

Bronchodilators are $\beta_2$-adrenergic receptor agonists and can be either short-acting (4-6h) or long-acting (>12h). They work by competitively binding to and activating the $\beta_2$-adrenergic receptor. As a result the smooth muscles in the lung relax, leading to the dilation of bronchial airways.[3]

Salmeterol is a long-acting bronchodilator that is usually prescribed to treat severe persistent asthma. Regular daily use of salmeterol in combination with corticosteroids reduces the frequency and severity of asthma attacks.[4] Since this therapy requires high doses, side effects such as increased heart rate and insomnia may occur. The long hydrophobic (oily) side chain (red) attached to the nitrogen of salmeterol serves to increase the solubility of the drug in the plasma membrane of the lung which leads to a slow and prolonged release. For this reason salmeterol is an effective long-acting drug.

The use of a short-acting $\beta_2$-adrenergic receptor agonist (e.g., salbutamol, *see* structure below) provides rapid relief of asthma symptoms. When inhaled, such faster-acting substances lessen sudden asthma symptoms, and when taken 15-20 minutes ahead of time, can prevent the onset of symptoms.

**Salbutamol**

The combination of salmeterol with the corticosteroid fluticasone (Flovent™, *see* page 51), known as Advair (GlaxoSmithKline), is now widely used, with annual sales of several billion dollars. Another application of Advair is in the treatment of chronic obstructive pulmonary disease (COPD, *see* page 53).

1. *Principles of Immunopharmacology (2nd Edition)* **2005**, 281-344; 2. *Prim. Care Resp. J. : J. Gen. Pract. Airways Group* **2006**, *15*, 326-331; 3. *Therapeutic Strategies in COPD* **2005**, 61-78; 4. *Drugs* **2005**, *65*, 1715-1734; Refs. p. 81

# FLUTICASONE PROPIONATE (FLOVENT™)

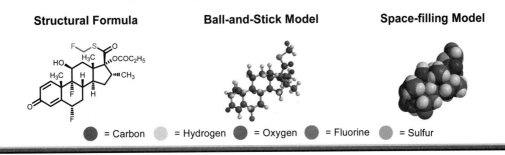

| **Structural Formula** | **Ball-and-Stick Model** | **Space-filling Model** |

● = Carbon   ● = Hydrogen   ● = Oxygen   ● = Fluorine   ● = Sulfur

*Year of discovery:* 1981 (Glaxo Wellcome); *Year of introduction:* 1994 (Flonase), 1996 (Flovent); *Drug category:* Anti-inflammatory glucocorticoid; *Main uses:* Treatment of asthma and allergic and non-allergic rhinitis; *Other brand names:* Flonase (for the treatment of rhinitis), Advair (in combination with salmeterol for the treatment of asthma and COPD); *Related drugs:* Beclomethasone (Vanceril), Triamcinolone acetonide (Azmacort), Flunisolide (Aerobid), Budesonide (Pulmicort).

Fluticasone propionate is a synthetic analog of cortisol that has been developed as an inhaled anti-inflammatory agent for the treatment of asthma, especially in combination with a vasodilator such as salmeterol (*see* page 50). Although cortisol and related glucocorticoids (*see* prednisone, page 44) are effective as inhaled antiasthmatics, they are not suitable for long term use because of cumulative side effects, such as osteoporosis and immunosuppression. Research over two decades led to improved antiasthmatic corticoid beclomethasone dipropionate, a prodrug of 16-methylprednisone.[1]

**Beclomethasone dipropionate (Vanceril™)**

This prodrug is advantageous for two reasons: (1) the residence time in the lung is prolonged relative to prednisone and (2) it is gradually released as local enzymes convert it into 16-methylprednisone, and these differences translate into greater efficacy and convenient dosing. However, because considerable amount of 16-methylprednisone still enters the circulation, the adverse side effects preclude the long-term use of beclomethasone in chronic asthma.

An intensive effort to solve the problem of side effects associated with prolonged use of inhaled corticosteroids led to the discovery of fluticasone propionate by researchers at Glaxo Wellcome (now GlaxoSmithKline). A key aspect of this research was the finding that the $C_2H_5OCOCH_2CO$ subunit attached to position 17 of beclomethasone could be replaced by the $FCH_2SCO$ group without loss of potency and with the great added benefit that the resulting steroid was very quickly deactivated by the action of enzymes in blood and liver. That deactivation occurred because of cleavage of the side chain at position 17 to a 17-carboxylic acid that had no glucocorticoid activity.[2]

liver, blood

**17-Carboxylic acid**

The systematic research leading to the structural changes from cortisol to fluticasone represents an excellent example of molecular engineering to optimize the properties of a therapeutic agent for a particular application. Although this process may look simple and straightforward in retrospect, in practice much time and effort are required to reach an optimized molecule. The combination of fluticasone propionate with salmeterol (preceding page) is sold as Advair.

1. *Adv. Ther.* **1997**, *14*, 153-159; 2. *J. Allergy Clin. Immunol.* **1998**, *101*, S434-S439; Refs. p. 81

# MONTELUKAST SODIUM (SINGULAIR™)

| Structural Formula | Ball-and-Stick Model | Space-filling Model |
|---|---|---|

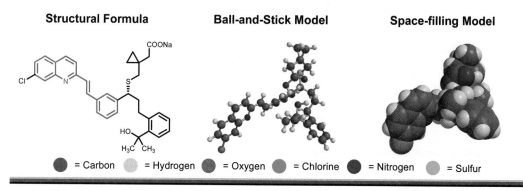

● = Carbon    ● = Hydrogen    ● = Oxygen    ● = Chlorine    ● = Nitrogen    ● = Sulfur

*Year of discovery:* 1991; *Year of introduction:* 1998 (Merck); *Drug category:* Oral leukotriene receptor antagonist; *Main uses:* Treatment of asthma and to relieve symptoms of seasonal allergies; *Related drugs:* Zafirlukast (Accolate), Pranlukast (Onon).

Montelukast sodium is a synthetic antagonist of the cysteinyl leukotriene receptor which has proven efficacious in the treatment of asthma.

The severity of asthma varies from mild to life-threatening. Narrowing of the airways is induced by the combination of inflammation of the bronchial tubes and muscular constriction. It can be triggered by hypersensitivity to allergens, irritants, infection or nerve signals.

Receptors on the surface of the airways bind mediators called cysteinyl leukotrienes (CysLT), small molecules produced in response to irritants. Upon binding to the target receptors, cysteinyl leukotrienes induce bronchial and smooth muscle contraction and enhance production of mucus in excess of the amount normally required to clear foreign particles. Leukotrienes have been shown to be inflammatory and to be involved in asthmatic and allergic reactions. A recent treatment of asthma is based on blocking the proinflammatory binding of leukotrienes to their receptors.

Cysteinyl leukotrienes were first detected as bronchoconstrictive factors in guinea pigs that had been sensitized with cobra venom (by Australian physiologist C. H. Kellaway in 1938). Since these factors were observed to cause a slow and prolonged contraction of muscles, they were termed as slow-reacting substances of anaphylaxis (SRS-A). The chemical nature of SRS-A was uncovered only years later, when Bengt Samuelsson at the Karolinska Institute discovered that SRS-A contained a lipid subunit derived from arachidonic acid. Because these substances could be formed *in vitro* from leukocytes (white blood cells), they were termed leukotrienes. The exact chemical structure of all the leukotrienes was determined by Samuelsson and the Corey group at Harvard. An efficient chemical synthesis of the leukotrienes was developed that was essential not only to the proof of structure but also to making these unstable and very rare compounds available for biological studies. With the availability of synthetic leukotrienes, investigation began to find molecules that either block leukotriene biosynthesis or receptor binding.

Among the innumerable CysLT antagonists made and investigated, the most widely used today is the sodium salt of montelukast, which is produced by Merck and Co. and marketed as Singulair. Given orally, it is bioavailable, potent and safe for many asthmatics and can be used with bronchodilating β-agonists (*see* salmeterol). Singulair is also effective in the treatment of allergic rhinitis. Since the approval of Singulair in 1998, sales have climbed to over $2 billion per year.[1,2]

An earlier leukotriene receptor antagonist, zafirlukast, introduced as Accolate in 1996 by AstraZeneca, is less widely used.

**Zafirlukast (Accolate™)**

1. *Journal of Drug Evaluation: Respiratory Medicine* **2002**, *1*, 53-88; 2. *Expert Opin. Pharmaco.* **2004**, *5*, 679-686; Refs. p. 81

# TIOTROPIUM BROMIDE (SPIRIVA™)

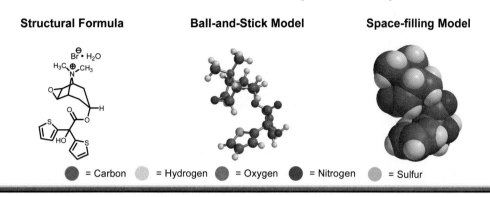

| Structural Formula | Ball-and-Stick Model | Space-filling Model |

● = Carbon    ○ = Hydrogen    ● = Oxygen    ● = Nitrogen    ● = Sulfur

*Year of discovery:* 1991; *Year of Introduction:* 2004 (Pfizer, Boehringer Ingelheim); *Drug category:* Long-acting antimuscarinic bronchodilator; *Main uses:* Treatment of chronic obstructive pulmonary disease (COPD); *Related drugs:* Ipratropium bromide (Atrovent).

Chronic obstructive pulmonary disease (COPD) is an umbrella term used to describe a set of chronic lung diseases in which there is obstruction or deficiency of airflow. Included are chronic bronchitis (inflamed and narrowed bronchial tubes) and emphysema, a condition in which the air sacs (alveoli) in the lungs are inelastic and weakened as a result of long-term damage. COPD is a serious illness that, according WHO estimates, affects 80 million people worldwide and over 10 million people in the US. It is the fourth leading cause of death (responsible for 2.75 million deaths globally and over 100 thousand deaths in the US in 2000) after heart disease, cancer and stroke, and is projected to be the third leading cause of death by 2020. The economic burden is also significant; COPD was estimated to cost about $32 billion in the US in 2002.

The most common cause of COPD is tobacco smoke. An estimated 15% of smokers develop the disease, and smoking is responsible for 90% of COPD cases. Other risk factors are intense and prolonged exposure to occupational dusts and chemicals, as well as high levels of urban air pollution. A history of severe childhood respiratory infection and low birth weight predispose to the development of the disease.

COPD is a progressive disease; the symptoms develop so gradually that they generally are not troublesome. Early telltale signs include shortness of breath, development of a chronic cough, increased sputum production, wheezing and a feeling of chest tightness. The first symptoms typically appear at age 40-50. Heavy smokers often loose 50% of their lung function by the age 50, and most of those with COPD die before reaching 65.

COPD is not curable, but it is possible to control the symptoms and reverse acute exacerbation. The most commonly used medications include inhaled bronchodilators that work by relaxing the muscles around the airways (*see* salmeterol, page 50) and inhaled anti-inflammatory corticosteroids (*see* fluticasone, page 51). Tiotropium bromide is a long-acting antimuscarinic agent that exerts its pharmacological effect by inhibiting the $M_3$ muscarinic receptors at the smooth muscles in the airways leading to bronchodilation. Tiotropium bromide capsules are comarketed by Boehringer Ingelheim and Pfizer under the trade name Spiriva and are self-administered by a commercial inhalation device.[1,2]

An earlier, structurally related bronchodilator, ipratropium, was introduced in 1986 by Boehringer Ingelheim under the name Atrovent. Ipratropium is a nonselective muscarinic receptor antagonist. It is also sold in combination with albuterol (Combivent, Duoneb).

**Ipratropium (Atrovent™)**

1. *Treatments in Resp. Med.* **2004**, *3*, 247-268; 2. *Nat. Rev. Drug Discov.* **2004**, *3*, 643-644; Refs. p. 82

# LORATADINE (CLARITIN™)

| Structural Formula | Ball-and-Stick Model | Space-filling Model |
|---|---|---|

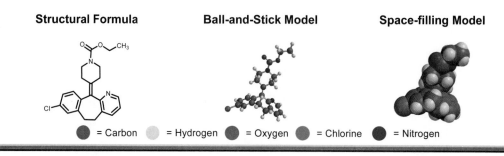

● = Carbon    ○ = Hydrogen    ● = Oxygen    ● = Chlorine    ● = Nitrogen

*Year of discovery:* 1981; *Year of introduction:* 1993 (Schering-Plough); *Drug category:* Long-acting second generation antihistamine; *Main uses:* Relieve the symptoms of seasonal allergies and rashes caused by food allergies; *Other brand names:* Claratyne (Schering-Plough), Alavert (Wyeth); *Related drugs:* Acrivastine (Semprex-D), Cetirizine (Zyrtec), Fexofenadine (Allegra).

Allergy, the immune system's misguided response to foreign substances that are normally harmless, affects about 50 million people and is the sixth leading cause of chronic illness in the United States. The most common type of allergy arises from hypersensitivity of the upper respiratory tract to airborne allergens, which can cause runny nose, itching of the nose and throat, and coughing. Certain foods such as tree nuts, peanut and shellfish as well as medicines can also act as allergens, leading to rash, vomiting, diarrhea, urticaria, and swelling. Other allergens occur in certain materials for instance in latex rubber. Extreme allergic reactions affect many organs. This response typically occurs when the allergen is eaten or injected (for example a bee sting) into an already hypersensitized person and is termed anaphylaxis.

In allergic reactions, the immune system that normally functions as the body's main line of defense against microbes, reacts to the allergens as harmful substances. Upon first contact with the allergen, the immune system generates large amounts of immunoglobulin E (IgE), a type of antibody that binds tightly to mast cells in the connective tissue and basophil cells in the blood. Upon further exposure, the allergen attaches to IgE on the surface of mast cells and basophils, triggering the release of powerful mediators including histamine, prostaglandins, and leukotrienes, which are inflammatory (*see* page 41).

In 1910, a factor in tissue extracts that stimulates the contraction of smooth muscle was discovered by H. Dale and P. Laidlaw. Seventeen years later, C. Best purified this factor from lung and liver samples and identified it as histamine, a molecule formed by loss of $CO_2$ from the amino acid histidine.

Histidine    $- CO_2$    Histamine

The first histamine blocker was described in 1937 by D. Bovet and A. Staub, who screened a series of synthetic amines for antihistamine activity. One of the substances effectively protected guinea pigs against lethal doses of histamine, but it was too toxic for clinical use. By 1944, Bovet and coworkers discovered pyrilamine maleate, which is one of the most effective antihistamines.

Pyrilamine

Although still commercially available, it is not generally used since it causes drowsiness, because it penetrates the brain, which has histamine receptors. In the following decades, several antihistamine drugs were developed that do not produce sedation. Loratadine (Claritin), developed by Schering-Plough in 1981, became the most widely used. Since 2002 it has been available without a prescription and as a generic drug. The antihistamines used to treat allergy bind to the H-1 histamine receptor whereas those used to reduce gastric acidity bind to another histamine receptor, H-2 (*see* page 103).[1,2]

1. *Clin. Exp. Allergy* **1998**, *28*, 15-19; 2. *Drugs* **1999**, *57*, 31-47; Refs. p. 82

# TYPE 2
# DIABETES

# AN OVERVIEW OF METABOLIC SYNDROME, A PRECURSOR OF DIABETES, HEART DISEASE AND STROKE

The conversion of food into a living body and into the energy that it needs depends on complex networks of biochemical reactions that must be exquisitely regulated for optimum function. Dietary carbohydrates are converted to glucose, which is used both as an energy source, through oxidation by $O_2$ to form $CO_2$ and $H_2O$, and as a source of intermediates for biosynthesis (e.g., of fatty acids). Fats are converted into fatty acids, which are likewise used both as fuel and to furnish building blocks for biosynthesis. Dietary proteins are broken down to their individual amino acids, which are then used for the biosynthesis of proteins and many other substances.

Energy is stored as adenosine triphosphate (ATP), which is produced in subcellular compartments called mitochondria. ATP activates biomolecules and drives, either directly or indirectly, much of the biosynthesis, development, regulation and repair required for life. Consequently, diseases arising from inefficiency (or misregulation) of energy production and metabolism have profound effects on health. Illness of this type was defined as metabolic syndrome by G.M. Reaven in 1988. This term encompasses a wide variety of molecular abnormalities which can result in:

(1) elevated blood glucose levels;

(2) inefficient utilization of glucose as an energy source;

(3) excessive metabolic conversion of glucose to fat tissue;

(4) lowered sensitivity to insulin;

(5) loss of mitochondrial function and energy production;

(6) elevated levels of circulating fatty acids, triglycerides and LDL;

(7) lowered levels of HDL in blood;

(8) excessive appetite and weight gain;

(9) hypertension;

(10) impaired circulation because of clogged arteries;

(11) inflammation;

(12) impairment of organ function and

(13) depressed immune response.

During the last 30 years metabolic syndrome has become increasingly common, and currently affects more than 50 million people in the US. Much of this increase is associated with higher caloric intake (especially of sugar and saturated fats), reduced physical activity and exercise, and an aging population.[1]

The earliest sign of metabolic syndrome is the presence of one or more of the following symptoms:

(1) excessive weight;

(2) hypertension;

(3) elevated levels of glucose in the blood and

(4) elevated levels of circulating lipids, especially triglycerides and LDL.

Individuals who have two or more of these symptoms are at higher risk of developing cardiovascular disease and type 2 diabetes (i.e., non-insulin dependent), especially if there is a family history of these illnesses. The symptoms of metabolic syndrome often occur concurrently because they arise from multiple, tightly-coupled biochemical abnormalities.[2]

One reason why excessive levels of circulating glucose are harmful to health is the potential for wide-ranging damage arising from the fact that glucose and some of its metabolites, for example methylglyoxal, can react with proteins and body tissues (see scheme below). The cross-linking process starts with the coupling reaction of the reactive carbonyl groups with the amino groups of proteins.

These reactions can occur spontaneously, i.e. without catalysis by enzymes, and can form covalently linked products which are harmful because they accumulate in the body, leading to diminished insulin sensitivity. Thus, although glucose is a critical fuel for the body, it must be tightly controlled.

Adipose cells function not only as energy storage vessels but also as sources of regulatory molecules. One such regulator is the small protein leptin, which after secretion by fatty tissue travels to the brain where it normally induces a signal to decrease food intake. Another regulator is adiponectin which increases insulin sensitivity, decreases glucose production in the liver, and increases glucose uptake by the muscle, all of which are antidiabetic activities. Adiponectin also decreases circulatory lipid and cholesterol and is anti-inflammatory. Excess body fat (especially abdominal fat) is associated with disregulation and abnormal function, including diminished adiponectin secretion and increased production of the proinflammatory TNF-$\alpha$. The dysfunctional behavior of adipocytes in excessively heavy or obese individuals may be a consequence of several factors, including inflammation due to macrophage infiltration, cumulative damage by reactive chemical fragments and disregulation of crucial enzymes. To sum up, the adipocyte is not just a storage cell for triglycerides but a complex cell that must be properly regulated.

Dislipidemia in metabolic syndrome elevates circulating fatty acids and their glycerol esters (triglycerides) and cholesterol-bearing LDL. It eventually leads to the deposition of arterial plaque, diminished circulation and oxygenation of tissues, with far reaching and adverse consequences to health.

$$HOH_2C-\overset{\overset{H}{|}}{\underset{\underset{OH}{|}}{C}}-CH_2OH$$

**Glycerol**

$$CH_3(CH_2)_{16}COOH$$

**Stearic Acid (a fatty acid)**

$$H_2C-O-CO(CH_2)_{16}CH_3$$
$$HC-O-CO(CH_2)_{16}CH_3$$
$$H_2C-O-CO(CH_2)_{16}CH_3$$

**A Fatty Acid Glycerol Ester (Triglyceride)**

Another aspect of metabolic syndrome is a reduction in the levels of the enzyme adenosine monophosphate-activated protein kinase (AMP-APK), which serves as a critical regulator of fatty acid synthesis, energy production from fatty acids by oxidation, and glucose uptake. The biosynthesis of fatty acids, which must be tightly regulated, is increased by a diet rich in carbohydrate. A key enzyme for the synthesis of fats is acetyl-CoA carboxylase, which catalyzes the first step in fatty acid synthesis from acetate (i.e.,

conversion of acetyl-CoA to malonyl-CoA). The activity of this enzyme is sensitive to the levels of substances such as insulin and ATP.

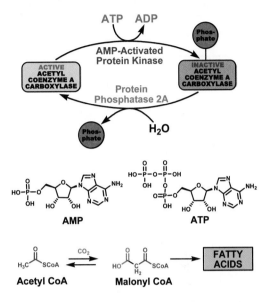

The enzyme AMP-activated protein kinase (AMP-APK) is a critical factor in metabolic syndrome and type 2 diabetes. There are two forms of AMP-activated protein kinase:

(1) a phosphorylated form which is active (AMP-APK*) and

(2) a non-phosphorylated form which is not active (AMP-APK$^0$)

It is only AMP-APK* that *inactivates* AcCoA by phosphorylation. In fact it does double duty, because it *activates* malonyl-CoA decarboxylase, which also downregulates fatty acid biosynthesis. There is an enzyme that converts AMP-APK$^0$ to AMP-APK* and one that catalyzes the reverse process; both of these enzymes are themselves regulated. The ratio of AMP-APK*/AMP-APK$^0$ is *increased* by exercise, pioglitazone, metformin and statins (such as atorvastatin), all of which are to some extent antidiabetic.

AMP-activated protein kinase is also present in the hypothalamic region of the brain, where it functions as a regulator of appetite and a controller of body weight. It does this in a way that runs counter to its antidiabetic effect in the rest of the body: an increase in the level of AMP-APK* actually causes an *increase* in appetite. The atypical antipsychotics olanzapine and clozapine, which cause weight gain as a serious side effect, *increase* AMP-APK* in the brain (*see* page 232)

The crystal structure of the core of the AMP-APK complex is shown below;[3] **Panel A** shows the three main subunits of the enzyme and **Panel B** shows the specific hydrogen-bonding interaction between AMP and the enzyme.

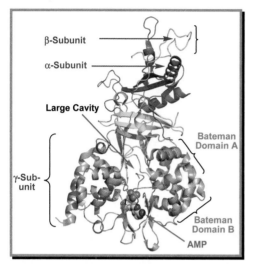

**Panel A:** AMP-activated protein kinase with AMP bound in the catalytic γ-subunit.

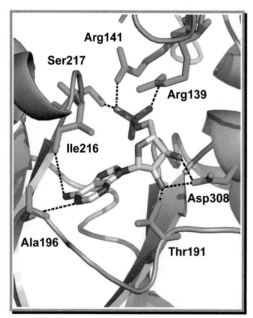

**Panel B:** Close-up view of AMP bound in the catalytic γ-subunit of the AMP-activated protein kinase. AMP is held in place by multiple hydrogen-bonding interactions.

The catalytic γ-subunit binds AMP which substantially increases kinase activity. The α- and β-subunits are regulatory subunits. When the α-subunit is phosphorylated, AMP-APK becomes approximately 100-fold more active as a kinase. Removal of the phosphate group of the α-subunit is inhibited when AMP binds to the catalytic γ-subunit.

Insulin plays a major role in the development of type 2 diabetes. The association of insulin with its receptor initiates a complex signaling sequence which, if dysfunctional at any stage, contributes to metabolic syndrome. Insulin is also anti-inflammatory.

The anti-inflammatory steroid cortisol is a factor in metabolic syndrome, stress and mental function (*see* also prednisone on page 44). At above-optimal levels it leads to elevated circulating glucose, weight gain, mental stress and anxiety.

**Cortisol**

Cushing's disease, which is caused by tumor-induced overproduction of cortisol in the adrenals and excessive circulating cortisol, leads to obesity, high blood pressure, osteoporosis, muscle weakness and several other problems. In contrast, patients with Addison's disease, who have a deficiency of cortisol, are both thin and non-diabetic.

Untreated metabolic syndrome generally progresses with age and leads to type 2 diabetes and the subsequent deterioration of health. In the more advanced stages, the pancreas becomes so damaged that the β-islet cells can no longer produce insulin, so that it must be administered to sustain life. Irreversible damage accumulates in the various organs of the body leading to loss of function that extends even to eyesight.

Although susceptibility to metabolic syndrome, mitochondrial dysfunction and type 2 diabetes may be inherited, only a few relevant gene mutations have been identified. The lines of defense against these conditions include:

(1) dietary weight control;

(2) physical activity and

(3) antidiabetic medication (*see* pages 59-63)

1. *Mayo Clin. Proc.* **2006**, *81*, 1615-1620; 2. *J. Clin. Endocrinol. Metab.* **2007**, *92*, 399-404; 3. *Science* **2007**, *315*, 1726-1729 (**2OOX**); Refs. p. 82

# ANTIDIABETIC

# AND

# CHOLESTEROL-LOWERING

# AGENTS

# METFORMIN (GLUCOPHAGE™)

| Structural Formula | Ball-and-Stick Model | Space-filling Model |
|---|---|---|

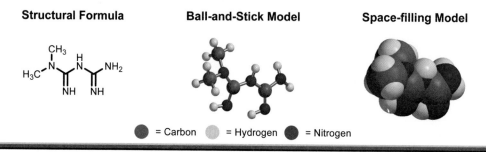

● = Carbon  ○ = Hydrogen  ● = Nitrogen

*Year of discovery:* 1957; *Year of introduction:* 1979 (BMS; FDA approval only in 1994); *Drug category:* Antidiabetic biguanide; *Main uses:* To lower blood glucose and to reduce cardiovascular complications; *Approximate number of people treated annually:* Over 6 million; *Other brand names:* Diabex, Diaformin and Fortamet; *Related drugs:* Troglitazone (Rezulin), Rosiglitazone (Avandia), Pioglitazone (Actos).

A key indicator of diabetes is persistent fasting blood glucose levels above a value of ca. 12.5 mg/mL which arise from defective conversion of glucose into energy. The utilization of glucose by organs and tissues is controlled by the hormone insulin which is produced in the β-islet cells of the pancreas. When this process becomes inefficient for any reason, blood glucose levels increase and cells are deprived of energy. Persistently high blood glucose levels (hyperglycemia) result in widespread damage in the body that can eventually lead to hypertension, heart disease, stroke, impaired circulation, nerve dysfunction, pain, infection, or organ failure.

There are two main categories of diabetes, type 1 and type 2. In type 1 diabetes, usually designated as insulin-dependent diabetes, the β-cells of the pancreas no longer produce insulin (e.g., due to autoimmune destruction of these cells) and administered insulin is required to sustain life. In type 2 diabetes, noninsulin-dependent or obesity related diabetes, insulin action is impaired (insulin resistance), blood glucose levels are elevated and both metabolism and energy production are compromised.

The onset of type 2 diabetes usually occurs in adults over the age of 40 and is often associated with excessive body weight (and especially abdominal fat).[1] The type 2 condition is multifactorial with genetic, nutritional, body weight, physical activity, and other components. Type 2 diabetes accounts for 90% of all cases of diabetes and affects over 60 million people in the US and Europe. Type 2 diabetes is a progressive disease for which there is currently no cure, although it can be managed. If not treated, people eventually develop the more serious combined type 1 and type 2 condition.[2]

Several classes of blood glucose lowering medications have been developed over the past half century to help normalize blood glucose in type 2 diabetes. They each work by a different mechanism. The most commonly used oral drug for type 2 diabetes, metformin (a biguanide), works by reducing glucose release from the liver, increasing the transport of glucose into muscle, hence increasing insulin sensitivity, and also decreasing the absorption of glucose from the gastrointestinal tract. Metformin acts biochemically to increase the level of adenosine monophosphate-activated protein kinase (AMP-APK), an enzyme that plays a control role in energy production and fatty acid metabolism. In turn, AMP-APK raises the levels of ATP, the source of power in cells.[3] Another beneficial effect of metformin is the downregulation of protein kinase C, a mediator of inflammation and, for example, a cause of macular degeneration.

Although metformin was discovered in 1957, it was not approved by the FDA until 1994 for the treatment of type 2 diabetes. Currently metformin is marketed both as a generic drug and under the brand name Glucophage. Metformin is also available in combination with the DPP-4 inhibitor sitagliptin (Janumet™, Merck; *see* page 63).

1. *Annual Review of Pathology: Mechanisms of Disease* **2007**, *2*, 31-56; 2. *Int. J. Vitam. Nutr. Res.* **2006**, *76*, 172-177; 3. *Diabetes, Obesity and Metabolism* **2006**, *8*, 591-602; Refs. p. 82

# GLIPIZIDE (GLUCOTROL™)

| Structural Formula | Ball-and-Stick Model | Space-filling Model |
|---|---|---|

● = Carbon    ● = Hydrogen    ● = Oxygen    ● = Nitrogen    ● = Sulfur

*Year of discovery:* ca. late 1970s; *Year of introduction:* 1984 (Pfizer); *Drug category:* Potassium channel blocker in pancreas beta cells (β-cells)/sulfonylurea class (sulfonylurea functional group is highlighted in red in the structures); *Main uses:* For treatment of type 2 diabetes; *Related drugs:* Tolazamide (Orinase), Tolbutamide (Tolinase), Glibenclamide (Glynase), Glimepiride (Amaryl).

The various medicines for type 2 diabetes take advantage of several different mechanisms of action (*see* metformin, sitagliptin phosphate and pioglitazone). The class of drugs known as sulfonylureas works by stimulating insulin secretion in pancreatic β-cells. This beneficial activity of sulfonylureas was discovered in 1942 by Marcel Janbon and coworkers who observed that these compounds lower blood glucose levels in experimental animals.[1]

Glipizide, a second generation sulfonylurea which is currently an important treatment of type 2 diabetes, has been widely used since it entered the market in the 1980s under the trade name of Glucotrol. Glipizide and other sulfonylureas bind to ATP-sensitive potassium ion (K⁺) channels traversing the membranes of β-cells and partially block these channels. Changes in the electric potential of the cell result and voltage-gated calcium ion ($Ca^{2+}$) channels open, allowing $Ca^{2+}$ ions into the cell. The higher levels of $Ca^{2+}$ in pancreatic β-cells stimulate insulin production and increase the secretion of insulin.

Ion channels are essential to mammalian life. They allow precise control of the levels of critical ions such as sodium ($Na^+$), potassium ($K^+$) and calcium ($Ca^{2+}$) in cells. Each ion flows through ion channels embedded in the cell membrane that are selective for it. Potassium channels are ubiquitous in cells and control a variety of cell functions (e.g., the shape of action potential in neurons, the duration of action potential in heart muscle, and the secretion of hormones). Potassium channel malfunction adversely effects the production of the hormone insulin by the β-cells of the pancreas. Lowered insulin levels result in elevated blood glucose levels and eventually in type 2 diabetes.[2]

Recent research suggests that in addition to increasing insulin production (which falls off with prolonged use), sulfonylureas also sensitize β-cells to glucose, limit excess glucose production in the liver and decrease metabolism of insulin in the liver. It has also been reported that sulfonylureas increase the levels of the protein which transports glucose into cells.

Glipizide can be used as a monotherapy or, at lower doses, in combination with other diabetes medications such as biguanides and thiazolidinediones. Side effects of glipizide include weight gain (which also can occur with insulin administration and various other antidiabetic therapies) and hypoglycemia (dangerously low concentration of blood glucose). Thus, the dose must be carefully chosen.[3] Profound or long-lasting hypoglycemia can produce permanent nerve damage and even death.

Glimepiride (Amaryl™, Sanofi Aventis), a third-generation sulfonylurea, was introduced in 1995. It is advantageous for those with impaired liver or kidney function.

**Glucose**    **Glimepiride (Amaryl™)**

1. *Drugs* **2004**, *64*, 1339-1358; 2. *Lancet* **2005**, *365*, 1333-1346; 3. *British Journal of Diabetes & Vascular Disease* **2006**, *6*, 159-165; Refs. p. 82

# PIOGLITAZONE (ACTOS™)

| Structural Formula | Ball-and-Stick Model | Space-filling Model |
|---|---|---|

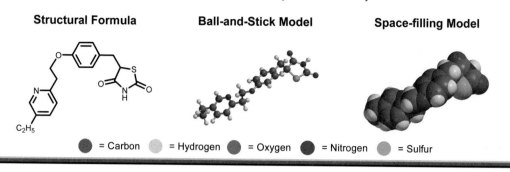

●= Carbon　　○= Hydrogen　　●= Oxygen　　●= Nitrogen　　●= Sulfur

*Year of discovery:* ca. 1990 by Takeda, Japan; *Year of introduction:* 1999 (Takeda & Eli Lilly); *Drug category:* Peroxisome proliferator-activated receptor-γ (PPARγ) activator/thiazolidinedione class of insulin sensitizers; *Main uses:* For treatment of type 2 diabetes; *Related drugs:* Rosiglitazone (Avandia, SKB).

Insulin resistance and elevated blood glucose predispose to type 2 diabetes and cardiovascular disease even when insulin levels are normal. A class of drugs known as thiazolidinediones was long known also to reduce blood glucose levels. However, these early agents generally were found to cause serious liver toxicity. Indeed, the first thiazolidinedione to be approved for human use in the US (troglitazone) had to be withdrawn from the market because it caused life-threatening liver damage in a tiny subset of patients. It was not until recently that safer thiazolidinediones were developed. Pioglitazone hydrochloride was approved in 1999 specifically for the treatment of type 2 diabetes and it is sold as Actos.[1] Rosiglitazone, another approved thiazolidinedione, is marketed as Avandia.

**Rosiglitazone (Avandia™)**

Pioglitazone is a highly selective and potent activator (agonist) for peroxisome proliferator-activated receptor-γ (PPARγ). PPARs are found in skeletal muscle and liver tissues that are the targets for insulin action. PPARs are ligand-activated transcription factors that belong to the nuclear hormone receptor family and are activated by dietary fatty acids and eicosanoids. Activation of PPAR nuclear receptors results in the transcription of a number of insulin-responsive genes involved in the control of glucose and lipid metabolism. Among the overall effects of PPARγ receptor activation are increased

glucose metabolism and insulin sensitivity, and downregulation of insulin in liver and blood.[2]

Recent evidence suggests that thiazolidinediones actually preserve β-cell function and protect cardiovascular and renal function in people with type 2 diabetes. Pioglitazone was also shown to reduce excess liver fat by over 50% in patients with nonalcoholic steatohepatitis, a fatty liver disease that can cause liver failure. This condition affects about 5% of Americans and occurs primarily as a result of obesity, insulin resistance, diabetes and elevated cholesterol. The progression of type 2 diabetes to more advanced stages may in many cases be halted by a combination of weight control, diet and antidiabetic medications such as pioglitazone.

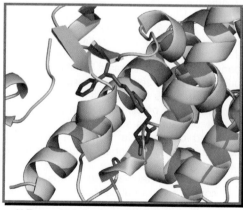

X-Ray crystal structure of rosiglitazone bound to PPAR-γ[3]

1. *Treatm. Endocrinol.* **2006**, *5*, 189-191; 2. *Curr. Diabet. Rev.* **2007**, *3*, 67-74; 3. *Nature* **1998**, *395*, 137-143 (**2PRG**); Refs. p. 82

# SITAGLIPTIN (JANUVIA™)

**Structural Formula**    **Ball-and-Stick Model**    **Space-filling Model**

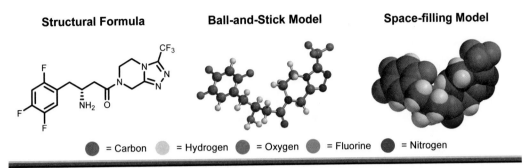

● = Carbon    ● = Hydrogen    ● = Oxygen    ● = Fluorine    ● = Nitrogen

---

*Year of introduction:* 2006 (Merck); *Drug category:* Dipeptidyl peptidase-4 (DPP-4) inhibitor; *Main uses:* For treatment of type 2 diabetes by enhancing the body's own ability to lower elevated blood sugar levels. It is used as a monotherapy as well as an add-on therapy along with other oral diabetes medications such as metformin or thiazolidinediones (e.g., pioglitazone); *Related drugs:* Vildagliptin (Galvus).

Since there are about 200 million people worldwide who have type 2 diabetes (*see* metformin on page 60), and no medicines for reversing or curing this disease, the discovery of molecules that correct the various underlying biochemical abnormalities is of great importance.

The incretins are naturally occurring peptide hormones that play a major role in the glucose-insulin pathway that is central to type 2 diabetes.[1] They signal the β-cells in the pancreas to release insulin and the α-cells to suppress glucose release from the liver, the latter by lowering the production of glucagon (also a peptide hormone). The most relevant incretins are glucagon-like peptide 1 (GLP-1) and glucose-dependent insulinotropic peptide (GIP). These incretins, which are produced by and released from the gut, do not last long in the body since they are rapidly inactivated by an enzyme called dipeptidyl peptidase-4 (DPP-4), which cleaves a two amino acid segment from the amino end (called the N-terminus) of the protein.[2] In a healthy person the synthesis of incretins peaks after each meal. The incretin process can be impaired in type 2 diabetics in two ways: (1) GLP-1 levels are decreased or (2) the response to GIP is diminished. It follows that blood glucose levels could be downregulated by blocking the enzymatic activity of DPP-4.

Several DPP-4 inhibitors are now in pharmaceutical development. Sitagliptin (Januvia, Merck) was approved by the FDA for human use as an antidiabetic agent in late 2006. This agent may be used as monotherapy or combined with metformin, glipizide, or

pioglitazone, as determined to be optimum for an individual patient.

An important aspect of the drug discovery process with these agents was the achievement of high selectivity for inhibition of DPP-4 over several other structurally related dipeptidyl peptidases that function to cleave other proteins in the body. High selectivity was essential to minimizing side effects. Of course, the optimized structure also had to jump over the usual hurdles of oral absorption, metabolic stability, attainment of an effective concentration of unbound drug in serum, and freedom from other undesired activities. It seems likely that DPP-4 inhibitors will become an important class of antidiabetic agents. The figure below shows the crystal structure of DPP-4 with sitagliptin (red) bound in the active site.[3]

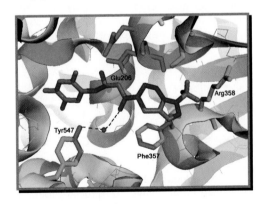

1. *Lancet* **2006**, *368*, 1696-1705; 2. *Ann. Pharmacother.* **2007**, *41*, 51-60; 3. *J. Med. Chem.* **2005**, *48*, 141-151 (1X70); Refs. p. 83

# ATORVASTATIN (LIPITOR™)

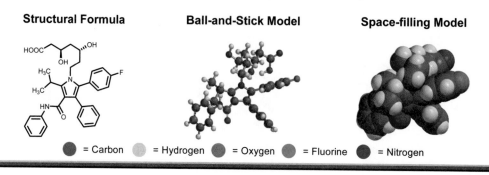

**Structural Formula**    **Ball-and-Stick Model**    **Space-filling Model**

● = Carbon    ○ = Hydrogen    ● = Oxygen    ● = Fluorine    ● = Nitrogen

*Year of discovery:* **1985**; *Year of introduction:* **1997 (Pfizer)**; *Drug category:* **Statins**; *Main uses:* **For the reduction of cholesterol levels (LDL) in blood and reducing the risk of cardiovascular disorders such as heart attack and stroke**; *Approximate number of people treated annually:* **Over 10 million**; *Related drugs:* **Lovastatin (Mevacor), Pravastatin (Pravachol), Simvastatin (Zocor), Fluvastatin (Lescol), Rosuvastatin (Crestor).**

Cholesterol is a critical component of cell membranes in mammals and the precursor of all the steroid hormones. Although essential for the maintenance of normal cell function, this important substance also plays a role in the development of cardiovascular disease since elevated cholesterol levels in the blood correlate with arterial plaque formation. If left untreated, growth of such plaques can eventually cause heart attack or stroke. About 80% of the cholesterol in the body is synthesized internally, mainly in the liver; the remainder is dietary.

Research over many years for drugs that decrease circulating cholesterol led to the discovery of compounds now known as statins. The first of these, compactin, a naturally occurring compound produced by fungi, was discovered in 1976 by Dr. Akira Endo of Japan. A compound of similar structure, lovastatin, was approved by the FDA in 1987 and marketed as Mevacor by Merck.

**Compactin**    **Lovastatin (Mevacor™)**

These drugs inhibit the enzyme HMG-CoA reductase that plays a crucial role in cholesterol biosynthesis. Since cholesterol is not soluble in water, it is carried in the blood by water-soluble lipoproteins of two types, LDL and HDL. Low density lipoprotein (LDL), is sometimes called "bad cholesterol" since

high levels of LDL in blood lead to plaque formation. On the other hand, high density lipoprotein (HDL), seems to protect the cardio-vascular system.[1,2] Since the early 1960s more than a dozen statins have been developed and commercialized, and currently many major pharmaceutical companies market a statin drug. The most widely used and potent statin, atorvastatin (Lipitor™), was first synthesized in 1985 by Dr. Bruce Roth in the US. This totally synthetic statin was introduced in 1997 and achieved sales over $13 billion in 2006.

The complete three-dimensional structure of the enzyme HMG-CoA reductase that regulates the production of cholesterol has been determined with the inhibitor atorvastatin bound to the critical catalytic site (the location for normal substrate binding); a close-up view is shown below.[3]

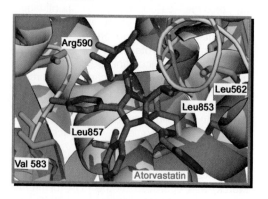

1. *Vascular Pharmacology* **2007**, *46*, 1-9; 2. *Current Atherosclerosis Reports* **2006**, *8*, 41-49; 3. *Science* **2001**, *292*, 1160-1164 (**1HWK**); Refs. p. 83

# EZETIMIBE (ZETIA™)

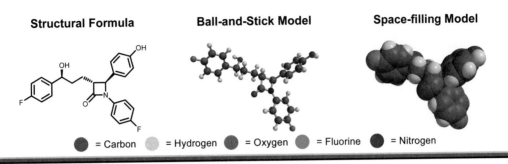

**Structural Formula**   **Ball-and-Stick Model**   **Space-filling Model**

● = Carbon   ● = Hydrogen   ● = Oxygen   ● = Fluorine   ● = Nitrogen

*Year of discovery:* 1998; *Year of introduction:* 2002 (Schering-Plough & Merck); *Drug category:* Antihypolipidemic agent/selective cholesterol absorption inhibitor; *Main uses:* For the reduction of high cholesterol levels (LDL) as an adjunct to dietary measures, as a monotherapy, or in combination with statins (e.g., atorvastatin, simvastatin); *Other brand names*: Ezetrol and Ezemibe; *Related drugs:* Atorvastatin (Lipitor), Lovastatin (Mevacor), Simvastatin (Zocor).

Tight control of the amount of cholesterol in the body is essential for long-term human health. Cholesterol regulation in humans is a complex process in which the level of cholesterol in cell membranes is sensed and coupled to the amount of cholesterol produced by *de novo* biosynthesis, principally in the liver, but also elsewhere (e.g., brain). The steady state amount of cholesterol is the result of internal biosynthesis, dietary intake, metabolism and clearance (mainly in the form of oxidized sterols as bile acids). The majority of dietary cholesterol is absorbed through the intestines. Since unregulated cholesterol levels, especially elevated LDL, may lead to the development of cardiovascular disease, lowering the absorption of dietary cholesterol is an attractive strategy for therapy with the overabundance of cholesterol typical of the western diet. Such an approach is complementary to the attenuation of cholesterol biosynthesis, e.g., by the use of HMG-CoA reductase inhibitors (*see* atorvastatin on page 64).

During the late 1990s, a research group at Schering-Plough identified a novel compound, ezetimibe, that selectively inhibits the absorption of cholesterol. In 2002 it was approved and marketed as Zetia. Although several naturally occurring plant sterols (e.g., sterols in soy) can inhibit the absorption of cholesterol, they are much less potent than ezetimibe.[1]

Ezetimibe acts in the small intestine where it localizes in the brush border (hair-like structures that facilitate the absorption of nutrients). It specifically binds to a protein that normally transports cholesterol through the intestinal wall into the circulation, and in this way reduces the absorption of dietary cholesterol. There is evidence that cholesterol is coupled to a sugar before transport. Ezetimibe apparently interferes either with this process, the transport itself or both. Only a minor fraction of administered ezetimibe is absorbed into the bloodstream, minimizing the amount in the body and lessening the chance of adverse reactions.

Ezetimibe has also been approved as combination therapy with statins (e.g., ezetimibe/simvastatin, which is known as Vytorin) since their cholesterol-lowering effects are additive.[2] When simvastatin (*see* below) is used alone, the reduction of LDL is about 35% whereas in combination with ezetimibe it is about 50%.

**Simvastatin (Zocor™)**

Ezetimibe is especially beneficial for those who do not tolerate statins because of muscle weakness or pain that can occur with these agents. In rare cases an extreme toxicity called rhabdomyolysis (a rapid breakdown of muscle fibers and release of toxins into the blood) can result from the use of statins.[3]

1. *Exp. Opin. Invest. Drugs* **2006**, *15*, 1337-1351; 2. *Nat. Clin. Prac. Card. Med.* **2006**, *3*, 664-672; 3. *Exp. Opin. Drug. Safety* **2006**, *5*, 651-666; Refs. p. 83

*Medicine brings hope, health, happiness, and extends life. There is hardly a better investment to be found than the purchase of an efficacious medicine, especially one that prevents disease from developing.*

*It was not always so. The famous 12th century physician and rabbinical sage Moses Maimonides recommended the consumption of chicken soup, prepared from fat hens, for the treatment of asthma, leprosy or migraine headaches.*

# CARDIOVASCULAR
AGENTS

# ATENOLOL (TENORMIN™)

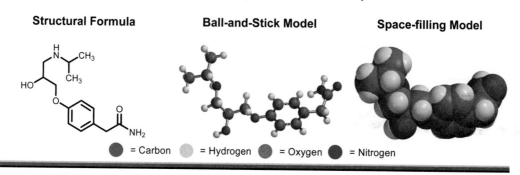

| Structural Formula | Ball-and-Stick Model | Space-filling Model |

● = Carbon    ● = Hydrogen    ● = Oxygen    ● = Nitrogen

*Year of discovery* of the first beta blocker (β-blocker): **1958**; *Year it entered the market:* **1976** (AstraZeneca); *Drug category:* β-blocker – a substance that affects the heart and the circulatory system (veins and arteries) by slowing the heart rate and relaxing the blood vessels. *Main uses:* To lower blood pressure, lower heart rate, reduce chest pain (angina), reduce risk of or attenuate heart attack. *Approximate number of people treated annually:* *Over* 20 million; *Related drugs:* Propranolol, Practolol, Nadolol, Oxprenolol, Penbutolol, Betaxolol, Metoprolol, Nebivolol, Carvedilol, Labetalol. Timolol (for glaucoma).

Hypertension, also known as high blood pressure, is a clinical condition in which the arterial blood pressure at rest exceeds on average 130/90 mm Hg. If left untreated, this condition promotes impaired circulation and cardiovascular disease, which can lead to heart attack or stroke. The incidence of hypertension increases with excess weight, age and family history of the disease. Statistics show that by 2005 more than 65 million people in the US had hypertension.

In the 1950s James Black, a British pharmacologist, and a team of chemists studied the possibility that molecules which could block the binding of β-adrenergic amines (e.g., adrenaline) to the β-adrenergic receptor might reduce blood pressure. This work led to the first successful antihypertensive β-blocker (and the 1988 Nobel Prize in Medicine).

The usual approach to drug discovery before the 1960s was to synthesize a large number of compounds and test them on animals. With Black's work and the growing knowledge of the biochemical basis of a given medical condition, the design of drugs became more rational. The development of β-blockers is an early example of this involving two important catecholamine hormones, adrenaline and noradrenaline (*see* structures on the right). The stimulation of the sympathetic nervous system (e.g., by fear) results in the increase of the adrenaline level that in turn increases blood pressure. Adrenaline and noradrenaline act by binding to the ubiquitous adrenergic receptors (adrenoceptors) of two types, α and β. Upon binding, the heart is stimulated while the blood vessels are constricted (this is the so-called fight or flight response).

**Adrenaline**      **Noradrenaline**

Atenolol is a compound that binds competitively to the β-adrenoceptor and blocks the action of adrenalin (thus the name β-blocker). This action slows the heart and lowers blood pressure. For patients with angina or congestive heart failure, atenolol is especially beneficial, since it reduces the oxygen demand of the heart muscle. However, β-blockers are contraindicated for asthmatics because they cause bronchoconstriction, but bronchodilating β-*agonists* are beneficial (*see* salmeterol, a β-agonist on page 50).[1]

The family of β-blocker compounds has been so successful that many major pharmaceutical companies have developed their own versions. Currently there are about 20 β-blocker drugs on the market.[2, 3]

1. *Curr. Pharm. Des.* **2007**, *13*, 229-239; 2. *Curr. Hypertens. Rep.* **2006**, *8*, 279-286; 3. *Curr. Hypertens. Rev.* **2007**, *3*, 15-20; Refs. p. 83

# ENALAPRIL (VASOTEC™)

**Structural Formula**     **Ball-and-Stick Model**     **Space-filling Model**

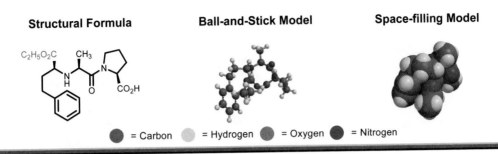

● = Carbon     ● = Hydrogen     ● = Oxygen     ● = Nitrogen

*Year of discovery:* 1981; *Year of introduction:* 1985 (Merck); *Drug category:* Angiotensin-converting enzyme (ACE) inhibitor; *Main uses:* Treatment of hypertension, congestive heart failure and atherosclerosis; *Related drugs:* Captopril (Captopril), Ramipril (Altace), Quinapril (Accupril), Perindopril (Aceon), Lisinopril (Prinivil, Zestoretic), Benazepril (Lotensin).

Angiotensin II is a very potent human blood pressure regulator that causes contraction of blood vessels and an increase in blood pressure. Lowering angiotensin II levels reduces blood pressure.

The history of angiotensin II dates back to 1898, when Robert Tigerstedt, a physiologist at the Karolinska Institute in Stockholm, discovered that crude extracts of the kidney contained a substance that led to increased blood pressure in laboratory animals. He named this substance renin.

During the following 100 years it was revealed that renin is an enzyme that is part of the complex renin-angiotensin system. Renin catalyzes the transformation of angiotensinogen, a protein produced in the liver, into angiotensin I, which has only a modest effect on blood pressure, but is the direct precursor of angiotensin II, a very potent pressor substance. This reaction is effected in the body by the angiotensin-converting enzyme (ACE). In 1967, John Vane and Sergio Ferreira at Oxford University found that the dried extract of Brazilian pit viper venom, which was known to reduce blood pressure, acts as a potent inhibitor of ACE. Vane, who was a consultant at the pharmaceutical company Squibb (now Bristol-Myers Squibb) at the time, suggested that the company investigates the venom's antihypertensive effect. Because the active peptides of the venom extract are unstable in the stomach, they were not orally active. However, chemists at the Squibb Co. succeeded in designing a non-peptide molecule that showed high inhibitory activity and excellent oral bioavailability. This substance, named captopril, entered the

market in 1978 under the brand name Capoten as the first commercially available ACE inhibitor for the treatment of hypertension. This drug became a huge therapeutic success, but because it had to be taken twice daily, it was overtaken by once-a-day relatives such as enalapril (Merck).

Enalapril is an orally active prodrug that is converted into the active drug, enalaprilat, in the gut. The prodrug to drug bioconversion involves the change of the carboxyethyl ($COOC_2H_5$) group of enalapril (shown in red in the above structure) to the carboxyl (COOH) of enalaprilat. Although enalaprilat is a very potent ACE inhibitor, it is not active when taken orally because of poor absorption.[1,2] Enalapril gained FDA approval in 1985 and became a billion-dollar drug a few years later. The specific mode of binding of enalapril to the ACE enzyme was determined using a high resolution X-ray crystal structure. The close-up image below shows enalaprilat directly interacting with the catalytic $Zn^{2+}$ ion at the active site of the enzyme.[3]

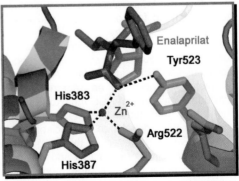

1. *Drugs* **1986**, *31*, 198-248; 2. *Drugs* **1992**, *43*, 346-381; 3. *Biochemistry* **2004**, *43*, 8718-8724 (1UZE); Refs. p. 83

# CANDESARTAN CILEXETIL (ATACAND™)

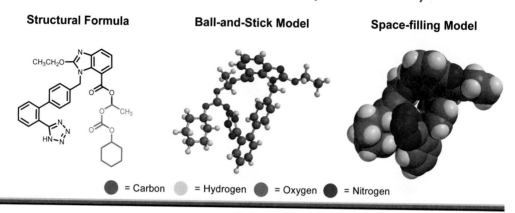

| **Structural Formula** | **Ball-and-Stick Model** | **Space-filling Model** |

● = Carbon   ● = Hydrogen   ● = Oxygen   ● = Nitrogen

*Year of discovery:* 1991; *Year of introduction:* 1998 (AstraZeneca); *Drug category:* Angiotensin receptor blocker (ARB); *Main uses:* Treatment of hypertension and chronic heart failure; *Related drugs:* Eprosartan (Teveten), Losartan (Cozaar), Olmesartan (Benicar), Telmisartan (Micardis), Valsartan (Diovan).

Candesartan cilexetil belongs to the newest class of pharmaceuticals called angiotensin receptor blockers (ARB-s) that are used for the treatment of high blood pressure and chronic heart failure.[1]

Angiotensin receptor blockers, like angiotensin-converting enzyme (ACE) inhibitors, act on the renin-angiotensin system (RAS), which plays a major role in regulating blood pressure. The main pressor substance in the renin-angiotensin system is angiotensin II. ACE inhibitors, as discussed in the case of enalapril (*see* page 69), attenuate the biosynthesis of angiotensin II. Angiotensin receptor blockers affect the last step of the RAS cascade by selectively binding to, but not activating, the angiotensin II receptors, thereby, blocking their interaction with angiotensin II. The result is reduction of blood pressure, as with ACE inhibition.

Although ACE inhibitors are first-line therapy in heart failure, they can cause side effects because they also inhibit the breakdown of bradykinin, which can lead to persistent dry cough in a sizeable fraction of the population (especially Asian). Angiotensin receptor blockade is an alternative for individuals intolerant of ACE inhibitor therapy. The first angiotensin receptor blocker, saralasin, was developed by scientists at Norwich Eaton. However, as this drug was a peptide, it was not orally active and was also quickly metabolized. The first non-peptide angiotensin receptor blocker, losartan, was

discovered at DuPont Merck Pharmaceuticals in 1990, and introduced in 1995 under the brand name Cozaar. The most prescribed ARB is Novartis' valsartan (Diovan), which generated more than $2.4 billion in sales in 2003. Valsartan was approved by the FDA in 1996. The competitor drug of AstraZeneca, candesartan cilexetil, entered the market in 1998 under the brand name Atacand.

**Losartan (Cozaar™)**     **Valsartan (Diovan™)**

Candesartan cilexetil is a prodrug that is converted to the active drug, candesartan, during absorbtion in the gastrointestinal tract by hydrolytic cleavage of the subunit at the lower right of the above structural formula (shown in red). Candesartan binds to the angiotensin II receptor more strongly than the other members of the sartan drug family allowing the use of lower doses and minimizing side effects.[2]

1. *Expert Opin. Pharmaco.* **2004**, *5*, 1589-1597; 2. *Am. J. Cardiol.* **1999**, *84*, 3S-8S; Refs. p. 84

# ALISKIREN (TEKTURNA™)

| Structural Formula | Ball-and-Stick Model | Space-filling Model |
|---|---|---|

● = Carbon ● = Hydrogen ● = Oxygen ● = Nitrogen

*Year of discovery:* 1995 (Ciba-Geigy, now Novartis); *Year of introduction:* March 2007; *Drug category:* Renin inhibitor; *Main uses:* For the treatment of hypertension.

Aliskiren is a recently approved (2007) antihypertensive agent that blocks the enzyme renin, an important regulator of blood pressure that is biosynthesized in the kidney. It cleaves the amino acid chain (453 amino acid subunits) of the protein angiotensinogen to form the decapeptide angiotensin I. Angiotensin I is further transformed by another enzyme, angiotensin-converting enzyme (ACE, *see* page 69), to angiotensin II, an octapeptide, with cleavage of two amino acids from the carboxyl terminus. Angiotensin I and angiotensin II both increase blood pressure, by promoting sodium retention and vaso-constriction.

Currently used antihypertensive medications (e.g., ACE inhibitors, angiotensin receptor blockers, β-blockers) and aldosterone antagonists) are not universally effective and do not reach the therapeutic goal (130/80 mm Hg) in about 70% of those treated. Some of them cause side effects. All ACE inhibitors, including enalapril (*see* page 69), cause serious airway inflammation, cough and angioedema in 15-20% of individuals due to the concurrent inactivation of bradykinin. ACE inhibition does not prevent the conversion of angiotensin I to angiotensin II by other proteases.

The inhibition of renin has been considered to be an attractive antihyper-tensive approach since 1957 because it might cause fewer side effects than the ACE inhibitors and also be more effective, since the levels of both angiotensins I and II would be reduced.[1] It is not surprising that the development of renin inhibitors was pursued simultaneously in many research laboratories over a period of several decades. However, success proved to be remarkably elusive.

The first safe and effective renin inhibitor, aliskiren, emerged from just such an effort at the Ciba-Geigy Co (now Novartis). Although a number of active inhibitors were discovered at Ciba-Geigy and elsewhere, none proved to be useful. Most of the therapeutic candidates were peptide-like in structure (peptidomime-tic) and were either too readily broken down in the stomach or non-bioavailable because of poor absorption. Some had unacceptable side effects. The search was further complicated by the fact that animal models (e.g., rat, dog, rabbit, pig) did not predict human activity. However, by the use of a doubly transgenic rat, containing both human renin and angiotensinogen genes, it was possible to develop a practical animal model for the evaluation of renin inhibition *in vivo* for candidate drugs.[2]

The extensive use of molecular modeling and X-ray structure analysis of various renin-inhibitor complexes guided the path to the discovery of aliskiren, an orally active, non-peptide, potent (low nM range) and specific human renin inhibitor.[3]

Aliskiren is a transition state analog. It mimics the structure of the high-energy, evanescent complex at the very top of the energy barrier that determines the rate of the biochemical reaction. It is more hydrophilic and more efficiently absorbed after oral administration than earlier renin inhibitors. In clinical trials, a single daily dose of aliskiren over 4 weeks reduced baseline blood pressure on average by over 30% and provided effective 24-hour renin inhibition with minimal side effects.[4]

The experience gained in the research on renin inhibition proved to be very helpful in the development of anti-AIDS HIV protease inhibitors.

1. *Adv. Drug. Del. Rev.* **2006**, *58*, 824-833; 2. *Lancet* **2006**, *368*, 1449-1456; 3. *Am. J. Hypertens.* **2007**, *20*, 11-20. 4. *Exp. Opin. Invest. Drugs* **2006**, *15*, 1269-1277; Refs. p. 84

# AMLODIPINE (NORVASC™)

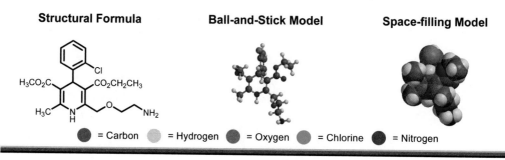

| Structural Formula | Ball-and-Stick Model | Space-filling Model |

● = Carbon ● = Hydrogen ● = Oxygen ● = Chlorine ● = Nitrogen

*Year of discovery:* 1970s; *Year of introduction:* 1990 (Pfizer); *Drug category:* Calcium channel blocker (or calcium antagonist) – a substance that affects the heart muscle, smooth muscles or neuron cells; *Main uses:* To treat angina (chest pain), hypertension (high blood pressure) and abnormally rapid heart beat (e.g., atrial fibrillation); *Approximate number of people treated annually:* Over 6 million; *Related drugs:* Felodipine (Plendil), Nifedipine (Adalat, Procardia), Lacidipine (Motens), Verapamil (Calan), Diltiazem (Cardizem), Nicardipine (Cardene).

Cardiovascular diseases are the leading cause of death worldwide. Two of these conditions, angina (chest pain) and hypertension, affect nearly 1 billion people globally. Angina occurs when the heart muscle receives insufficient oxygen (termed ischemia or anoxia). Blood pressure (force on the walls of arteries), when excessive, increases the workload of the heart, stresses arteries, and eventually causes damage to vital organs (e.g., kidneys and brain). There are multiple biochemical/physiological causes of hypertension.

During the past 50 years, many drugs have been introduced for the treatment of angina and hypertension that work by different mechanisms. These drugs fall into five main categories: (1) β-blockers; (2) ACE (angiotensin-converting enzyme) biosynthesis inhibitors; (3) angiotensin receptor antagonists (4) calcium channel blockers and (5) diuretics, which cause excretion of sodium chloride and a decrease in blood volume and pressure.[1]

Calcium channel blockers are small molecules that block the transport of calcium ions into the muscle cells of the heart and the arteries. Muscle contraction is regulated by the entry of calcium ions into the muscle cells via specific calcium ion channels. Calcium channel blockers decrease the contraction of the heart, dilate the arteries, lower blood pressure, and reduce chest pain.[2] There are certain differences between the currently used calcium channel blocker drugs. They can be either short-acting or long-acting and may or may not have an effect on the heart rate. Amlodipine is a long-acting calcium channel blocker that was developed during the 1980s and marketed in 1990 under the trade name of Norvasc.

Amlodipine is a member of the 1,4-dihydropyridine chemical family which has high bioavailability, a long half-life in plasma (once-a-day dosage), minimal effect on the heart rate, and smooth onset of action with only gradual decrease of blood pressure. The first member of the class was nifedipine (Adalat) which was developed by Bayer in Germany. It was followed by nicardipine (Cardene) which was also widely used for a time.[3]

**Nifedipine (Adalat™)**

**Nicardipine (Cardene™)**

Since its introduction, amlodipine has become the most prescribed antihypertensive agent in the world with annual sales of $4.7 billion in 2005.

1. *Curr. Hypertens. Rev.* **2007**, *3*, 9-13; 2. *Exp. Opin. Pharmacother.* **2006**, *7*, 2385-2401; 3. *Curr. Hypertens. Rev.* **2006**, *2*, 103-111; Refs. p. 84

# NITROGLYCERIN (NITROGLYCERIN™)

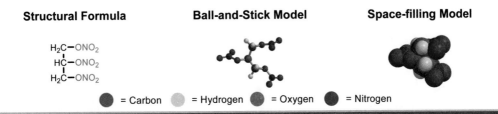

**Structural Formula**  **Ball-and-Stick Model**  **Space-filling Model**

● = Carbon ● = Hydrogen ● = Oxygen ● = Nitrogen

*Year of discovery:* 1847 (Ascanio Sobrero at University of Torino); *Year of introduction:* 1878 (William Murrell at Westminster Hospital); *Drug category:* Vasodilator; *Main uses:* Treatment of angina and congestive heart failure; *Other brand names:* Minitran, Nitro-dur, Nitrolingual, Nitromist, Nitrostat; *Related drugs:* Isosorbide dinitrate (Bidil, Dilatrate, Isordil), Isosorbide mononitrate (Imdur, Ismo, Monoket).

Nitroglycerin, an organic nitrate also known as glyceryl trinitrate, is a vasodilator used for the treatment of angina (chest pain caused by hypoxia in the heart) and congestive heart failure. It is most commonly applied as a sublingual tablet or transdermal patch.

Nitroglycerin, a highly unstable and explosive substance, was first prepared by the Italian chemist Ascanio Sobrero in 1847. It was later utilized by Alfred Nobel, who became acquainted with Sobrero in Paris, as the key component of dynamite. From this discovery, Nobel made the fortune that was used to establish the Nobel Prize.

In the 1860s, the English surgeon, T. L. Brunton reported that the inhalation of isoamyl nitrite vapor rapidly alleviates angina and reduces blood pressure. However, relief lasted only a few minutes and dosing was uncertain.

**Isoamyl nitrite**

This observation prompted William Murrell at Westminster Hospital to test the possibility that nitroglycerin might relieve chest pain. He found that nitroglycerin effectively controls angina and that the beneficial effect lasts for about one hour. Nitroglycerin was introduced into clinical practice in 1878 and remains a principal medication for angina. It is interesting that, years later, Nobel was pre-scribed nitroglycerin as therapy for chest pain.

Nitroglycerin efficiently relieves angina because it dilates the coronary blood vessels and improves oxygenation of the heart. Nitroglycerin is converted *in vivo* into the molecule nitric oxide (NO) that serves as a trigger of vasodilation.

Nitric oxide is produced naturally in the body from the amino acid arginine with the enzyme nitric oxide synthase (NOS) as catalyst.

**L-Arginine**  **L-Citrulline**

Nitric oxide has many functions, one of which is regulation of arterial tone and blood pressure. NO is synthesized by the endothelial cells that line blood vessels. It diffuses from the endothelia to the vascular smooth muscle and activates the enzyme guanylate cyclase by binding to the heme iron in the enzyme. (Heme is the structural unit of hemoglobin that carries oxygen in blood.) This enzyme promotes the formation of cyclic guanosine monophosphate (cGMP). The elevation of cGMP levels decreases intracellular calcium and causes relaxation and vasodilation. For their role in the discovery of nitric oxide as a key regulator in the cardiovascular system, Robert Furchgott, Louis Ignarro and Ferid Murad were awarded the Nobel Prize in Medicine in 1998.

NO plays a role in penile erection, in controlling blood clotting by inhibition of platelet aggregation, and in the immune response, wherein it is produced by macrophages. NO is a very reactive, short-lived species which is deactivated with a half-life of less than a minute.[1]

1. *Chem. Eng. News* **1993**, *71*, 26-38; Refs. p 84.

# CLOPIDOGREL BISULFATE (PLAVIX™)

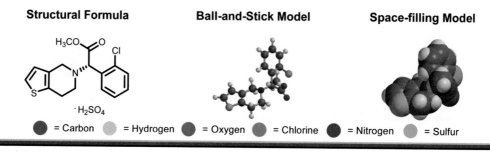

**Structural Formula**   **Ball-and-Stick Model**   **Space-filling Model**

· $H_2SO_4$

● = Carbon   ● = Hydrogen   ● = Oxygen   ● = Chlorine   ● = Nitrogen   ● = Sulfur

*Year of discovery:* 1990; *Year of introduction:* 1997 (Sanofi Aventis/Bristol-Myers Squibb); *Drug category:* Antiplatelet agent — a substance that prevents the formation of blood clots; *Main uses:* reduction of thrombotic events for persons with recent stroke, myocardial infarction or established peripheral artery disease. *Approximate number of individuals treated since its introduction:* Over 20 million; *Related drugs:* Acetylsalicylic acid (Aspirin).

Blood clots cause about 90% of heart attacks and 80% of strokes. Therefore, substances that are capable of preventing platelets from sticking together can be life-saving. Low-dose acetylsalicylic acid – aspirin – (*see* page 38) is currently used to decrease the risk of heart attack and occlusive stroke by inhibiting the biosynthesis of thromboxane, a type of prostaglandin that promotes clotting (*see* below).

**Arachidonic Acid**

Enzymes

**Thromboxane (TXA$_2$)**

The use of aspirin for extended periods carries a small risk of stomach irritation or ulceration even at low doses. The development of clopidogrel was motivated by the need for a safer medicine for aspirin intolerant people.[1]

The Sanofi company (now Sanofi Aventis) identified clopidogrel bisulfate as a very potent oral antiplatelet agent and introduced it in 1997 under the trade name Plavix. It works by a different biochemical mechanism than aspirin.[2] Clopidogrel bisulfate inhibits the binding of adenosine diphosphate (ADP) to its platelet receptor and the subsequent ADP-mediated activation of glycoprotein GPIIb/IIIa,

a proclotting complex, with high selectivity. This modification of the platelet ADP receptor is irreversible, and the platelets that are exposed to clopidogrel are affected for their remaining lifespan (about 2 weeks), thus inhibiting platelet aggregation.

An earlier product by Sanofi/Roche with the same ring system as clopidogrel, ticlopidine, was found to have similar antiplatelet activity but to be toxic to white blood cells (*see* structure below). Clopidogrel is not the biologically active agent (it is a prodrug) as it must undergo certain biotransformations to become active. To date the active metabolite has not been identified.[3]

**Ticlopidine**

Recently, the FDA has approved the use of clopidogrel for individuals who have suffered a common form of heart attack in which a sudden blockage of the coronary artery occurs, a condition affecting an estimated 500,000 Americans each year. It has been shown that treatment with clopidogrel after a heart attack improves blood flow in the damaged coronary artery and reduces the risk of new heart attack. Clopidogrel is also used to reduce the risk of stroke. Annual sales of clopidogrel have exceeded three billion dollars.

1. *Future Cardiol.* **2006**, 2, 343-366; 2. *Exp. Opin. Pharmacother.* **2006**, 7, 1109-1120; 3. *Proc. Natl. Acad. Sci. U. S. A.* **2006**, 103, 11069-11074; Refs. p. 84

# DIGOXIN (LANOXIN™)

**Structural Formula**      **Ball-and-Stick Model**      **Space-filling Model**

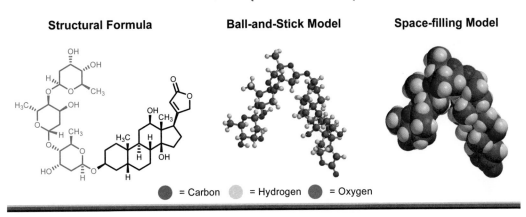

● = Carbon      ● = Hydrogen      ● = Oxygen

---

*Other brand names:* Digitek; *Therapeutic use:* Approximately since 1500 B.C.; *Drug category:* Cardiac glycoside - $Na^+/K^+$ ATPase pump inhibitor; *Main uses:* To treat congestive heart failure, increase cardiac output, and decrease heart rate. *Approximate number of people treated annually:* 550,000 New patients in the US; *Related drugs:* Ouabain.

Cardiac failure, which occurs when the heart can not deliver the required amount of blood to the tissues and organs, is along with ischemia and arrhythmia (irregular heartbeat), a common and major medical problem. This condition is usually caused by damaged and weakened heart muscles (especially the ventricles) that do not contract efficiently. The net effect is reduced circulation, accumulation of blood in the heart, lowered blood pressure and inadequate blood flow to the lungs and kidneys. Digoxin is efficacious in people with heart failure because it increases the contractile force of the heart and improves the blood flow to vital organs. Digoxin remains a useful treatment for this condition despite the fact that the efficacious dose is not far below the toxic dose.[1]

Cardiac glycosides have a long history in medical practice since their early use in the form of plant extracts as diuretics and heart stimulants which dates back to 1500 B.C. The first modern account of the beneficial effects of cardiac glycosides was made in the late 1700s by a British doctor, William Withering, who described the effect of digitalis (extract from foxglove plants) for the treatment of edema (also known as dropsy). Withering noted that the foxglove extract was both beneficial and toxic and that correct dosing was critical to successful treatment.

We now know that the foxglove extract contains two main components, digoxin and digitoxin that are easily separated. Digoxin proved to be the more effective and less toxic

substance. Digoxin works by binding to the α-subunit of the $Na^+/K^+$ ATPase pump in the membranes of heart cells (mycocytes). With the inhibition of the $Na^+/K^+$ pump, the concentration of $Na^+$ increases in the cells. This also leads to an increase of intracellular $Ca^{2+}$ concentration and ultimately to an increase in the force of heart muscle contraction (positive inotropic effect). Until recently digoxin was the first-line treatment for individuals with congestive heart failure. However, it is now administered mainly to those who remain symptomatic despite treatment with diuretics and ACE inhibitors. Digoxin can only be used safely with strict medical monitoring since severe side effects, especially heart rhythm disturbances, may occur.[2, 3]

The structure of digoxin is made up of two distinct moieties. The carbohydrate part (the three six-membered rings at the left of the above structural formula) and the steroidal part (the right side). The carbohydrate part improves aqueous solubility, but it is not essential for activity. Numerous cardioactive glycosides occur in plants which share the steroid framework and differ mainly in the carbohydrate units and in the number and location of OH groups attached to the steroid part. The development of safer versions of digoxin would be medically useful.

1. *Prog. Cardiovasc. Dis.* **2006**, *48*, 372-385; 2. *Exp. Opin. Drug. Safety* **2006**, *5*, 453-467; 3. *Am. J. Cardiovasc. Drugs* **2006**, *6*, 77-86; **Refs. p. 85**

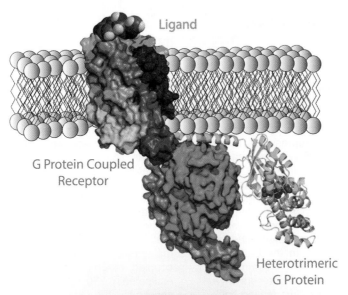

Ligand

G Protein Coupled
Receptor

Heterotrimeric
G Protein

# RECEPTORS
# AND
# SIGNALING

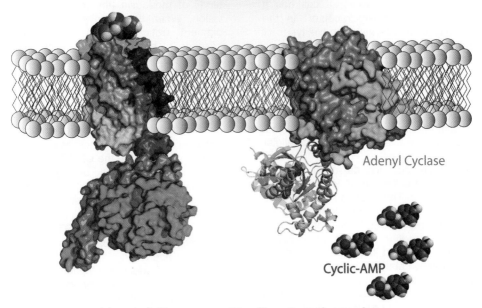

Adenyl Cyclase

Cyclic-AMP

Ligand-Receptor Binding Leads to the
Generation of Many Molecules of
Cyclic-AMP

# INFORMATION FLOW INTO THE CELL BY CHEMICAL SIGNALING

Information is essential to life, even at the unicellular level of a microorganism. A microbe needs to be able to sense the level of nutrients in its environment and migrate toward locations that are progressively richer in nourishment. With multicellular creatures such as man, the flow of information is many orders of magnitude more complicated, diverse, and crucial. At the level of individual cells, each cell must respond to changes around it, and at the level of an organ, each individual cell must be coordinated with the others to optimize function. In the brain, for example, each of the ca. $10^{11}$ neurons are interconnected with each of the others in a network that not only carries information back and forth but also responds to messages from the rest of the body.

A human receives vast amounts of external information via the senses and processes an even greater mass of complex, internally generated information. Chemical signaling is central to both. The information transfer is generally initiated by a specific signaling molecule (e.g., a hormone, a neurotransmitter, or a cell activator) for which a cell has a discriminating, high-affinity receptor. In the human body there are innumerable types of signal-initiating molecules and of their cognate receptors. Generally, a part of the receptor is accessible at the surface of the cell, a part spans the cell membrane and a third section protrudes into the cell. When the receptor is occupied by its ligand, its three-dimensional shape (conformation) changes in a way that switches on an associated intracellular enzyme. The activated enzyme, in turn, catalyzes numerous rounds of a reaction which allows the cell to adjust to the changes in its environment. The incoming signal can be greatly amplified not only because the activated enzyme catalyzes the formation of many molecules of its product but because the product itself can trigger a cascade of chemical change catalyzed by other enzymes. At an appropriate point, the signal is turned off. Thus, a change in the cell's environment is communicated to a cell which makes the appropriate molecular adjustments internally.

There are other paths of communication into a cell. For example, the signaling molecule-receptor complex may be internalized so as to carry the signaling molecule directly into the cell. Once there, it can affect the cell in various ways. For instance, it can bind to a specific protein which then carries it into the nucleus where the carrier protein-signaling molecule complex can regulate gene transcription. There are about 50 such human nuclear hormone receptors which are activated by ligands such as, androgens, estrogens, glucocorticoids, vitamin $D_3$, thyroid hormone, dopamine, serotonin and peptide growth factors and regulators.

The most common signaling mechanism that is targeted by current medicinal molecules involves receptors called G protein-coupled receptors (GPCRs). There are an estimated 950 human GPCRs which constitute the largest family of integral, membrane-spanning proteins. Many of the molecules in this book (roughly 40%) act on GPCRs, so it is important to examine in more detail how these molecular machines function in chemical signaling. There are also a large number of signaling pathways that involve receptors other than GPCRs.

Much of our understanding of the structure of G protein-coupled receptors came from the study of rhodopsin, the receptor involved in vision. When light is absorbed by a visual pigment (retinal) bound to rhodopsin, a cascade of reactions takes place within the cell. These reactions generate nerve signals to the brain that result in a visual image.[1]

Generally, the membrane-spanning part of GPCRs consists of seven $\alpha$-helical domains in a bundle that parallels the fatty acid chains of the membrane. The ligand-binding region lies between 3 or 4 of the $\alpha$-helices of this bundle. The conformation of the GPCR changes upon ligand binding to an arrangement which causes the replacement of guanosine diphosphate (GDP) on an attached G protein $\alpha$-subunit (at the inner membrane) by guanosine triphosphate (GTP). The $\alpha$-subunit–GTP complex then activates the enzyme adenyl cyclase to convert many molecules of adenosine triphosphate (ATP) to cyclic adenosine monophosphate (c-AMP), which is essentially an amplified and transduced version of the original signal. c-AMP plays a major role in triggering the ensuing cascade of events. The various ligands for GPCRs similarly cause a change in GPCR conformation that leads to the production of c-AMP. Termination of the signal can be effected by GTPase which converts GTP to GDP.[2]

c-AMP activates a kinase - protein kinase A (PKA) - that then can affect many cellular proteins by phosphorylation with profound consequences for the cell. These changes can affect metabolism, cell function or signaling. PKA can also alter gene expression by phosphorylating transcription factors. The level of c-AMP can be modulated or reduced by the action of phosphodiesterases (PDEs), which catalyze the hydrolysis of c-AMP to 5-AMP. There are at least 11 known classes of PDEs and many subtypes and splice variants.

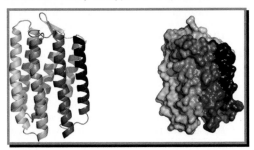

**Panel A**: G protein-coupled receptor in ribbon and surface (space-filling) representation. The seven helical domains are shown in seven different colors.[3]

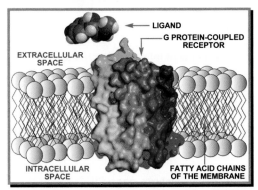

**Panel B**: The seven helical domains of the GPCR are bundled with an axis parallel to the fatty acid chains of the membrane.

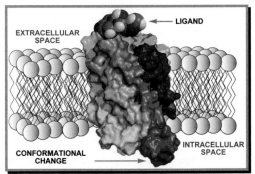

**Panel C**: Ligand-induced conformational change. For clarity, the ligand is shown at the surface, although it becomes embedded between helices.

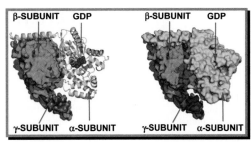

**Panel D**: Two different representations of the heterotrimeric G protein. GDP is bound in the α-subunit. The α-subunit is shown on the left with the ribbon diagram in order for the GDP to be visible.[4]

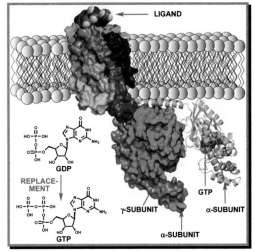

**Panel E**: Upon binding of the G protein to the transmembrane receptor, the GDP molecule in the α-subunit is replaced with a GTP molecule.

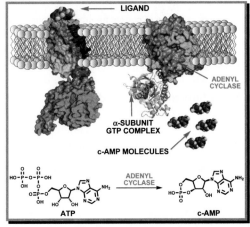

**Panel F**: The α-subunit-GTP complex separates from the G protein heterotrimer and moves to adenyl cyclase which it activates to convert ATP into c-AMP.[5]

1. Stryer, L. et al. Biochemistry, 6th Revised Edition (W.H. Freeman, **2007**); 2. Weinberg, R. A. The Biology of Cancer (Garland Science, **2007**); 3. *Structure* **2002**, *10*, 473-482. (**1GUE**); 4. *Cell* **1995**, *83*, 1047-1058. (**1GG2**); 5. *Biochemistry* **2000**, *39*, 14464-14471. (**1CUL**; Refs. p. 85

# REFERENCES FOR PART II.

### Acetylsalicylic Acid (Aspirin, page 38)[1-4]

1. Higgs, G. After 100 years of aspirin, the anticytokine drugs offer a new approach to anti-inflammatory therapy. *Curr. Opin. Invest. Drugs (Thomson Curr. Drugs)* **2003**, *4*, 517-518.

2. Colwell, J. A. Aspirin for the primary prevention of cardiovascular events. *Drugs of Today* **2006**, *42*, 467-479.

3. Kroetz, F., Hellwig, N., Schiele, T. M., Klauss, V. & Sohn, H.-Y. Prothrombotic potential of NSAID in ischemic heart disease. *Mini-Rev. Med. Chem.* **2006**, *6*, 1351-1355.

4. Morris, T., Stables, M. & Gilroy, D. W. New perspectives on aspirin and the endogenous control of acute inflammatory resolution. *TheScientificWorld* **2006**, *6*, 1045-1068.

### Naproxen (Aleve, page 39)[1-5]

1. Sutton, L. B. Naproxen sodium. *J. Am. Pharm. Assoc.* **1996**, *NS36*, 663-667.

2. Rainsford, K. D. Discovery, mechanisms of action and safety of ibuprofen. *Int. J. Clin. Pract., Suppl.* **2003**, *135*, 3-8.

3. Graham, G. G. & Williams, K. M. Metabolism and pharmacokinetics of ibuprofen. *Aspirin and Related Drugs* **2004**, 157-180.

4. Hawkey, C. J. NSAIDs, Coxibs, and the Intestine. *J. Cardiovasc. Pharmacol.* **2006**, *47*, S72-S75.

5. Naesdal, J. & Brown, K. NSAID-associated adverse effects and acid control aids to prevent them: a review of current treatment options. *Drug Saf.* **2006**, *29*, 119-132.

### How Do Anti-inflammatory Drugs Work? (page 40)[3, 4]
### X-Ray Structures (1DIY and 1EQG)[1, 2]

1. Malkowski, M. G., Ginell, S. L., Smith, W. L. & Garavito, R. M. The productive conformation of arachidonic acid bound to prostaglandin synthase. *Science* **2000**, *289*, 1933-1938.

2. Selinsky, B. S., Gupta, K., Sharkey, C. T. & Loll, P. J. Structural Analysis of NSAID Binding by Prostaglandin H2 Synthase: Time-Dependent and Time-Independent Inhibitors Elicit Identical Enzyme Conformations. *Biochemistry* **2001**, *40*, 5172-5180.

3. Calder, P. C. Polyunsaturated fatty acids and inflammation. *Prostaglandins, Leukotrienes Essent. Fatty Acids* **2006**, *75*, 197-202.

4. Capone, M. L. et al. Pharmacodynamic of cyclooxygenase inhibitors in humans. *Prostaglandins & Other Lipid Mediators* **2007**, *82*, 85-94.

### Other Eicosanoids in Inflammation (page 41)[1-5]

1. Calder, P. C. Polyunsaturated fatty acids and inflammation. *Prostaglandins, Leukotrienes Essent. Fatty Acids* **2006**, *75*, 197-202.

2. Kojima, F., Kapoor, M., Kawai, S. & Crofford, L. J. New insights into eicosanoid biosynthetic pathways: implications for arthritis. *Exp. Rev. Clin. Immun.* **2006**, *2*, 277-291.

3. Biscione, F., Pignalberi, C., Totteri, A., Messina, F. & Altamura, G. Cardiovascular effects of omega-3 free fatty acids. *Curr. Vasc. Pharmacol.* **2007**, *5*, 163-172.

4. Gonzalez-Periz, A. & Claria, J. New approaches to the modulation of the cyclooxygenase-2 and 5-lipoxygenase pathways. *Curr. Top. Med. Chem.* **2007**, *7*, 297-309.

5. Khanapure, S. P., Garvey, D. S., Janero, D. R. & Letts, L. G. Eicosanoids in inflammation: biosynthesis, pharmacology, and therapeutic frontiers. *Curr. Top. Med. Chem.* **2007**, *7*, 311-340.

### An Overview of Inflammation (page 42)[1, 2]

1. Gallin, J. I. & Snyderman, R. Inflammation: Basic Principles and Clinical Correlates (Lippincott Williams & Wilkins, **1999**).

2. Cytokines and Joint Injury (eds. van der Berg, W. D. & Miossec, P.) (Birkhauser Basel, **2004**).

### Celecoxib (Celebrex, page 43)[2-5]
### X-Ray Structure (1CX2)[1]

1. Kurumbail, R. G. et al. Structural basis for selective inhibition of cyclooxygenase-2 by anti-inflammatory agents. *Nature (London)* **1996**, *384*, 644-648.

2. Liew, D. & Krum, H. The cardiovascular safety of celecoxib. *Future Cardiol.* **2005**, *1*, 709-722.

3. Aronson, J. K. The NSAID roller coaster: more about rofecoxib. *Br. J. Clin. Pharmacol.* **2006**, *62*, 257-259.

4. Furberg, C. D. The COX-2 inhibitors - an update. *Am. Heart J.* **2006**, *152*, 197-199.

5. McGettigan, P. & Henry, D. Cardiovascular risk and inhibition of cyclooxygenase. A systematic review of the observational studies of selective and nonselective inhibitors of cyclooxygenase 2. *JAMA* **2006**, *296*, 1633-1644.

## Prednisone (Deltasone, page 44)[1-5]

1. Conn, D. L. & Lim, S. S. New role for an old friend: prednisone is a disease-modifying agent in early rheumatoid arthritis. *Curr. Opin. Rheumatol.* **2003**, *15*, 193-196.

2. Rhen, T. & Cidlowski, J. A. Antiinflammatory action of glucocorticoids - new mechanisms for old drugs. *N. Engl. J. Med.* **2005**, *353*, 1711-1723.

3. Barnes, P. J. How corticosteroids control inflammation: Quintiles Prize Lecture 2005. *Br. J. Pharmacol.* **2006**, *148*, 245-254.

4. Buckingham, J. C. et al. Annexin 1, glucocorticoids, and the neuroendocrine-immune interface. *Ann. N. Y. Acad. Sci.* **2006**, *1088*, 396-409.

5. Goecke, A. & Guerrero, J. Glucocorticoid receptor β in acute and chronic inflammatory conditions: clinical implications. *Immunobiology* **2006**, *211*, 85-96.

## Methotrexate (Trexall, page 46)[2-5]
## X-Ray Structure (1RG7)[1]

1. Sawaya, M. R. & Kraut, J. Loop and subdomain movements in the mechanism of Escherichia coli dihydrofolate reductase: crystallographic evidence. *Biochemistry* **1997**, *36*, 586-603.

2. Pile, K. Methotrexate. *Antirheumatic Therapy: Actions and Outcomes* **2005**, 175-197.

3. Stamp, L. et al. The use of low dose methotrexate in rheumatoid arthritis-are we entering a new era of therapeutic drug monitoring and pharmacogenomics? *Biomed. Pharmacother.* **2006**, *60*, 678-687.

4. Swierkot, J. & Szechinski, J. Methotrexate in rheumatoid arthritis. *Pharmacol. Rep.* **2006**, *58*, 473-492.

5. van der Heijden, J. W., Dijkmans, B. A. C., Scheper, R. J. & Jansen, G. Drug insight: resistance to methotrexate and other disease-modifying antirheumatic drugs-from bench to bedside. *Nat. Clin. Prac. Rheumatol.* **2007**, *3*, 26-34.

## Allopurinol (Zyloprim, page 47)[1-5]

1. Becker, M. A. The biochemistry of gout. *Crystal-Induced Arthropathies* **2006**, 189-212.

2. Martinon, F. & Glimcher, L. H. Gout: new insights into an old disease. *J. Clin. Invest.* **2006**, *116*, 2073-2075.

3. Niel, E. & Scherrmann, J.-M. Colchicine today. *Joint, Bone, Spine* **2006**, *73*, 672-678.

4. Pacher, P., Nivorozhkin, A. & Szabo, C. Therapeutic effects of xanthine oxidase inhibitors: renaissance half a century after the discovery of allopurinol. *Pharmacol. Rev.* **2006**, *58*, 87-114.

5. Teng, G. G., Nair, R. & Saag, K. G. Pathophysiology, clinical presentation and treatment of gout. *Drugs* **2006**, *66*, 1547-1563.

## Salmeterol (Serevent, page 50)[1-5]

1. Barnes, P. J. Drugs for the treatment of asthma and COPD. *Principles of Immunopharmacology (2nd Edition)* **2005**, 281-344.

2. Cazzola, M. & Matera, M. G. Beta-adrenoceptor agonists: Clinical use. *Therapeutic Strategies in COPD* **2005**, 61-78.

3. Reynolds, N. A., Lyseng-Williamson, K. A. & Wiseman, L. R. Inhaled salmeterol/fluticasone propionate: a review of its use in asthma. *Drugs* **2005**, *65*, 1715-1734.

4. Crompton, G. A brief history of inhaled asthma therapy over the last fifty years. *Prim. Care Resp. J.: J. Gen. Pract. Airways Group* **2006**, *15*, 326-331.

5. Meinke, L., Chitkara, R. & Krishna, G. Advances in the management of chronic obstructive pulmonary disease. *Exp. Opin. Pharmacother.* **2007**, *8*, 23-37.

## Fluticasone Propionate (Flonase, page 51)[1-6]

1. Johnson, M. Anti-inflammatory properties of fluticasone propionate *Int. Arch. Allergy Immunol.* **1995**, *107*, 439-440.

2. Storms, W. W. Treatment of seasonal allergic rhinitis with fluticasone propionate aqueous nasal spray: review of comparator studies *Allergy* **1995**, *50*, 25-29.

3. Harding, S. M. Fluticasone propionate: pharmacology and implications for clinical practice *Adv. Ther.* **1997**, *14*, 153-159.

4. Johnson, M. Development of fluticasone propionate and comparison with other inhaled corticosteroids *J. Allergy Clin. Immunol.* **1998**, *101*, S434-S439.

5. Fenton, C.; Keating, G. M. Inhaled salmeterol/fluticasone propionate: a review of its use in chronic obstructive pulmonary disease *Drugs* **2004**, *64*, 1975-1996.

6. Reynolds, N. A.; Lyseng-Williamson, K. A.; Wiseman, L. R. Inhaled salmeterol/fluticasone propionate: a review of its use in asthma *Drugs* **2005**, *65*, 1715-1734.

## Montelukast Sodium (Singulair, page 52)[1-3]

1. Jones, T. R.; Metters, K.; Evans, J. Preclinical pharmacological studies with montelukast (Singulair), a selective cysteinyl leukotriene

receptor (CysLT1) antagonist *Clin. Exp. Allergy Rev.* **2001**, *1*, 205-209.

2. Diamant, Z.; Sampson, A. P. Montelukast *Journal of Drug Evaluation: Respiratory Medicine* **2002**, *1*, 53-88.

3. Nayak, A. A review of montelukast in the treatment of asthma and allergic rhinitis *Expert Opin. Pharmaco.* **2004**, *5*, 679-686.


## Tiotropium Bromide (Spiriva, page 53) [1-5]

1. Keam, S. J.; Keating, G. M. Tiotropium bromide: A review of its use as maintenance therapy in patients with COPD *Treatments in Resp. Med.* **2004**, *3*, 247-268.

2. Mundy, C.; Kirkpatrick, P. Fresh from the pipeline: Tiotropium bromide *Nat. Rev. Drug Discov.* **2004**, *3*, 643-644.

3. Somand, H.; Remington, T. L. Tiotropium: a bronchodilator for chronic obstructive pulmonary disease *Ann. Pharmacother.* **2005**, *39*, 1467-1475.

4. Gross, N. J. Anticholinergic agents in asthma and COPD *Eur. J. Pharmacol.* **2006**, *533*, 36-39.

5. Lipson, D. A. Tiotropium bromide *International Journal of Chronic Obstructive Pulmonary Disease* **2006**, *1*, 107-114.


## Loratadine (Claritin, page 54) [1-5]

1. Bachert, C. Histamine - a major role in allergy? *Clin. Exp. Allergy* **1998**, *28*, 15-19.

2. Kay, G. G.; Harris, A. G. Loratadine: a nonsedating antihistamine. Review of its effects on cognition, psychomotor performance, mood and sedation *Clin. Exp. Allergy* **1999**, *29*, 147-150.

3. Slater, J. W.; Zechnich, A. D.; Haxby, D. G. Second-generation antihistamines: a comparative review *Drugs* **1999**, *57*, 31-47.

4. Baena-Cagnani, C. E. Safety and tolerability of treatments for allergic rhinitis in children *Drug Saf.* **2004**, *27*, 883-898.

5. Unal, M.; Hafiz, G. The role of antihistamines in the management of allergic rhinitis *Curr. Med. Chem: Anti-Inflammatory & Anti-Allergy Agents* **2005**, *4*, 477-480.


## An Overview of Metabolic Syndrome (page 56) [1-6]

1. Despres, J.-P. & Lemieux, I. Abdominal obesity and metabolic syndrome. *Nature* **2006**, *444*, 881-887.

2. Johnson, L. W. & Weinstock, R. S. The metabolic syndrome: concepts and controversy. *Mayo Clin. Proc.* **2006**, *81*, 1615-1620.

3. Grundy, S. M. Metabolic syndrome: a multiplex cardiovascular risk factor. *J. Clin. Endocrinol. Metab.* **2007**, *92*, 399-404.

4. Towler, M. C. & Hardie, D. G. AMP-activated protein kinase in metabolic control and insulin signaling. *Circ. Res.* **2007**, *100*, 328-341.

5. Townley, R. & Shapiro, L. Crystal Structures of the Adenylate Sensor from Fission Yeast AMP-Activated Protein Kinase. *Science* **2007**, *315*, 1726-1729.

6. Waki, H. & Tontonoz, P. Endocrine functions of adipose tissue. *Annual Review of Pathology: Mechanisms of Disease* **2007**, *2*, 31-56.


## Metformin (Glucophage, page 60) [1-5]

1. Strowig, S. M. & Raskin, P. Combination therapy using metformin or thiazolidinediones and insulin in the treatment of diabetes mellitus. *Diabetes, Obesity and Metabolism* **2005**, *7*, 633-641.

2. Deeks, E. D. & Scott, L. J. Pioglitazone/metformin. *Drugs* **2006**, *66*, 1863-1877.

3. Keller, U. From obesity to diabetes. *Int. J. Vitam. Nutr. Res.* **2006**, *76*, 172-177.

4. Schimmack, G., DeFronzo, R. A. & Musi, N. AMP-activated protein kinase: role in metabolism and therapeutic implications. *Diabetes, Obesity and Metabolism* **2006**, *8*, 591-602.

5. Waki, H. & Tontonoz, P. Endocrine functions of adipose tissue. *Annual Review of Pathology: Mechanisms of Disease* **2007**, *2*, 31-56.


## Glipizide (Glucotrol, page 61) [1-5]

1. Korytkowski, M. T. Sulfonylurea treatment of type 2 diabetes mellitus: focus on glimepiride. *Pharmacotherapy* **2004**, *24*, 606-620.

2. Rendell, M. The role of sulfonylureas in the management of type 2 diabetes mellitus. *Drugs* **2004**, *64*, 1339-1358.

3. Matthews, D. R. & Wallace, T. M. Sulphonylureas and the rise and fall of beta-cell function. *British Journal of Diabetes & Vascular Disease* **2005**, *5*, 192-196.

4. Stumvoll, M., Goldstein, B. J. & van Haeften, T. W. Type 2 diabetes: principles of pathogenesis and therapy. *Lancet* **2005**, *365*, 1333-1346.

5. Valensi, P. & Slama, G. Sulphonylureas and cardiovascular risk: facts and controversies. *British Journal of Diabetes & Vascular Disease* **2006**, *6*, 159-165.


## Pioglitazone (Actos, page 62) [2-5]
## X-Ray Structure (2PRG) [1]

1. Nolte, R. T. et al. Ligand binding and co-activator assembly of the peroxisome proliferator-activated receptor-γ. *Nature (London)* **1998**, *395*, 137-143.

2. Isley, W. L. Hepatotoxicity of thiazolidinediones. *Exp. Opin. Drug. Safety* **2003**, *2*, 581-586.

3. Pfuetzner, A., Schneider, C. A. & Forst, T. Pioglitazone: an antidiabetic drug with cardiovascular therapeutic effects. *Exp. Rev. Cardiovasc. Ther.* **2006**, *4*, 445-459.

4. Waugh, J., Keating, G. M., Plosker, G. L., Easthope, S. & Robinson, D. M. Spotlight on pioglitazone in type 2 diabetes mellitus. *Treatm. Endocrinol.* **2006**, *5*, 189-191.

5. Stojanovska, L., Honisett, S. Y. & Komesaroff, P. A. The anti-atherogenic effects of thiazolidine-diones. *Curr. Diabet. Rev.* **2007**, *3*, 67-74.

## Sitagliptin (Januvia, page 63)[2-5]
## X-Ray Structure (1X70)[1]

1. Kim, D. et al. (2R)-4-Oxo-4-[3-(Trifluoromethyl)-5,6-dihydro[1,2,4]triazolo[4,3-a]pyrazin-7(8H)-yl]-1-(2,4,5-trifluorophenyl)butan-2-amine: A Potent, Orally Active Dipeptidyl Peptidase IV Inhibitor for the Treatment of Type 2 Diabetes. *J. Med. Chem.* **2005**, *48*, 141-151.

2. Barnett, A. DPP-4 inhibitors and their potential role in the management of type 2 diabetes. *Int. J. Clin. Pract.* **2006**, *60*, 1454-1470.

3. Drucker, D. J. & Nauck, M. A. The incretin system: glucagon-like peptide-1 receptor agonists and dipeptidyl peptidase-4 inhibitors in type 2 diabetes. *Lancet* **2006**, *368*, 1696-1705.

4. Miller, S. A. & St. Onge, E. L. Sitagliptin: a dipeptidyl peptidase IV inhibitor for the treatment of type 2 diabetes. *Ann. Pharmacother.* **2006**, *40*, 1336-1343.

5. Campbell, R. K. Rationale for dipeptidyl peptidase 4 inhibitors: a new class of oral agents for the treatment of type 2 diabetes mellitus. *Ann. Pharmacother.* **2007**, *41*, 51-60.

## Atorvastatin (Lipitor, page 64)[2-5]
## X-Ray Structure (1HWK)[1]

1. Istvan, E. S. & Deisenhofer, J. Structural mechanism for statin inhibition of HMG-CoA reductase. *Science* **2001**, *292*, 1160-1164.

2. Ramos, J., Khan, Q. A., Thoenes, M. & Khan, B. V. Atorvastatin: beyond lipid-lowering effects in the diabetic population. *Future Cardiol.* **2006**, *2*, 527-533.

3. Schaefer, E. J. & Asztalos, B. F. The effects of statins on high-density lipoproteins. *Curr. Atheroscl. Rep.* **2006**, *8*, 41-49.

4. Ii, M. & Losordo, D. W. Statins and the endothelium. *Vasc. Pharm.* **2007**, *46*, 1-9.

5. Lahera, V. et al. Endothelial dysfunction, oxidative stress, and inflammation in atherosclerosis: beneficial effects of statins. *Curr. Med. Chem.* **2007**, *14*, 243-248.

## Ezetimibe (Zetia, page 65)[1-5]

1. Sudhop, T., Luetjohann, D. & von Bergmann, K. Sterol transporters: targets of natural sterols and new lipid lowering drugs. *Pharmacol. Ther.* **2005**, *105*, 333-341.

2. Ahmed, M. H., Saad, R. A. & Osman, M. M. Ezetimibe: effective and safe treatment for dyslipidemia associated with nonalcoholic fatty liver disease. *Exp. Opin. Drug. Safety* **2006**, *5*, 487-488.

3. Burnett, J. R. & Huff, M. W. Cholesterol absorption inhibitors as a therapeutic option for hypercholesterolemia. *Exp. Opin. Invest. Drugs* **2006**, *15*, 1337-1351.

4. Gotto, A. M., Jr. & Farmer, J. A. Drug insight: the role of statins in combination with ezetimibe to lower LDL cholesterol. *Nat. Clin. Prac. Card. Med.* **2006**, *3*, 664-672.

5. Tiwari, A., Bansal, V., Chugh, A. & Mookhtiar, K. Statins and myotoxicity: a therapeutic limitation. *Exp. Opin. Drug. Safety* **2006**, *5*, 651-666.

## Atenolol (Tenormin, page 68)[1-4]

1. Pedersen, M. E. & Cockcroft, J. R. The latest generation of beta-blockers: new pharmacologic properties. *Curr. Hypertens. Rep.* **2006**, *8*, 279-286.

2. Burnier, M., Bullani, R. & Vogt, B. Beta-blockers for the treatment of essential hypertension: what are the arguments against their use as first line therapy? *Curr. Hypertens. Rev.* **2007**, *3*, 15-20.

3. Chrysant, S. G. Beta-blockers as first line treatment of hypertension: a proponent's view. *Curr. Hypertens. Rev.* **2007**, *3*, 21-28.

4. Karagiannis, A., Mikhailidis, D. P., Kakafika, A. I., Tziomalos, K. & Athyros, V. G. Atenolol: differences in mode of action compared with other antihypertensives. An opportunity to identify features that influence outcome? *Curr. Pharm. Des.* **2007**, *13*, 229-239.

## Enalapril (Vasotec, page 69)[1-5]
## X-Ray Structure (1UZE)[6]

1. Patchett, A. A. The chemistry of enalapril *Br. J. Clin. Pharmacol.* **1984**, *18*, 201-207.

2. Perry, R. S. Enalapril maleate (MK-421, Reniten): an improved ACE inhibitor *Drugs of Today* **1985**, *21*, 31-40.

3. Todd, P. A.; Heel, R. C. Enalapril. A review of its pharmacodynamic and pharmacokinetic properties, and therapeutic use in hypertension and congestive heart failure *Drugs* **1986**, *31*, 198-248.

4. Bauer, J. H. Angiotensin converting enzyme inhibitors *Am. J. Hypertens.* **1990**, *3*, 331-337.

5. Todd, P. A.; Goa, K. L. Enalapril. A reappraisal of its pharmacology and therapeutic use in hypertension *Drugs* **1992**, *43*, 346-381.

6. Natesh, R.; Schwager, S. L. U.; Evans, H. R.; Sturrock, E. D.; Acharya, K. R. Structural Details on the Binding of Antihypertensive Drugs Captopril and Enalaprilat to Human Testicular Angiotensin I-Converting Enzyme *Biochemistry (Mosc).* **2004**, *43*, 8718-8724.

## Candesartan Cilexetil (Atacand, page 70)[1-5]

1. Siragy, H. Angiotensin II receptor blockers: review of the binding characteristics *Am. J. Cardiol.* **1999**, *84*, 3S-8S.

2. Stoukides, C. A.; McVoy, H. J.; Kaul, A. F. Candesartan cilexetil: an angiotensin II receptor blocker *Ann. Pharmacother.* **1999**, *33*, 1287-1298.

3. Gleiter, C. H.; Jaegle, C.; Gresser, U.; Moerike, K. Candesartan *Cardiovasc. Drug Rev.* **2004**, *22*, 263-284.

4. Oestergren, J. Candesartan for the treatment of hypertension and heart failure *Expert Opin. Pharmaco.* **2004**, *5*, 1589-1597.

5. Ross, A.; Papademetriou, V. Candesartan cilexetil in cardiovascular disease *Expert Rev. Cardiovasc. Ther.* **2004**, *2*, 829-835.

## Aliskiren (Tekturna, page 71)[1-5]

1. Kang, J. J., Toma, I., Sipos, A., McCulloch, F. & Peti-Peterdi, J. Imaging the renin-angiotensin system: An important target of anti-hypertensive therapy. *Adv. Drug. Del. Rev.* **2006**, *58*, 824-833.

2. O'Brien, E. Aliskiren: a renin inhibitor offering a new approach for the treatment of hypertension. *Exp. Opin. Invest. Drugs* **2006**, *15*, 1269-1277.

3. Staessen, J. A., Li, Y. & Richart, T. Oral renin inhibitors. *Lancet* **2006**, *368*, 1449-1456.

4. O'Brien, E. et al. Aliskiren Reduces Blood Pressure and Suppresses Plasma Renin Activity in Combination With a Thiazide Diuretic, an Angiotensin-Converting Enzyme Inhibitor, or an Angiotensin Receptor Blocker. *Hypertension* **2007**, *49*, 276-284.

5. Pool, J. L. et al. Aliskiren, an Orally Effective Renin Inhibitor, Provides Antihypertensive Efficacy Alone and in Combination With Valsartan. *Am. J. Hypertens.* **2007**, *20*, 11-20.

## Amlodipine (Norvasc, page 72)[1-5]

1. Blank, R. A single-pill combination of amlodipine besylate and atorvastatin calcium. *Drugs of Today* **2006**, *42*, 157-175.

2. Godfraind, T. Calcium-channel modulators for cardiovascular disease. *Exp. Opin. Emerg. Drugs* **2006**, *11*, 49-73.

3. Liebson, P. R. Calcium channel blockers in the spectrum of antihypertensive agents. *Exp. Opin. Pharmacother.* **2006**, *7*, 2385-2401.

4. Ozawa, Y., Hayashi, K. & Kobori, H. New generation calcium channel blockers in hypertensive treatment. *Curr. Hypertens. Rev.* **2006**, *2*, 103-111.

5. de la Sierra, A. & Ruilope, L. M. Are calcium channel blockers first-line drugs for the treatment of hypertension and cardiovascular disease? *Curr. Hypertens. Rev.* **2007**, *3*, 9-13.

## Nitroglycerin (Nitroglycerin, page 73)[1-6]

1. Feldman, P. L., Griffith, O. W. & Stuehr, D. J. The surprising life of nitric oxide. *Chem. Eng. News* **1993**, *71*, 26-38.

2. Ignarro, L. J. Nitric oxide: a unique endogenous signaling molecule in vascular biology (Nobel lecture). *Angew. Chem., Int. Ed. Engl.* **1999**, *38*, 1882-1892.

3. Murad, F. Discovery of some of the biological effects of nitric oxide and its role in cell signaling (Nobel lecture). *Angew. Chem., Int. Ed. Engl.* **1999**, *38*, 1856-1868.

4. Marsh, N. & Marsh, A. A short history of nitroglycerine and nitric oxide in pharmacology and physiology. *Clin. Exp. Pharmacol. Physiol.* **2000**, *27*, 313-319.

5. Blaise, G. A., Gauvin, D., Gangal, M. & Authier, S. Nitric oxide, cell signaling and cell death. *Toxicology* **2005**, *208*, 177-192.

6. Miller, M. R. & Megson, I. L. Recent developments in nitric oxide donor drugs. *Br. J. Pharmacol.* **2007**, *151*, 305-321.

## Clopidogrel Bisulfate (Plavix, page 74)[1-5]

1. Savi, P. et al. The active metabolite of clopidogrel disrupts P2Y12 receptor oligomers and partitions them out of lipid rafts. *Proc. Natl. Acad. Sci. U. S. A.* **2006**, *103*, 11069-11074.

2. Tantry, U. S., Etherington, A., Bliden, K. P. & Gurbel, P. A. Antiplatelet therapy: current strategies and future trends. *Future Cardiol.* **2006**, *2*, 343-366.

3. Walsh, S. J., McClelland, A. J. & Adgey, J. A. Clopidogrel in the treatment of ischaemic heart disease. *Exp. Opin. Pharmacother.* **2006**, *7*, 1109-1120.

4. Wang, T. H., Bhatt, D. L. & Topol, E. J. Aspirin and clopidogrel resistance: an emerging clinical entity. *Eur. Heart J.* **2006**, *27*, 647-654.

5. Weimar, C. & Diener, H.-C. Antiplatelet therapy and oral anticoagulation for prevention of ischemic stroke. *Curr. Drug Ther.* **2006**, *1*, 249-256.

## Digoxin (Lanoxin, page 75)[1-5]

1.  Schmidt, T. A. & Kjeldsen, K. Regulation of digitalis glycoside receptors in digoxin treatment. *Prog. Exp. Cardiol.* **2003**, *5*, 501-510.

2.  Wasserstrom, J. A. & Aistrup, G. L. Digitalis: New actions for an old drug. *Am. J. Physiol.* **2005**, *289*, H1781-H1793.

3.  Bauman, J. L., DiDomenico, R. J. & Galanter, W. L. Mechanisms, manifestations, and management of digoxin toxicity in the modern era. *Am. J. Cardiovasc. Drugs* **2006**, *6*, 77-86.

4.  Bukharovich, I. F. & Kukin, M. Optimal medical therapy for heart failure. *Prog. Cardiovasc. Dis.* **2006**, *48*, 372-385.

5.  Tamargo, J., Delpon, E. & Caballero, R. The safety of digoxin as a pharmacological treatment of atrial fibrillation. *Exp. Opin. Drug. Safety* **2006**, *5*, 453-467.

## Information Flow into the Cell by Chemical Signaling (page 78)[2, 5]
## X-Ray Structures (1GUE, 1GG2, 1CUL)[1, 3, 4]

1.  Wall, M. A. et al. The structure of the G protein heterotrimer Giα1β1γ2. *Cell (Cambridge, Mass.)* **1995**, *83*, 1047-1058.

2.  Berg, J.M., Tymoczko, J.L. and Stryer, L. Biochemistry (W.H. Freeman, 6th Revised Edition, **2007**).

3.  Tesmer, J. J. G. et al. Molecular Basis for P-Site Inhibition of Adenylyl Cyclase. *Biochemistry* **2000**, *39*, 14464-14471.

4.  Edman, K. et al. Early Structural Rearrangements in the Photocycle of an Integral Membrane Sensory Receptor. *Structure* **2002**, *10*, 473-482.

5.  Weinberg, R. A. The Biology of Cancer (Garland Science, **2007**).

*The fate of humanity depends on its reproductive health and on its ability to align the population of this planet with the resources available for the present and future.*

# PART III.

PART III

# REPRODUCTIVE MEDICINE

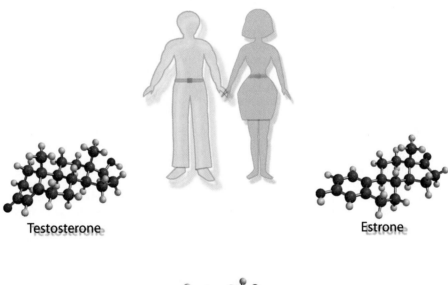

Testosterone

Estrone

Progesterone

# ORAL CONTRACEPTIVES

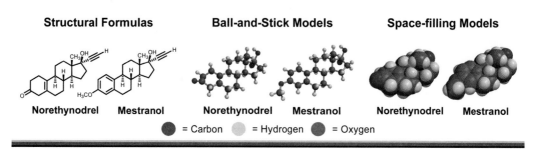

| Structural Formulas | Ball-and-Stick Models | Space-filling Models |
|---|---|---|
| Norethynodrel   Mestranol | Norethynodrel   Mestranol | Norethynodrel   Mestranol |

● = Carbon   ◯ = Hydrogen   ● = Oxygen

*Year of discovery:* 1950s; *Year of introduction:* 1960 (GD Searle & Company); *Drug category:* Oral contraceptives; *Main uses:* Prevention of unplanned pregnancy; *Commercial names:* Ortho Tri-Cyclen, Triphasil, Alesse, Cyclessa, Ortho-Cept, Yasmin, etc,.

Estrogens and progestins are natural steroidal hormones that are vital in sexual development and in the regulation of the physiological process associated with repro-ductive fertility in women. The combination of such biological knowledge with chemistry led to the development of effective contraceptives for prevention of unplanned pregnancy.

The fertility cycle in women is controlled by a complex cascade of events involving signaling by protein and steroid hormones between the hypothalamus, the pituitary, and the ovaries. In the first phase of the cycle, maturation of the Graafian follicle is stimulated by the follicle stimulating hormone (FSH), which is secreted in the pituitary gland along with the luteinizing hormone (LH) in response to pulsative release of gonadotropin-releasing hormone from the hypothalamus. Luteinizing hormone stimulates estrogen secretion from the ovaries that inhibits pituitary release of FSH and initiates thickening of the inner lining of the uterus (endometrium). Increased estrogen levels spur an acute release of LH that triggers ovulation. The residual follicle (corpus luteum) produces progesterone, causing further changes in the endometrium that prepare it for implantation of the fertilized egg. Progesterone is also essential for suppressing ovulation during pregnancy.

**Estrone   Progesterone   Ethinyl estradiol**

The role of progesterone as an inhibitor of ovulation was first observed in the 1930s, but was not further investigated as a mechanism of contraception until the 1950s (by Gregory Pincus). Since the dose of progesterone required to prevent fertilization was high,

Pincus had to screen more then 200 progestin analogs to find a useful contraceptive. The search culminated in the discovery of norethynodrel. Although subsequent clinical trials were successful, some of those treated experienced breakthrough bleeding. Investigation of the synthetic drug revealed that when it was contaminated with ~2% of mestranol, an estrone derivative, the occurrence of bleeding was decreased. Based on this finding, mestranol was later deliberately incorporated into the drug, leading to the development of the first combined oral contraceptive, Enovid, which gained approval in 1960. Oral contraceptives are now used by about 100 million women worldwide. The introduction of contraceptives has had a major impact on society.

Currently used contraceptives are taken at considerably lower dosages than Enovid. The estrogen component of current drugs most frequently is ethinyl estradiol, which is combined with any of several available potent progestin derivatives. Monophasic oral contraceptives contain the same dose of the progestin component in each tablet, whereas bi- or triphasic drugs contain variable amounts depending on the stage of the fertility cycle.

In addition to the traditional formulation, newer delivery methods have been developed including transdermal patch, intrauterine device, or injectable hormonal agents. Extended-cycle contraceptives consisting of 84 active pills and 7 inactive pills are also available. Progestin only pills (POP) are mainly used by women who are breastfeeding or cannot take estrogen. Emergency contra-ceptives are available to prevent pregnancy after unprotected intercourse.[1,2]

1. *N. Engl. J. Med.* **2003**, *349*, 1443-1450; 2. *Mayo Clin. Proc.* **2006**, *81*, 949-955; **Refs. p 105**

# TESTOSTERONE (TESTOSTERONE™)

| **Structural Formula** | **Ball-and-Stick Model** | **Space-filling Model** |
|---|---|---|

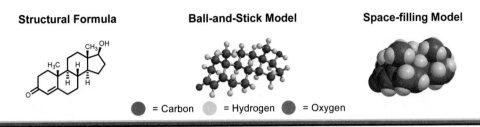

●  = Carbon      ○ = Hydrogen      ● = Oxygen

*Year of discovery:* 1935; *Drug category:* Anabolic steroid hormone; *Main uses:* Treatment of male hypogonadism; *Related drugs:* Testosterone enanthate (Delatestryl), Testosterone cypionate (Depotestosterone); *Other brand names:* Androderm, Androgel, Testim, Striant.

The steroid testosterone is the principal male sex hormone. The main approved therapeutic use of testosterone is the treatment of testosterone deficiency (hypogonadism). Although testosterone and other anabolic hormones enhance muscle development and improve athletic performance, risks are involved with their use.

Testosterone in males is produced in the testes. Testosterone secretion is stimulated by the release of the luteinizing hormone (LH) from the pituitary gland, which is regulated by the hypothalamic gonadotropin-releasing hormone (GnRH). Testosterone is transported in the body by specific plasma proteins. A fraction of testosterone is metabolized to active steroids, such as estradiol (by the enzyme complex aromatase), and dihydrotestosterone, which exhibits even greater androgenic activity (by the enzyme 5α-reductase).

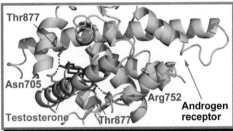

Testosterone and its active metabolites have many physiological effects. Testosterone and dihydrotestosterone are responsible for the differentiation of internal and external genitalia in the fetus and maturation during puberty. Testosterone exhibits anabolic effects; it stimulates muscle protein synthesis that leads to larger muscle mass, increased muscle strength and resistance. Estradiol in the fetus is associated with the sexual differentiation of the brain. During puberty, estradiol accelerates the maturation of cartilage into bone leading to epiphysis and termination of bone growth.

Testosterone and dihydrotestosterone act via the androgen receptor, which belongs to the nuclear receptor superfamily.[1] Binding of testosterone or dihydrotestosterone to the receptor initiates conformational changes that allow translocation of the complex into the nucleus, where it acts as a transcription factor and stimulates the expression of certain genes.

Binding of testosterone to the androgen receptor.

Testosterone deficiency in men can result from either inadequate function of the testes (primary hypogonadism), or in the hypothalamus and pituitary (secondary hypogonadism). Testosterone levels in men generally decrease with age. Although administration of testosterone to elderly men produces beneficial results, such as increased muscle mass, bone mineral density, and mental function, its use is limited due to the possible promotion of prostatic malignancy.[2]

Since the oral bioavailability of testosterone is low due to rapid metabolism in the liver, various alternative formulations have been developed including subcutaneous injection, transdermal skin patch (Androderm), topical gel (Androgel, Testim), and buccal tablet (Striant).[3]

1. *Protein Sci.* **2006**, *15*, 987-999 (2AM9); 2. *Nat. Clin. Pract. Endoc.* **2006**, *2*, 146-159; 3. *Nat. Clin. Pract. Urol.* **2006**, *3*, 653-665; **Refs. p. 105**

# MIFEPRISTONE (MIFEPREX™)

| Structural Formula | Ball-and-Stick Model | Space-filling Model |
|---|---|---|

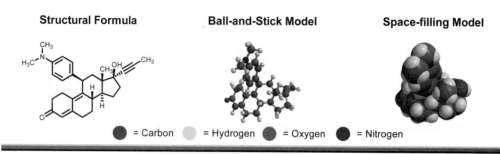

● = Carbon   ○ = Hydrogen   ● = Oxygen   ● = Nitrogen

*Year of discovery:* 1980 (Roussel Uclaf); *Year of introduction:* 1987 (France); 2000 (US, Danco Laboratories); *Drug category:* Progesterone receptor modulator; *Main uses:* Termination of pregnancy up to 49 days of gestation; *Other brand names:* Mifegyne.

Mifepristone, a synthetic steroid, is used for termination of intrauterine pregnancy up until about the eighth week. Other possible applications include preoperative dilation of the cervix prior to surgical abortion, emergency contraception, and induction of labor in case of fetal death.

Mifepristone was discovered at the French company Roussel Uclaf in 1980 in a program to find antagonists of glucocorticoids. In addition to inhibiting glucocorticoid action, mifepristone was found to exhibit anti-progestational properties. Concurrently, it was developed as an abortion-inducing agent. It was first approved for use in France in 1987, and then in other European countries. It was introduced in the US in 2000 under the trade name Mifeprex.

Progesterone is vital in pregnancy because it prepares the uterus for implantation of the fertilized ovum. After fertilization, it plays an essential role in sustaining pregnancy by suppressing ovulation, decreasing the contractility of the uterine smooth muscle, and decreasing the maternal immune response. Mifepristone acts by binding tightly to the progesterone receptor.[1] Although it blocks the action of progesterone, it is a partial agonist in the absence of the natural hormone. Administration of mifepristone during pregnancy blocks the action of progesterone that leads to degeneration of the uterine lining (decidua) and softening of the cervix. This causes detachment of the blastocyst, which leads to decreased production of the peptide hormone, human chorionic gonadotropin (hCG). The peptide hCG inhibits the disintegration of the corpus luteum, which produces progesterone. Decreased progesterone production results in

further breakdown of the decidua. In addition to these effects, prostaglandin levels are increased and the sensitivity of the uterus toward prostaglandins is enhanced leading to uterine contraction that ultimately causes expulsion of the detached blastocyst.[2,3]

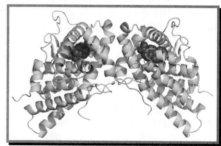
Progesterone receptor dimer.

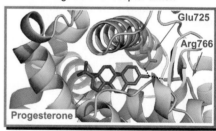
Close-up view of the binding of progesterone.

Mifepristone is administered orally in a 600 mg dose. If complete abortion does not occur within two days, misoprostol, a synthetic prostaglandin derivative is administered to induce contraction of the uterus.

**Misoprostol (Cytotec™)**

1. *Nature (London)* **1998**, *393*, 392-396 (**1A28**); 2. *Contraception* **2003**, *68*, 471-476; 3. *Contraception* **2006**, *74*, 66-86; **Refs. p. 105**

# OXYTOCIN (OXYTOCIN™)

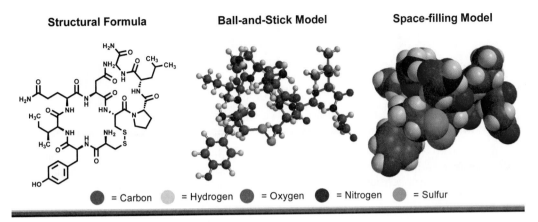

**Structural Formula**     **Ball-and-Stick Model**     **Space-filling Model**

= Carbon     = Hydrogen     = Oxygen     = Nitrogen     = Sulfur

*Year of discovery:* 1952; *Year of introduction:* 1950s; *Drug category:* Neuropeptide hormone; *Main uses:* Induction of labor; *Other brand names:* Pitocin; *Related drugs:* Dinoprostone (Cervidil, Prostin E$_2$), Misoprostol (Cytotec).

Oxytocin is a natural peptide hormone that stimulates uterine contraction and induces labor. It is intravenously administered to facilitate the progression of labor or to reduce bleeding after expulsion of the placenta. The infusion of oxytocin is adjusted to control the intensity and frequency of uterine contraction.[1]

Oxytocin is produced mainly in the hypothalamus and stored in the pituitary gland, but it is also synthesized in the ovaries, endometrium and placenta. The powerful contractile action of pituitary extracts on uterine smooth muscle first was recognized in the early 1900s. Oxytocin was isolated in 1928 at Parke Davis and Company along with a structurally closely related peptide hormone, vasopressin, which is a pressor and antidiuretic. The structures of these hormones were determined by Vincent du Vigneaud at Cornell University in 1953. He showed that each consists of nine amino acids and that they differ only in the amino acids in position 3 and 8. Six of the amino acid units are contained in a disulfide ring. du Vigneaud was awarded the Nobel Prize in Chemistry in 1955 for this work.

During childbirth, oxytocin secretion is stimulated by dilation of the cervix and vagina. During the last stage of pregnancy, uterine responsiveness to oxytocin is elevated because of increased expression of oxytocin receptors, which is stimulated by estrogen. Oxytocin also increases local prostaglandin production, which further stimulates uterine contraction. Oxytocin acts by binding to specific G protein-coupled receptors which initiate the physiological process leading to uterine contraction. Oxytocin is also implicated in milk ejection during breastfeeding.

In addition to these effects, oxytocin plays a role in social behavior. Animal studies have revealed that it enhances maternal behavior and reduces stress.[2]

The induction of labor can also be stimulated by prostaglandins, natural substances synthesized from fatty acids, that are involved in many physiological processes (*see* page 40). In addition to causing uterine contraction, prostaglandins assist in ripening of the cervix. Currently, only PGE$_2$ (dinoprostone) is approved for the induction of labor (Cervidil, Forest Laboratories; Prostin E$_2$, Pfizer), but PGF$_{2\alpha}$ (dinoprost), another natural prostaglandin used earlier, is also effective. Dinoprostone is administered locally. Misoprostol, an orally active synthetic PGE$_1$ derivative (see page 92), is also efficacious for induction of labor, but it is used in the US mainly for the prevention of NSAID-induced gastric ulcer.

**Dinoprost**     **Dinoprostone (Prostin E$_2$™)**

1. *Reproduction* **2006**, *131*, 989-998; 2. *Nat. Neurosci.* **2004**, *7*, 1048-1054; **Refs. p. 105**

# SILDENAFIL (VIAGRA™)

| Structural Formula | Ball-and-Stick Model | Space-filling Model |
|---|---|---|

● = Carbon   ○ = Hydrogen   ● = Oxygen   ● = Nitrogen   ● = Sulfur

*Year of discovery:* 1989; *Year of introduction:* 1998 (Pfizer); *Drug category:* Selective inhibitor of cGMP specific type 5 phosphodiesterase; *Main uses:* Treatment of male erectile dysfunction and pulmonary arterial hypertension; *Related drugs:* Tadalafil (Cialis), Vardenafil (Levitra); *Other brand names:* Revatio.

Sildenafil, the first selective type 5 phosphodiesterase inhibitor (PDE-5), is an effective oral treatment for erectile dysfunction, a condition that affects about 30 million men in the US. It is also efficacious for pulmonary artery hypertension caused by narrowed arteries in the lung. This condition results in a reduced supply of oxygen in the body (ischemia) and multiple consequences including damage to the heart.[1]

The development of sildenafil as a potent therapy for erectile dysfunction was the result of a serendipitous discovery during research to identify phosphodiesterase (PDE) enzyme inhibitors for the treatment of high blood pressure and angina at Pfizer. One of the candidate molecules, UK92480, later called sildenafil, exhibited highly selective inhibitory activity against type 5 phosphodiesterase. It was observed clinically that although sildenafil is not efficacious for the treatment of angina, it markedly increases erectile function. Clinical trials of sildenafil for the treatment of erectile dysfunction began in 1993 and led to approval in 1998. Sildenafil, marketed as a citrate salt under the name Viagra, soon became a billion-dollar drug. In the late 1990s it was recognized that sildenafil is beneficial for the treatment of pulmonary artery hypertension. It is available for this indication since 2005 under the name Revatio.[2]

The physiological process of erection is initiated by the release of nitric oxide (NO), a short-lived neurotransmitter that activates the enzyme guanylate cyclase, leading to increased production of cyclic guanosine monophosphate (cGMP). cGMP triggers a cascade of biochemical processes that results in vasodilation in the corpus cavernosum, increased penile blood flow and erection. This process is reversed by type 5 phospho-

diesterase, which catalyzes the conversion of cGMP to guanosine monophosphate (5'-GMP).

Sildenafil increases cGMP levels and improves erectile function. Sildenafil does not affect the production of cGMP and has no effect in the absence of sexual stimulation.

The beneficial effect of sildenafil in pulmonary arterial hypertension is due to the elevated levels of type 5 phosphodiesterase in the pulmonary arterial walls of hypertensive individuals.[3,1]

Binding of sildenafil to type 5 phosphodiesterase.

Two additional PDE-5 inhibitors, vardenafil (Levitra™, Bayer) and tadalafil (Cialis™, Eli Lilly) were introduced in 2003.

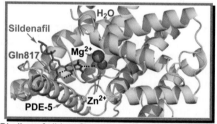

**Vardenafil (Levitra™)**    **Tadalafil (Cialis™)**

1. *Nature* **2003**, *425*, 98-102 (**1UDT**); 2. *Nat. Rev. Drug Discov.* **2006**, *5*, 689-702; 3. *N. Engl. J. Med.* **2000**, *342*, 1802-1813; **Refs. p. 105**

# OSTEOPOROSIS

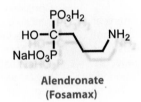

**Alendronate
(Fosamax)**

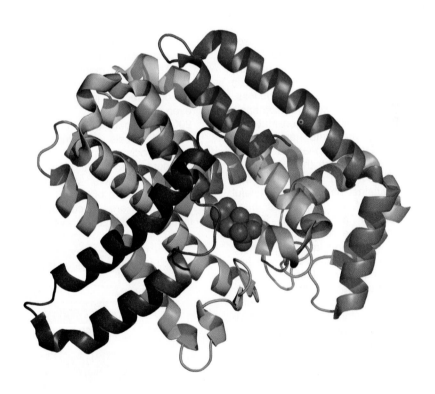

Alendronate (magenta) Inhibition of the Enzyme Farnesyl Pyrophosphate Synthase

# SOME ASPECTS OF OSTEOPOROSIS

Osteoporosis, the result of abnormal loss of bone, leads to increased risk of bone fracture.[1] It is a major health problem in postmenopausal woman and the elderly, who lose 3% of their bone mass per year. About 40% of women experience an osteoporotic fracture during life. Osteoporosis is the result of an imbalance in the normal process of bone remodeling, the continuous formation of new bone and removal of existing bone. The deposition of new bone is effected by the osteoblast cell line, and the dissolution of bone is caused by osteoclast cells. The populations and activities of these cells, which are attached to the surface of bone, are coupled and tightly regulated by multiple factors.

One such factor is simply the mechanical stress on bone. It has been long known that prolonged immobilization due to illness or injury results in substantial loss of bone, as much as 15% over a three month period. Astronauts in the zero gravity environment of space lose bone mass at a rate of 2% per month (more than 20% per year). This "disuse" form of osteoporosis is a serious concern in planning for extended human exploration of space and an additional reason to understand bone homeostasis more deeply. Disuse osteoporosis is the result of both reduced osteoblast-mediated bone formation and enhanced osteoclast-mediated bone loss.

Osteoclasts secrete an enzyme, cathepsin K, which breaks down the collagen matrix of bone. They also generate acid which dissolves the mineral component, a form of calcium phosphate. Although there has been much research to find molecules that selectively inhibit cathepsin K, so far no useful therapeutic agent of this type has been discovered. Another approach to limiting osteoclast-induced bone resorption depends on downregulating the population of this cell line, which has a lifetime *in vivo* of only two weeks (vs. 4 months for osteoblasts). This development of new osteoclasts is strongly stimulated by a protein called receptor activator of nuclear factor-KB ligand (RANKL), a member of the TNF-$\alpha$ class. Other mediators that enhance osteoclast formation, activation or lifespan are (1) parathyroid hormone, (2) prostaglandin $E_2$, (3) calcitriol, and (4) interleukin-11. A monoclonal antibody against RANKL, denosumab (Amgen), which is currently in Phase 3 clinical trials, has been shown to promote the formation of strong and dense bone. Inflammatory cytokines upregulate the levels of RANKL and, as a consequence, increase osteoclast action and bone loss. This is an important cause of the joint destruction that accompanies arthritis. On the other hand, the levels of RANKL are lowered with increasing concentrations of estrogen, which can explain the beneficial effect of estrogen on bone.

Osteoblast proliferation or development is stimulated by parathyroid hormone and by calcitriol. The balance between osteoblast and osteoclast populations is further maintained by the capacity of osteoblasts to secrete RANKL and stimulate the development of osteoclasts (and also by the reverse of this effect). Corticosteroids induce osteoporosis in part by inhibiting osteoblast function as well as by downregulating calcium levels.

There are rare human disorders of genetic origin that involve either abnormally high or low bone mass or strength. One family, which had been found to possess strikingly high bone mass, turned out to have a single amino acid mutation in a protein on the Wnt signaling pathway that reduced its function and increased osteoblast proliferation. Genetic studies offer great promise in the fundamental understanding of osteoporosis.

The annual medical cost of osteoporosis in the US is roughly $15 billion for the 10 million affected inividuals. The problem is even more serious in Asia. Worldwide, at least 200 million people are affected. The need for early diagnosis and improved treatment of osteoporosis will only increase as world demographics shift toward older ages. Increased muscle weakening with aging exacerbates the effects of osteoporosis. Both conditions are accelerated by physical inactivity (or the zero gravity of space).

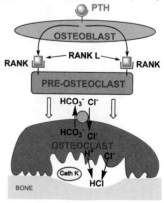

1. Orwoll, E. S. & Bliziotes, M. Osteoporosis: Pathophysiology and Clinical Management (2003); Refs. p. 106

# ALENDRONATE (FOSAMAX™)

| Structural Formula | Ball-and-Stick Model | Space-filling Model |
| --- | --- | --- |

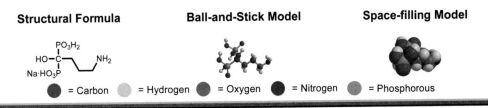

● = Carbon  ○ = Hydrogen  ● = Oxygen  ● = Nitrogen  ○ = Phosphorous

*Year of discovery:* 1977; *Year of introduction:* 1995 (Merck); *Drug category:* Inhibitor of bone resorption/farnesyl pyrophosphate synthase inhibitor; *Main uses:* Prevention and treatment of osteoporosis and Paget's disease; *Other brand names:* Fosavance (with vitamin D); *Related drugs:* Etidronate (Didronel), Risedronate (Actonel), Ibandronate (Boniva), Zoledronic acid (Zometa).

Alendronate, a simple bisphosphonate structure, in which two phosphate fragments are linked by carbon, is widely used for the treatment and prevention of osteoporosis. Osteoporosis can be caused by aging, physical immobility, inflammation, lowered estrogen levels in postmenopausal women or extended use of glucocorticoids (e.g., prednisone). Alendronate has also been applied to the treatment of Paget's disease, a chronic condition, in which defective bone remodeling leads to soft, brittle and deformed bones.

Bisphosphonates are ionized, water-soluble compounds that mimic the pyrophosphate ion, $(HOPO_2-O-PO_2OH)^{2-}$, which is produced biochemically in the body by several different processes (e.g., the conversion of ATP to AMP). In contrast to pyrophosphate, which is prone to cleavage in water with formation of phosphate ions, the bisphosphonates are resistant to hydrolysis. As with pyrophosphate, the bisphosphonates have an affinity for calcium ions, which is a major constituent of bone along with phosphate and collagen type proteins. Because of this affinity, bisphosphonates were investigated in the 1960s for the amelioration of atherosclerosis, which involves the deposition of calcium as a part of atherosclerotic plaque. However, bisphosphonates were found not to inhibit plaque formation. They were also studied unsuccessfully as inhibitors of dental plaque deposition.

Somewhat surprisingly, it was discovered that bisphosphonates are not only taken up by bone, but also inhibit bone resorption. Because bone is a dynamic structure in living organisms, constantly being dissolved (by cells called osteoclasts) and replaced (by cells called osteoblasts), this discovery led to the application of bisphosphonates to the treat-

ment and prevention of osteoporosis, and even for palliative treatment of metastatic bone cancer .

The antiresorptive potency of the first generation bisphosphonates such as etidronate (Didronel) was improved by adding a nitrogen-containing side chain to the bisphosphonate unit. This led to the development of alendronate (Fosamax), which achieved annual sales over $3 billion.[1] Even more potent antiresorptive agents are now available including ibandronate (Boniva), and zoledronic acid (Zometa). Once-a-week dosing with alendronate or once a month dosing with ibandronate or a single annual infusion of zoledronic acid (5 mg over 15 min) are just as effective as once daily alendronate.

Etidronate
(Didronel™)

Ibandronate
(Boniva™)

Zoledronic acid
(Zometa™)

After administration, bisphosphonates accumulate at the site of bone resorption, where they are taken up by osteoclasts. There is evidence that alendronate inhibits farnesyl pyrophosphate synthase (*see* below),[2] which inhibits signaling proteins that activate osteoclasts.[3]

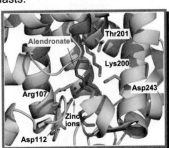

1. *Ann. N. Y. Acad. Sci.* **2006**, *1068*, 367-401; 2. *ChemMedChem* **2006**, *1*, 267-273 (2F92); 3. *Endocr. Rev.* **1998**, *19*, 80-100; Refs. p. 106

# CALCITRIOL (ROCALTROL™)

**Structural Formula**    **Ball-and-Stick Model**    **Space-filling Model**

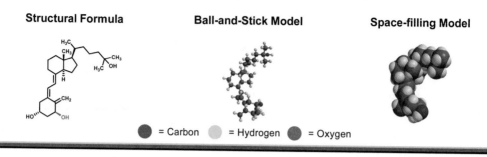

⬤ = Carbon    ⚪ = Hydrogen    ⬤ = Oxygen

*Year of discovery:* 1968; *Year of introduction:* 1978 (Roche); *Drug category:* Hormone, responsible for the regulation of calcium and phosphate metabolism; *Main uses:* Treatment of hypocalcemia, rickets, and osteoporosis; *Other brand names:* Calcijex.

During the 19th century, the childhood disease, rickets, characterized by soft, deformed bones and stunted growth, spread throughout Europe. By the 1920s it was discovered that the disease can be prevented by exposure to sunlight or by consuming cod liver oil or foods irradiated with ultraviolet light. The nature of the substances responsible for the curative effect, collectively called vitamin D, was determined during the 1930s. For the work on vitamin D, the german organic chemist, Adolf Windaus, who also prepared the compound synthetically from 7-dehydrocholesterol, was awarded the Nobel Prize in Chemistry.[1]

It was discovered in the 1970s that the active agent is not vitamin $D_3$, but the $1\alpha,25$-dihydroxy derivative, now known as calcitriol. It is formed from the precursor vitamin $D_3$ by an enzyme-catalyzed oxydative process in which an oxygen atom is introduced at position $1\alpha$ and also 25, as shown below. Also shown is the chemical reaction in which light promotes the transformation of 7-dehydrocholesterol to vitamin $D_3$.

**7-Dehydrocholesterol** → UV light → **Pre-vitamin $D_3$**

isomerization ↓

**$1\alpha,25$-Dihydroxyvitamin $D_3$ (Calcitriol)** ← liver kidney ← **Vitamin $D_3$**

Calcitriol is crucial in the maintenance of healthy bones, because it controls circulating calcium and phosphate ion levels. This hormone promotes the absorption of calcium ions from food in the intestines and the reabsorption of these ions in the kidneys. It also inhibits parathyroid hormone secretion from the parathyroid glands (*see page 100*).

Vitamin D deficiency can cause rickets and osteoporosis. When the amount of vitamin $D_3$ produced in the body is insufficient, it needs to be obtained from food sources such as fatty fish or from dietary supplements. Another member of the vitamin D family, vitamin $D_2$, which is derived from fungal and plant sources, is often used instead of vitamin $D_3$, although it is less effective. Calcitriol is a useful treatment for osteoporosis in older patients with an impaired capacity to make it from vitamin $D_3$.

Calcitriol acts by binding to the vitamin D receptor (VDR), a member of the steroid/thyroid hormone receptor superfamily. The receptor-hormone complex then translocates to the nucleus, where it acts as a transcription factor that modulates gene expression of the transport proteins involved in calcium absorption in the intestine.[2]

In addition to controlling calcium and phosphate metabolism, calcitriol also has an immunomodulating effect that may be useful for the treatment of autoimmune diseases. It has already been approved for the topical treatment of immune-mediated skin disease, such as psoriasis. Calcitriol also has been found to induce apoptosis (biochemically-programmed cell death) in some cancer cells.

1. *J. Cell. Biochem.* **2003**, *88*, 296-307; 2. *Osteoporos. Int.* **1997**, *7*, 24-29; Refs. p. 106

# RALOXIFENE (EVISTA™)

**Structural Formula**  **Ball-and-Stick Model**  **Space-filling Model**

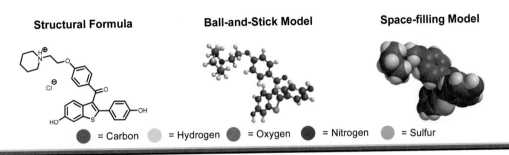

● = Carbon  ● = Hydrogen  ● = Oxygen  ● = Nitrogen  ● = Sulfur

*Year of discovery:* 1982; *Year of introduction:* 1997 (Eli Lilly); *Drug category:* Selective estrogen receptor modulator (SERM); *Main uses:* Prevention and treatment of osteoporosis.

Raloxifene is a synthetic compound which acts as an estrogen on bone, but as an antiestrogen in other tissues. It is a member of a class known as selective estrogen receptor modulators (SERM).[1]

Reduced levels of estrogen in postmenopausal women is a major risk factor for osteoporosis. Unfortunately, the use of natural estrogens for the treatment and prevention of osteoporosis is accompanied by an increased risk of uterine and breast cancer. Although the combination of estrogen with progestin in hormone replacement therapy (HRT) is protective against uterine cancer, it leads to an increased risk of coronary heart disease, invasive breast cancer, stroke and pulmonary embolism.

Estrogens exert a wide range of effects by binding to estrogen receptors (ER) that occur in various tissues. The two known types of estrogen receptors, ERα and ERβ are ligand-activated transcription factors that activate gene expression. Different tissues express the two receptor subtypes in different proportions. Binding of estrogen to these receptors initiates conformational changes and dimerization leading to structures that have an increased ability to complex with specific segments of DNA, termed estrogen response elements. The estrogen receptor/DNA complex recruits further coactivators and other proteins leading to the formation of an organized molecular assembly that initiates transcription. The conformation that the estrogen receptor assumes when bound to an antagonist is different from that occupied by an agonist. The antagonist conformation results in a reduced ability to initiate transcription.

The first estrogen receptor modulators were discovered because these compounds were found to produce a tissue-specific response upon binding to estrogen receptors. This discovery stimulated the search for substances that have beneficial estrogenic effects in certain tissues (e.g., bone, brain, liver), but antagonistic or no activity in other tissues (e.g., breast and endometrium), where estrogenic action is undesirable. The tissue-specific activation produced by these substances arises because the geometry of the SERM-ER complexes varies. These variations lead to the recruitment of different co-activators and differences in gene activation.[2]

Raloxifene, a selective estrogen receptor modulator developed by Eli Lilly, is an *agonist at bone* estrogen receptors and decreases bone resorption. It was approved for the prevention and treatment of osteoporosis in 2002. Raloxifene also decreases the level of LDL cholesterol without affecting HDL cholesterol. Raloxifene is an estrogen receptor *antagonist in breast* tissue and is effective in reducing the occurrence of breast cancer in high risk patients, in common with tamoxifen (Nolvadex), another selective estrogen receptor modulator developed for the treatment of breast cancer. The binding of raloxifene to the estrogen receptor is depicted in the figure below.[3]

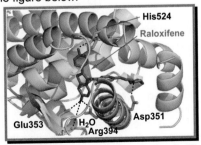

1. *Drugs* **2000**, *60*, 370-411; 2. *Reproduction, Fertility and Development* **2001**, *13*, 331-336; 3. *Nature* **1997**, *389*, 753-758 (1ERR); Refs. p. 106

# TERIPARATIDE (FORTEO™)

**Structural Formula**  **Ball-and-Stick Model**  **Space-filling Model**

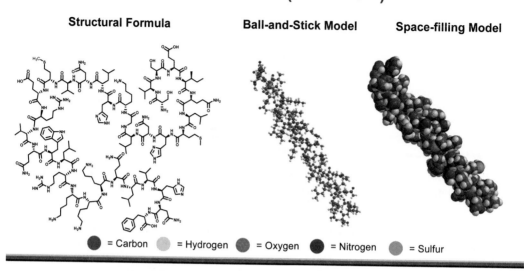

● = Carbon   ○ = Hydrogen   ● = Oxygen   ● = Nitrogen   ● = Sulfur

*Year of discovery:* 1970s; *Year of introduction:* 2002 (Eli Lilly); *Drug category:* Anabolic agent; *Main uses:* Treatment of osteoporosis for patients with high risk of fractures; *Drug category:* First drug in its class, directly affecting osteoblasts and causing bone growth.

The maintenance of strong bones requires optimum calcium ($Ca^{2+}$) and phosphate ($PO_4^{3-}$) levels. The key regulators of these ions in the human body are parathyroid hormone and 1α,25-dihydroxy-vitamin $D_3$ (*see* page 98). Parathyroid hormone, an 84 amino acid polypeptide secreted by the parathyroid glands, controls the concentration of calcium ions in extracellular fluid by regulating the supply of these ions from bone, and from reabsorption in the kidneys. Parathyroid hormone also stimulates the formation of calcitriol from its precursor, vitamin $D_3$, which enhances absorption of calcium ions in the intestines. The secretion of parathyroid hormone varies inversely with plasma calcium ion concentration.

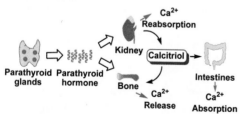

Sustained elevation of parathyroid hormone levels in the circulation causes bone loss. This is one consequence of hyperpara-thyroidism, a disease, in which parathyroid hormone is produced in excess due to the overactivation of the thyroid glands.

Paradoxically, intermittent administration of parathyroid hormone stimulates bone formation. The timing of administration of the drug is critical, since exposure for a short period (< 2 h) leads to bone growth (anabolic response) whereas exposure for more than two hours induces bone loss (catabolic response). The beneficial effect of brief exposure to parathyroid hormone on bone formation may be the result of its stimulation of transformation of stem cells in bone marrow into osteoblasts. Only the 27 amino acid N-terminal segment of parathyroid hormone is required for efficacy.[1]

Teriparatide is a synthetic parathyroid hormone segment containing the first 34 amino acid of the natural protein. This drug, developed by Eli Lilly and marketed under the name Forteo, was approved in 2002 for the treatment of osteoporosis. It is the first parathyroid hormone derivative on the market and also the first agent to act by affecting osteoblasts, thereby directly causing bone growth and an increase of bone density.[2] Teriparatide is used as second-line therapy mainly in cases involving individuals with a history of osteoporosis-related fracture, or intolerant of other therapies. Since teriparatide is not orally absorbed, it is administered by injection.

1. *N. Engl. J. Med.* **2001**, *344*, 1434-1441; 2. *Nat. Rev. Drug Discov.* **2003**, *2*, 257-258; Refs. p. 106

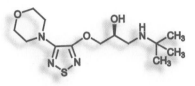

Latanoprost

Timolol

# GLAUCOMA AND ANTIULCER AGENTS

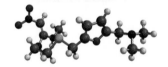

Ranitidine
(Zantac)

# LATANOPROST (XALATAN™)

| Structural Formula | Ball-and-Stick Model | Space-filling Model |
|---|---|---|

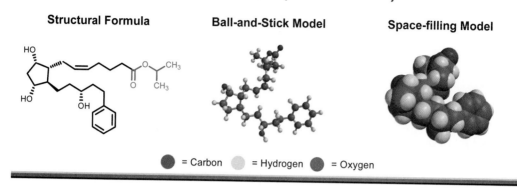

● = Carbon    ○ = Hydrogen    ● = Oxygen

*Year of discovery:* Early 1990s; *Year of introduction:* 1996 (Pharmacia & Upjohn; now marketed by Pfizer); *Drug category:* Prostaglandin $F_{2\alpha}$ analog/Prostanoid selective FP receptor agonist; *Main uses:* To reduce eye-pressure in those with glaucoma or elevated eye-pressure; *Related drugs:* Bimatoprost (Lumigan), Travoprost (Travatan), Timolol (Timoptic).

Glaucoma is an eye condition that damages the optic nerve. Although initially asymptomatic, glaucoma leads to visual field loss and eventually blindness, if left untreated. Damage to the optic nerve is irreversible and irreparable. Glaucoma is the second leading cause of vision loss worldwide. It is estimated that over 70 million people have the disease, but less than 50% are aware of it. There are several factors that may lead to glaucoma, including increased intraocular pressure (IOP), a family history of glaucoma, high blood pressure, diabetes, female gender, and possibly Asian or African-American lineage. Although increased IOP has not positively been linked to the onset of the disease, it is nonetheless a risk factor. Studies have shown that the reduction of IOP can delay glaucomatous nerve or visual field damage. Primary open angle glaucoma is the most prevalent form and it is caused by chronic obstruction of the outflow of aqueous humour within the trabecular meshwork (the primary aqueous drainage system of the eye).

Current treatment for glaucoma includes the use of drugs, surgery, and argon laser trabeculoplasty when drug treatment alone fails. Pharmacotherapies for glaucoma target the reduction of IOP by: (1) reduction of aqueous humour production and (2) elevation of the outflow of aqueous humour.[1] The reduction of aqueous humour production can be achieved by use of topical β-blockers such as timolol (*see* structure at right), but the efficacy of this treatment decreases over time.

(The oral form of timolol is used to treat high blood pressure and prevent heart attack.)

Prostaglandin analogs such as latanoprost (Pharmacia & Upjohn) are now the most successful drugs for decreasing IOP by raising the outflow of aqueous humour.[2] Latanoprost is a prodrug which is hydrolyzed in the cornea by esterases which cleave the isopropyl ester moiety (red). The resulting biologically active carboxylic acid binds to ocular PG receptors, leading to an increase in the outflow of aqueous humour and a lowering of IOP. The structures of two other efficacious ocular prostaglandin analogs, bimatoprost (Lumigan™, Allergan) and travoprost (Travatan™, Alcon), are shown below.[3]

**Timolol (Timoptic™)**

**Bimatoprost (Lumigan™)**    **Travoprost (Travatan™)**

1. *Ther. Clin. Risk Manag.* **2006**, *2*, 193-205; 2. *Drugs & Aging* **2003**, *20*, 597-630; 3. *Curr. Med. Res. Opin.* **2005**, *21*, 1875-1883; **Refs. p. 106**

# RANITIDINE (ZANTAC™)

| Structural Formula | Ball-and-Stick Model | Space-filling Model |
|---|---|---|

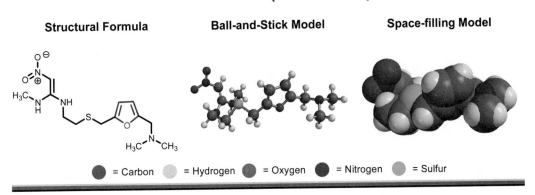

● = Carbon  ○ = Hydrogen  ● = Oxygen  ● = Nitrogen  ○ = Sulfur

*Year of discovery:* early 1970s (Glaxo); *Year of introduction:* 1981; *Drug category:* Histamine $H_2$-receptor antagonist; *Main uses:* For the treatment of dyspepsia, peptic ulcer disease, gastroesophageal reflux disease or prophylaxis; *Related drugs:* Cimetidine (Tagamet), Famotidine (Pepcid), Nizatidine (Axid) and Roxatidine.

Erosion of the mucosal lining of the stomach by an excess of the digestive enzyme pepsin and hydrochloric acid produces gastric (peptic) ulcers. Gastric ulcers occur as a consequence of insufficient protection against irritation by foreign agents (e.g., certain spices, histamine, medicines or microbes) and the corrosive action of stomach acid and pepsin. Ulcers affect more than 5 million people in the US alone and cause approximately 15,000 fatalities per year due to ulcer-related complications, such as perforation of the stomach wall, internal bleeding and infection.[1]

The search for substances that control the secretion of stomach acid began in the 1950s, but it was not until the early 1970s that James Black and coworkers at SmithKline used pharmacological evidence that the histamine $H_2$-receptor is the important regulator of gastric acid production to guide the search for an antiulcer drug. Their research led to the discovery of cimetidine, an effective histamine $H_2$-receptor antagonist.[2] Cimetidine competitively inhibits the binding of histamine to $H_2$-receptors on the surface of the acid-releasing parietal cells of the stomach. It selectively targets $H_2$-receptors, and has almost no effect on $H_1$-receptors. The suppression of gastric acid secretion is dose-dependent. $H_2$-receptor antagonists reduce both the concentration of gastric acid and the amount produced. Consequently, $H_2$-receptor antagonists protect against gastric ulceration and help to promote healing.

Cimetidine (Tagamet™, *see* structure below) became the first billion-dollar drug during the 1980s.

Cimetidine is not an ideal antiulcer agent because it has a half-life of only 2 hours and can cause minor skin rash. It was succeeded by the longer-acting and more potent $H_2$-receptor antagonist, ranitidine (Zantac™). Ranitidine is a very safe, twice-a-day medication (half-life 3 hours) that became the world's best-selling prescription drug by 1988. Both cimetidine and ranitidine are now available without a prescription in the US.

Other potent $H_2$-receptor antagonists have been developed including nizatidine (Axid™, Eli Lilly) and famotidine (Pepcid™, Merck & Co.). The structures of these drugs are shown below.

**Cimetidine (Tagamet™)**

**Nizatidine (Axid™)**

**Famotidine (Pepcid™)**

1. *Nat. Clin. Prac. Gastro. Hepat.* **2006**, *3*, 80-89; 2. *Drug Disc. Dev.* **2006**, *1*, 295-311; Refs. p. 107

# OMEPRAZOLE (PRILOSEC™)

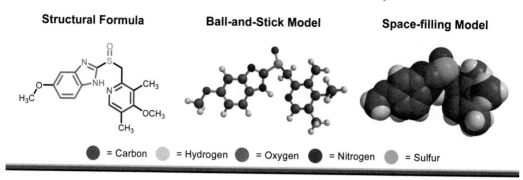

| Structural Formula | Ball-and-Stick Model | Space-filling Model |

● = Carbon  ○ = Hydrogen  ● = Oxygen  ● = Nitrogen  ○ = Sulfur

*Year of discovery:* late 1970s; *Year of introduction:* 1987 (Astra Pharmaceuticals, now AstraZeneca) *Drug category:* Proton pump inhibitor; *Main uses:* For the treatment of dyspepsia, peptic ulcer disease, and gastroesophageal reflux disease. In combination with antibiotics such as clarithromycin and amoxicillin it is used to eradicate the bacterium *Helicobacter pylori* that can lead to gastric ulcers; *Related drugs:* Lansoprazole (Prevacid), Rabeprazole (Aciphex), Pantoprazole (Protonix).

During the late 1970s researchers at Astra Pharmaceuticals in Sweden (now Astra-Zeneca) developed omeprazole (Prilosec™), the first gastric proton pump inhibitor. Proton pump inhibitors irreversibly block the hydrogen/potassium adenosine triphosphate enzyme system ($H^+/K^+$ ATPase) that is responsible for the secretion of acid into the stomach. Omeprazole allows the reduction of gastric acid secretion to very low levels in a dose-dependent way. A key structural feature of omeprazole (and other proton pump inhibitors) is the presence of a sulfoxide functional group (shown above in green) between the benzimidazole (red) and pyridine (blue) rings.

Omeprazole is administered orally. At neutral pH, it is both stable and devoid of proton pump inhibitory activity. However, it is activated once it reaches the parietal cells of the stomach via the bloodstream. In the acidic environment of the stomach omeprazole undergoes an acid-catalyzed rearrangement (*see* scheme at right) to form a reactive species (sulfenamide), which attacks and inactivates the enzyme $H^+/K^+$ ATPase.[1] As a result, the proton pump is shut off and gastric acidity is lowered.

Due to the overall lipophilicity of omeprazole, it readily crosses cell membranes and has high bioavailability. In terms of both potency and bioavailability, proton pump inhibitors have proved superior to $H_2$-receptor antagonists in the reduction of gastric acidity.[2,3]

Annual sales of omeprazole approached $7 billion in 2001.

**Omeprazole** → Rearrangement → **Sulfenic Acid**

Loss of water ($H_2O$)

**Inactive Enzyme-Omeprazole Complex** ← HS⟍ (Enzyme) ← **Sulfenamide**

Omeprazole is a mixture of enantiomers. The (*S*)-enantiomer, the active form, is now available as a proprietary product under the brand name Nexium™.

1. *Trends Pharmacol. Sci.* **1987**, *8*, 399-402; 2. *Therapy* **2006**, *3*, 227-236; 3. *Aliment. Pharmacol. Ther.* **2006**, *24*, 743-750; Refs. p. 107

# REFERENCES FOR PART III.

## Oral Contraceptives (page 90)[1-4]

1. Forinash, A. B.; Evans, S. L. New hormonal contraceptives: a comprehensive review of the literature *Pharmacotherapy* **2003**, *23*, 1573-1591.

2. Petitti, D. B. Combination estrogen-progestin oral contraceptives *N. Engl. J. Med.* **2003**, *349*, 1443-1450.

3. Mishell, D. R. State of the art in hormonal contraception: an overview *Am. J. Obstet. Gynecol.* **2004**, *190*, S1-S4.

4. David, P. S.; Boatwright, E. A.; Tozer, B. S.; Verma, D. P.; Blair, J. E.; Mayer, A. P.; Files, J. A. Hormonal contraception update *Mayo Clin. Proc.* **2006**, *81*, 949-955.

## Testosterone (Testosterone, page 91)[1-4,6]
## X-Ray Structure (2AM9)[5]

1. Lamberts, S. W. J.; van den Beld, A. W.; van der Lely, A.-J. The endocrinology of aging *Science (Washington, D. C.)* **1997**, *278*, 419-424.

2. Riggs, B. L.; Khosla, S.; Melton, L. J., III Primary osteoporosis in men. Role of sex steroid deficiency *Mayo Clin. Proc.* **2000**, *75*, S46-S50.

3. Bhasin, S.; Calof, O. M.; Storer, T. W.; Lee, M. L.; Mazer, N. A.; Jasuja, R.; Montori, V. M.; Gao, W.; Dalton, J. T. Drug insight: testosterone and selective androgen receptor modulators as anabolic therapies for chronic illness and aging *Nat. Clin. Pract. Endoc.* **2006**, *2*, 146-159.

4. Cunningham, G. R. Testosterone replacement therapy for late-onset hypogonadism *Nat. Clin. Pract. Urol.* **2006**, *3*, 260-267.

5. De Jesus-Tran, K. P.; Cote, P.-L.; Cantin, L.; Blanchet, J.; Labrie, F.; Breton, R. Comparison of crystal structures of human androgen receptor ligand-binding domain complexed with various agonists reveals molecular determinants responsible for binding affinity *Protein Sci.* **2006**, *15*, 987-999.

6. Srinivas-Shankar, U.; Wu, F. C. W. Drug insight: testosterone preparations *Nat. Clin. Pract. Urol.* **2006**, *3*, 653-665.

## Mifepristone (Mifeprex, page 92)[2-6]
## X-Ray Structure (1A28)[1]

1. Williams, S. P.; Sigler, P. B. Atomic structure of progesterone complexed with its receptor *Nature (London)* **1998**, *393*, 392-396.

2. Sanders Wanner, M.; Couchenour, R. L. Hormonal emergency contraception *Pharmaco-therapy* **2002**, *22*, 43-53.

3. Gemzell-Danielsson, K.; Mandl, I.; Marions, L. Mechanisms of action of mifepristone when used for emergency contraception *Contraception* **2003**, *68*, 471-476.

4. Spitz, I. M. Progesterone antagonists and progesterone receptor modulators: an overview *Steroids* **2003**, *68*, 981-993.

5. Westhoff, C. Emergency contraception *N. Engl. J. Med.* **2003**, *349*, 1830-1835.

6. Fiala, C.; Danielsson, K.-G. Review of medical abortion using mifepristone in combination with a prostaglandin analogue *Contraception* **2006**, *74*, 66-86.

## Oxytocin (Oxytocin, page 93)[1-4]

1. Norwitz, E. R.; Robinson, J. N.; Challis, J. R. G. The control of Labor *N. Engl. J. Med.* **1999**, *341*, 660-666.

2. Insel, T. R.; Young, L. J. The neurobiology of attachment *Nat. Rev. Neurosci.* **2001**, *2*, 129-136.

3. Young, L. J.; Wang, Z. The neurobiology of pair bonding *Nat. Neurosci.* **2004**, *7*, 1048-1054.

4. MacKenzie, I. Z. Induction of labour at the start of the new millennium *Reproduction* **2006**, *131*, 989-998.

## Sildenafil (Viagra, page 94)[1,2,4-7]
## X-Ray Structure (1UDT)[3]

1. Lue, T. F. Erectile dysfunction *N. Engl. J. Med.* **2000**, *342*, 1802-1813.

2. Rotella, D. P. Phosphodiesterase 5 inhibitors: current status and potential applications *Nat. Rev. Drug Discov.* **2002**, *1*, 674-682.

3. Sung, B.-J.; Hwang, K. Y.; Jeon, Y. H.; Lee, J. I.; Heo, Y.-S.; Kim, J. H.; Moon, J.; Yoon, J. M.; Hyun, Y.-L.; Kim, E.; Eum, S. J.; Park, S.-Y.; Lee, J.-O.; Lee, T. G.; Ro, S.; Cho, J. M. Structure of the catalytic domain of human phosphodiesterase 5 with bound drug molecules *Nature* **2003**, *425*, 98-102.

4. Osterloh, I. H. The discovery and development of Viagra (sildenafil citrate) *Sildenafil* **2004**, 1-13.

5. Doggrell, S. A. Comparison of clinical trials with sildenafil, vardenafil and tadalafil in erectile dysfunction *Expert Opin. Pharmaco.* **2005**, *6*, 75-84.

6. Cirino, G.; Fusco, F.; Imbimbo, C.; Mirone, V. Pharmacology of erectile dysfunction in man *Pharmacol. Ther.* **2006**, *111*, 400-423.

7. Ghofrani, H. A.; Osterloh, I. H.; Grimminger, F. Sildenafil: from angina to erectile dysfunction to

pulmonary hypertension and beyond *Nat. Rev. Drug Discov.* **2006**, *5*, 689-702.

## Some Aspects of Osteoporosis (page 96)[1-3]

1. Orwoll, E. S. & Bliziotes, M. Osteoporosis: Pathophysiology and Clinical Management (Humana Press, **2003**).

2. Mulder, J. E., Kolatkar, N. S. & LeBoff, M. S. Drug insight: existing and emerging therapies for osteoporosis. *Nat. Clin. Prac. Endocr. Met.* **2006**, *2*, 670-680.

3. Rosen, C. J. & Bouxsein, M. L. Mechanisms of disease: is osteoporosis the obesity of bone? *Nat. Clin. Prac. Rheumatol.* **2006**, *2*, 35-43.

## Alendronate (Fosamax, page 97)[1-4,6]
## X-Ray Structure (2F92)[5]

1. Fleisch, H. Bisphosphonates: mechanisms of action *Endocr. Rev.* **1998**, *19*, 80-100.

2. Reid, I. R.; Siris, E. Alendronate in the treatment of Paget's disease of bone *International Journal of Clinical Practice, Supplement* **1999**, *101*, 62-66.

3. Sambrook, P. Once weekly alendronate *Drugs of Today* **2003**, *39*, 339-346.

4. Brandi, M. L. The use of alendronate in osteoporosis treatment: 10-year experience *Postmenopausal Osteoporosis* **2006**, 289-301.

5. Rondeau, J.-M.; Bitsch, F.; Bourgier, E.; Geiser, M.; Hemmig, R.; Kroemer, M.; Lehmann, S.; Ramage, P.; Rieffel, S.; Strauss, A.; Green, J. R.; Jahnke, W. Structural basis for the exceptional in vivo efficacy of bisphosphonate drugs *ChemMedChem* **2006**, *1*, 267-273.

6. Russell, R. G. G. Bisphosphonates: from bench to bedside *Ann. N. Y. Acad. Sci.* **2006**, *1068*, 367-401.

## Calcitriol (Rocaltrol, page 98)[1-5]

1. DeLuca, H. F. 1,25-Dihydroxyvitamin D3 in the pathogenesis and treatment of osteoporosis *Osteoporos. Int.* **1997**, *7*, 24-29.

2. Nuti, R.; Bonucci, E.; Brancaccio, D.; Gallagher, J. C.; Gennari, C.; Mazzuoli, G.; Passeri, M.; Sambrook, P. The role of calcitriol in the treatment of osteoporosis *Calcif. Tissue Int.* **2000**, *66*, 239-240.

3. Holick, M. F. Vitamin D: a millenium perspective *J. Cell. Biochem.* **2003**, *88*, 296-307.

4. Suda, T.; Ueno, Y.; Fujii, K.; Shinki, T. Vitamin D and bone *J. Cell. Biochem.* **2003**, *88*, 259-266.

5. Nakamura, K.; Iki, M. Efficacy of optimization of vitamin D in preventing osteoporosis and osteoporotic fractures: a systematic review *Environmental Health and Preventive Medicine* **2006**, *11*, 155-170.

## Raloxifene (Evista, page 99)[2-6]
## X-Ray Structure (1ERR)[1]

1. Brzozowski, A. M.; Pike, A. C. W.; Dauter, Z.; Hubbard, R. E.; Bonn, T.; Engstrom, O.; Ohman, L.; Greene, G. L.; Gustafsson, J.-A.; Carlquist, M. Molecular basis of agonism and antagonism in the estrogen receptor *Nature* **1997**, *389*, 753-758.

2. Balfour, J. A.; Goa, K. L. Raloxifene *Drugs & Aging* **1998**, *12*, 335-341.

3. Clemett, D.; Spencer, C. M. Raloxifene-A review of its use in postmenopausal osteoporosis *Drugs* **2000**, *60*, 370-411.

4. Kellen, J. A. Raloxifene *Current Drug Targets* **2001**, *2*, 423-425.

5. Thiebaud, D.; Secrest, R. J. Selective estrogen receptor modulators: Mechanism of action and clinical experience. Focus on raloxifene *Reproduction, Fertility and Development* **2001**, *13*, 331-336.

6. Diez-Perez, A. Selective estrogen receptor modulators *Postmenopausal Osteoporosis* **2006**, 203-207.

## Teriparatide (Forteo, page 100)[1-5]

1. Neer, R. M.; Arnaud, C. D.; Zanchetta, J. R.; Prince, R.; Gaich, G. A.; Reginster, J.-Y.; Hodsman, A. B.; Eriksen, E. F.; Ish-Shalom, S.; Genant, H. K.; Wang, O.; Mitlak, B. H. Effect of parathyroid hormone (1-34) on fractures and bone mineral density in postmenopausal women with osteoporosis *N. Engl. J. Med.* **2001**, *344*, 1434-1441.

2. Berg, C.; Neumeyer, K.; Kirkpatrick, P. Fresh from the pipeline: Teriparatide *Nat. Rev. Drug Discov.* **2003**, *2*, 257-258.

3. Eriksen, E. F.; Robins, D. A. Teriparatide: A bone formation treatment for osteoporosis *Drugs of Today* **2004**, *40*, 935-948.

4. Potts, J. T. Parathyroid hormone: past and present *J. Endocrinol.* **2005**, *187*, 311-325.

5. Bogado, C. E.; Massari, F. E.; Zanchetta, J. R. Parathyroid hormone (1-84) and teriparatide in the treatment of postmenopausal osteoporosis *Women's Health* **2006**, *2*, 447-457.

## Latanoprost (Xalatan, page 102)[1-5]

1. Perry, C. M., McGavin, J. K., Culy, C. R. & Ibbotson, T. Latanoprost: an update of its use in glaucoma and ocular hypertension. *Drugs & Aging* **2003**, *20*, 597-630.

2. Holmstrom, S., Buchholz, P., Walt, J., Wickstrom, J. & Aagren, M. Analytic review of bimatoprost, latanoprost and travoprost in primary open angle glaucoma. *Curr. Med. Res. Opin.* **2005**, *21*, 1875-1883.

3. Ishida, N., Odani-Kawabata, N., Shimazaki, A. & Hara, H. Prostanoids in the therapy of glaucoma. *Cardiovasc. Drug. Rev.* **2006**, *24*, 1-10.

4. Noecker, R. J. The management of glaucoma and intraocular hypertension: current approaches and recent advances. *Ther. Clin. Risk Manag.* **2006**, *2*, 193-205.

5. Hollo, G. The side effects of the prostaglandin analogues. *Exp. Opin. Drug. Safety* **2007**, *6*, 45-52.

## Ranitidine (Zantac, page 103)[1-4]

1. Wallace, J. L. Recent advances in gastric ulcer therapeutics. *Current Opinion in Pharmacology* **2005**, *5*, 573-577.

2. Ganellin, C. R. Discovery of the antiulcer drug Tagamet. *Drug Disc. Dev.* **2006**, *1*, 295-311.

3. Moenkemueller, K. & Malfertheiner, P. Drug treatment of functional dyspepsia. *World Journal of Gastroenterology* **2006**, *12*, 2694-2700.

4. Yuan, Y., Padol, I. T. & Hunt, R. H. Peptic ulcer disease today. *Nat. Clin. Prac. Gastro. Hepat.* **2006**, *3*, 80-89.

## Omeprazole (Prilosec, page 104)[1-5]

1. Lindberg, P., Braendstroem, A. & Wallmark, B. Structure-activity relationships of omeprazole analogs and their mechanism of action. *Trends Pharmacol. Sci.* **1987**, *8*, 399-402.

2. Raghunath, A. S., O'Morain, C. & McLoughlin, R. C. Review article: the long-term use of proton-pump inhibitors. *Aliment. Pharmacol. Ther.* **2005**, *22*, 55-63.

3. Edwards, S. J., Lind, T. & Lundell, L. Systematic review: proton pump inhibitors (PPIs) for the healing of reflux esophagitis - a comparison of esomeprazole with other PPIs. *Aliment. Pharmacol. Ther.* **2006**, *24*, 743-750.

4. Moenkemueller, K. & Malfertheiner, P. Drug treatment of functional dyspepsia. *World Journal of Gastroenterology* **2006**, *12*, 2694-2700.

5. Wong, G. L.-H. & Sung, J. J. Y. Esomeprazole: a new proton pump inhibitor for NSAID-associated peptic ulcers and dyspepsia. *Therapy* **2006**, *3*, 227-236.

*The immune system is essential for human survival.*

*It is an exceedingly complex collection of many types of special cells that must be exquisitely regulated and coordinated.*

# PART IV.

*There are at least one hundred illnesses involving abnormal function of the immune system. They range from rheumatoid arthritis to psoriasis, a severe skin disease, to multiple sclerosis, which affects the nervous system, to lupus erythematosus, a severe systemic disease.*

*Immune function declines with age which causes serious illnesses in the elderly.*

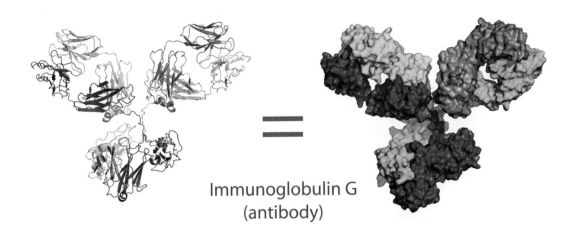

Immunoglobulin G
(antibody)

# AUTOIMMUNE DISEASE
# AND
# ORGAN TRANSPLANT

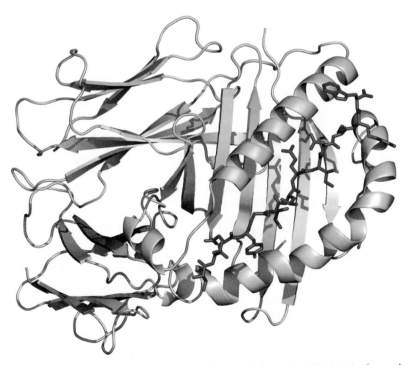

Antigenic Peptide (red) Bound to a Class I HLA Molecule

# A BRIEF SURVEY OF THE IMMUNE SYSTEM

## Introduction

The immune system plays a crucial role in human health, but one that has two sides. On the one hand, it is absolutely indispensable for protection against pathogens; on the other, malfunctions in the immune system can result in serious disease. Earlier sections of this book have mentioned the involvement of the immune system in inflammatory diseases such as arthritis and asthma, and also in autoimmune diseases such as type 1 diabetes (diabetes mellitus) and Addison's disease. As medicine becomes more molecular in character, the cellular and molecular details of immunity must be clarified if truly effective treatments are to be discovered.

Human life and health depend on the ability of the immune system to distinguish between self and non-self and selectively to destroy viruses, bacteria, fungi, multicellular parasites, or other foreign invaders. That discrimination involves the sensing of foreign matter by receptors on the surface of various immune cells. Such receptors can detect foreign matter that exhibits telltale, non-self chemical features. The body also uses outer defenses wherever it is directly exposed to the environment. These include antimicrobial enzymes and molecules in skin and in the respiratory and digestive tracts, and also sentinel immune cells that effectively destroy microorganisms.[1-4]

## Innate immunity

The inner defense mounted by the immune system when foreign invaders enter the body is broadly of two types, depending on the kind of immune cell that is involved. The first response is a predetermined defensive system that deals with a broad assortment of invaders. These cells make up the innate arm of the immune response. Members of this class include monocytes, macrophages, dendritic cells, granulocytes, and natural killer cells. They all originate in bone marrow from a common hemopoietic stem cell precursor. These cells often carry surface receptors that bind surface components of microbes, a form of pattern recognition selected for by evolution.

## Adaptive Immunity

The second line of cells form the "adaptive arm" of the immune response which includes B- and T-lymphocytes, and the various subtypes derived from them. The response of these cells is highly specific and exquisitely tailored to be effective against the foreign agent. In addition, some of those cells are unusually long-lived and remember a previous encounter with a particular invading microorganism. They are primed for later action.

## Complexity of Immunity

The many component cells of the immune system are linked together in a functional network by chemical messengers that influence their activation or suppression, their migration to the site of infection, and also their proliferation or demise. The various immune cells express a large number of surface receptors that receive and transmit a huge variety of informational signals from the cell's environment. The direction, coordination and regulation of the immune reaction to infection involves multiple levels of chemical signaling and redundancy that add up to an enormously complex system. For this reason, it has taken many decades to attain our present understanding of its operation and the molecular basis for the exquisite distinction between the cells of our own body and those of a foreign organism. Although there is still much to learn, we are beginning to appreciate the ways in which the immune system malfunctions in inflammatory and autoimmune diseases.

A brief summary of the working elements of the immune system is challenging because of the various types of cells that are involved and the large cast of biochemical agents that serve as weapons, messengers, activators, deactivators or regulators. The interactions of these components are complex and dynamic. For these reasons we shall consider first the various mediators that guide cell action and their general properties, and then discuss the actions of the different classes of immune cells and the role of the mediators in these actions.

## Mediators

*1. Chemotactic agents.* These attract immune cells to the site where the immune defense needs to be focused. The attraction arises from sensing receptors on a target cell that stimulate its movement toward increasing concentrations of the agent.

*1A. Chemokines.* These are small signaling proteins (8-10 kDa molecular weight) secreted by several types of immune cells, especially to attract monocytes, macrophages, and dendritic (phagocytic) cells of the innate response, and the T- and B-cells of the adaptive response. There are roughly 50 chemokines and numerous receptors (about 20) for them.

*1B. Products of the complement system.* "Complement" is a large family of proteins, made in liver and in certain immune cells, that has several functions, for example:

(1)   to effect proteolytic cleavage of proteins,

(2)   to provide reactive fragments that can tag foreign or abnormal host cells and

(3)   to generate small peptide fragments (e.g., complement peptides C3a, C5a and N-formyl-Met-Leu-Phe) that serve as chemotactic agents.

*1C.* Fragments from degradation of microbes or viruses or from blood clotting or fibrinolysis.

*1D.* Other small molecules, e.g., leukotrienes $B_4$ and $C_4$ (*see* page 41), which are powerfully attractive to immune cells.

*2. Cytokines.* These are proteins secreted by immune cells that regulate and coordinate the immune response, but also play a role in inflammation and cell proliferation. Subtypes include monokines (made by monocytes), lymphokines (made by lymphocytes), chemokines (*see* above) and interleukins (made by leukocytes to act on other leukocytes). Examples include:

(1)   TNF–$\alpha$ that activates cells such as macrophages and T-cells,

(2)   IL–1 that activates T-cells,

(3)   IL–2 that activates T, B and also natural killer (NK) cells and

(4)   interferons, produced by a variety of cells and serving to upregulate antiviral defenses. There are about 30 different interleukins now known.

*3. C-reactive protein (CRP).* The production of this protein which circulates in blood is increased by about two orders of magnitude after an infection. CRP binds to

bacteria synergistically with the C3b fragment of complement and stimulates monocytes and macrophages to attack the cells so tagged.

*4. Antibodies (Ab).* Antibodies are proteins of the immunoglobulin type that recognize and bind to foreign agents including microbes, cell fragments or foreign molecules, effectively tagging these species for destruction or elimination. There are five classes of immunoglobulins: IgG (most abundant in blood, ca. 75%), IgM (ca. 10%), IgA (15-20%), IgD (<1%) and IgE (<1%), all synthesized by B-cells. IgE mainly targets mast cells and basophils. As if this is not complex enough, there are 4 subclasses of IgG and 2 of IgA in humans.

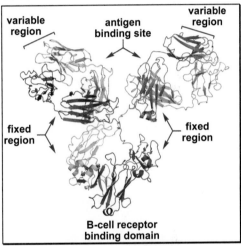

X-Ray structure of IgG showing key structural units.

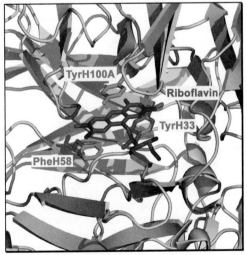

A close-up X-ray structure of the antigen binding site of an IgG molecule with riboflavin as the bound antigen. The crystal used for this structural analysis was obtained from the serum of a patient with the cancerous disease multiple myeloma. (**2FL5**)

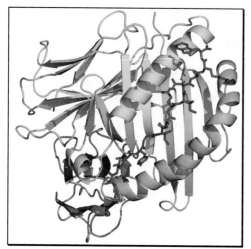

An X-ray view of a small peptide (red) bound to a class I HLA molecule. (**1IAK**)

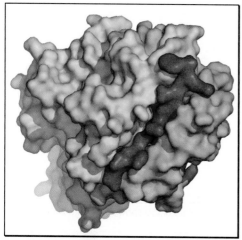

A space-filling representation of the class I HLA molecule-peptide binary complex shown above. The peptide is shown in red. (**1IAK**)

The structures of the various classes have been determined by X-ray crystallographic studies. These proteins are roughly Y-shaped with the stem being a fixed structural domain that binds to a specific cell surface receptor for each class.

Their receptors occur, for example, on B-cells (for IgG) and mast cells (for IgE). The uppermost part of the slanting Y arms of the immunoglobulin are highly variable in amino acid sequence and contain the antigen binding region. One particular B-cell clone produces only one antibody species, specific for that clone. Every day millions of non-identical immunoglobulins are produced in the body by the total population of ca. $10^{12}$ cells. When an antigen is bound to an Ig molecule

attached to the surface of the B-cell, it is internalized and broken down. Each B-cell clone deals with only one particular antigen.

## Human Leukocyte Antigen Molecules (HLA)

Human leukocyte antigens, also known as major histocompatibility complex (MHC) are proteins that, at one end, function as ligands which bind to cognate receptors and, at the other end, possess groves that always carry one specific small peptide in a stretched-out condition, and only that particular peptide. There are two broad classes of human leukocyte antigens, class I HLAs and class II HLAs. The peptides which fit class I HLAs are 8-11 amino acid long, whereas those that are recognized by class II HLAs are 10-30 amino acid long. The peptides may be derived from the host's own proteins (self peptides) or from foreign proteins (non-self or antigenic peptides). The HLA-peptide complexes are expressed by antigen presenting cells (APC). When these HLA-peptide complexes are activated by pathogens, they migrate to lymph nodes. Once there, the APC-HLA-peptide complex can bind to a T-cell causing its activation. An X-ray structure of a binary complex of class I HLA and a small peptide is shown on the left.

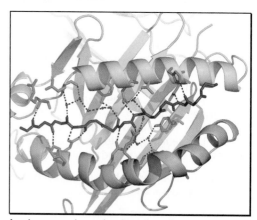

A close-up view showing the direct and water-mediated hydrogen bonding between the class I HLA molecule and peptide ligand. The peptide is shown in red (without its side chains), the $\alpha$ and $\beta$ subunits of the HLA molecule are shown in yellow and cyan, the water molecules are shown as small spheres in magenta, and the hydrogen bonds are depicted as blue dotted lines. (**1IAK**)

The binding of class II HLA molecules to peptides is generally similar to that shown for class I HLA-peptide complex.

## Families of Immune Cells

*1.* Phagocytes are immune cells that recognize, engulf and digest viruses, microbes, and foreign cells or organic materials. There are three principle subtypes, which are part of the innate response: monocytes, macrophages and dendritic cells. In addition, there is another family of phagocytic cells, the granulocytes, that contain subcellular enzyme packets.

*1A. Monocytes.* Roughly 5% of the circulating white blood cells in the body are monocytes. These cells have lifetimes of just a few days, after which they develop into macrophages, which reside in body tissues. Monocytes selectively destroy cells or materials that are tagged by having surface labels such as antibodies or a complex of complement fragment C3b and C-reactive protein. Monocytes are attracted to sites of infection or damage by the various chemotactic factors mentioned above.

*1B. Macrophages.* The name macrophage derives from the Greek for "big eater", which is appropriate since they devour foreign cells or fragments and since one macrophage can ingest dozens of microorganisms over their lifetime, which can vary from months to years. Because they produce a variety of noxious enzymes and chemicals, macrophages cause significant collateral damage to surrounding cells and tissues, thereby resulting in inflammation and, also, in their own demise.

Macrophages are activated by mediators secreted by other immune cells; they also produce their own mediators to stimulate the immune response of other cells.

Macrophages are guided to their prey by a variety of attractant molecules. They play a role as a viral reservoir in HIV infection since the human immunodeficiency virus can gain surreptitious entry by binding to a surface receptor to form a complex which is then taken up by the macrophage (endocytotic internalization). Once inside, the HIV guest enjoys safe haven since it is protected from attack by the immune system.

*1C. Dendritic cells* migrate from blood to tissues that are exposed to the environment (e.g., airways, gastrointestinal tract and skin). They are attracted to sites of infection, and if they sense a tagged foreign invader, they are activated to ingest and break down the various constituents.

Dendritic cells, in common with ordinary human cells, contain a cylindrically shaped assembly of 28 protein subunits (the "proteasome") that can cleave any protein which carries a polyubiquitin chain (a general tag for disposal) to small fragments (mostly 8-12 amino acids). These then are transported to the surface of the dendritic cells where they are held by a human leukocyte antigen (HLA) protein. The dendritic cell so equipped next migrates to a lymph node where it awaits an encounter with a T-cell that possesses a surface receptor, which provides a matching fit with the HLA-peptide complex. The dendritic cell then functions as an antigen presenting cell which activates the T-cell. Most dendritic cells carry small peptides derived from self that do not activate T-cells. There are actually three types of dendritic cells with respect to origin, but they all function in the same general way as antigen presenting cells.

*2. Granulocytes.* This class of white blood cells, also known as polymorphonuclear leukocytes, comprises the subclasses *neutrophils, eosinophils*, *basophils* and *mast* cells. These cells contain, in addition to nuclei, granules that hold enzymes which break down DNA, RNA, triglycerides and other lipids, complex carbohydrates and proteins. These enzymes (e.g., $\alpha$-amylase, elastase and collagenase) can break down not only organic materials but also bacterial cell walls (using the enzyme lysozyme), and tissues. Other granule products are (1) lactoferrin which binds iron strongly and deprives bacteria of this element which is essential for their proliferation and (2) an array of enzymes that break down bacterial cells and their contents, and powerful bacteriocidal oxidants such as peroxide-derived oxygen radicals and hypochlorite (NaOCl). The latter (the active ingredient in chlorine bleach) is especially lethal to bacteria. One milliliter of bleach will kill 1 gram of *E. coli* bacteria in seconds.

*2A. Neutrophils.* The most abundant of the phagocytic white blood cells (about 50-70%, roughly $10^{10}$ cells in the average human), neutrophils circulate in blood and are drawn to sites of infection by cytokines, chemokines and other mediators. They arrive rapidly (ca. 1 hour) and in force at the site of infection because they are widely distributed and mobile. Neutrophils have half lives of roughly 7-8 hours, which is fortunate because they are not only deadly to invading microorganisms, but very destructive to the body's own cells and tissues.

*2B. Eosinophils.* Although they comprise less than 5% of the white blood cell population, eosinophils can effectively deal with parasites and viruses because they possess not only protein digesting enzymes and oxidants but an abundance of DNAase and RNAase enzymes. An elevated eosinophil blood count can accompany various disease states, e.g., parasitic infection, Hodgkin's malignancy and rheumatoid arthritis (in which they contribute to tissue and joint damage).

*2C. Basophils and mast cells.* These immune cells have many common features including surface receptors for IgE and granules containing inflammatory mediators such as histamine, leukotrienes B and C, $PGD_2$ and various cytokines. These mediators are released when the cells are activated. Mast cells are found in most tissues and are considerably more abundant than basophils. Full activation of these cells occurs when IgE-bound surface receptors are cross-linked by an antigen. The IgE is provided by a B-cell which can interact with the basophil or mast cell. Mast cells appear to be important in asthma and rhinitis and in various allergies.

*3. Natural killer (NK) cells.* NK cells of the immune system are very cytotoxic, and consequently, highly regulated weapons against infection. Their outer surfaces are studded with regulatory receptors that bind either activating or deactivating mediators. Activation can be effected by cytokines such as TNF-$\alpha$ (released, for example, by macrophages which are stimulated by microbial lipopolysaccharides), viral double stranded RNA and interferons (especially IF-$\alpha$ and IF-$\beta$, which are secreted by virally infected cells). These bind to activation receptors on the surface of the NK cells. NK cells contain granules that release at least two types of attack mediators: (1) perforin proteins which enter the membranes of target cells and form pores, and (2) proteolytic enzymes that then invade the target cell through these pores. The activity of NK cells is eventually downregulated by mediators, including the body's own HLA proteins that bind to inhibitory receptors on the outer surface of the NK cells.

NK cells are especially well suited to the removal of the body's own cells that become malignant or infected by viruses, or any other cells whose surfaces are tagged by IgG.

*4. Lymphocytic T-cells.* The white blood cells known as T- and B-cells, originating in the thymus and bone marrow, respectively, are major components of the adaptive immune response. The daily human production of these cells is about $10^8$ each per day, and each population consists of many millions of distinctly different clones. This enormous diversity is combinatorial and is the result of gene-segment shuffling during cell formation. The diversity is manifested in T-cells by the huge variety of binding domains of immunoglobulin-like T-cell receptors (TCR) on the surface of the T-cell. These receptors recognize and bind to specific HLA-peptide complexes on the antigen presenting cells. The diversity-producing gene shuffling occurs in Ig/TCR genes of T-cells (and for B-cells in genes coding for Igs). In general, T-cells and B-cells require two different signals for full activation. One of these is directly related to interaction with an antigen. The maturation and proliferation of T-cells is greatly increased in those cells that are bound to HLA-peptide complexes and also receive a co-stimulatory signal from the activated APC's. As a result, there is an exponential expansion of those highly competent cells and an elimination of any incompetent T-cells.

A negative selection also occurs in the thymus during T-cell generation for any subset of T-cells that happen to bind strongly with HLA molecules bearing a "self" nonapeptide. Those T-cells which could be toxic to the body's own cells are caused to undergo bio-chemically programmed cell death (apoptosis) and thus depart from the immune system or are rendered innocuous. (*See* section below on discrimination between self and non-self.)

T-cells can be distinguished by two differentiating surface proteins, CD4 and CD8. CD4-bearing T-cells (CD4$^t$) bind class II HLAs and CD8-bearing T-cells (CD8$^t$) bind class I HLAs. CD8$^t$ cells are cytotoxic T-cells (T$_c$) whereas CD4$^t$ cells are helper T-cells which are involved in activating other immune cells or cytokine release (T$_h$). Cytotoxic T-cells are powerful destroyers of foreign cells, but also sources of inflammation. There are two sub-sets of T$_h$-cells, T$_h$1 and T$_h$2, which produce different cytokines and affect different components of the immune response. T$_h$1-cells produce IF-$\gamma$ and are highly stimulating to phago-cytes. T$_h$2-cells produce IL-4, IL-5, and IL-13 and stimulate the immune response of mast cells, basophils and eosinophils, effectively upregulating antiparasitic and allergic responses. Another type of $T_h$-cell, $T_h17$, makes IL-17, and seems to play a role in autoimmune diseases such as multiple sclerosis and inflammatory bowel disease.

There are T-cells that carry both CD4 and CD25 surface protein markers and serve as regulatory cells ($T_r$). One important function of $T_r$-cells, which contributes to the ability of the immune system to distinguish between self and non-self, is to cause apoptosis (biochemically programmed cell death) of self-reactive $T_c$-cells. A second critical role of $T_r$ – cells is to hasten the death of activated $T_c$-cells after an infection has been cleared. This process supplements another control on $T_c$-cells in which apoptosis of the $T_c$-cell occurs spontaneously after it has been activated multiple times.

Finally, there are long-lived T-cells that are termed "memory cells" because they live for several years and are able to deal rapidly with an agent from an earlier infection. Normally $T_c$-cells survive only for a few weeks, and even less if they have been activated several times.

*5. Lymphocytic B-cells.* B-cells recognize foreign organic matter including microbes and viruses, produce high affinity immunoglobulin antibodies that attach to these foreign entities, and cooperate with other immune cells to eliminate such threats. B-cells originate from stem cells in bone marrow and progress through a multi-stage maturation process, which only a small fraction survive. Humans have at any one time about $10^{12}$ B-cells and a huge diversity of B-cell clones, approximately $10^9$. Each one produces a different antibody which may be of the IgG, IgM, IgA, IgD or IgE class, as mentioned in the section on antibodies above. The enormous number of different B-cell clones, and Igs arises because of very extensive, random shuffling of certain segments of the genes that code for the variable, antigen-binding, portion of the immunoglobulins. This genetic shuffling is all the more remarkable because the genes coding for the Igs consist of many segments that are separated by intervening, non-coding DNA and are even on different chromosomes. It is clear that great B-cell diversity is required because there are $20^9$ or $5.12 \times 10^{11}$ possible nonapeptides with the 20 genetically coded amino acids.

Another reason for the effectiveness of the B-cells in combating infection is the existence of a mechanism for expanding the population of those clones that produce Igs which recognize antigens and bind with high affinity to the HLAs of antigen presenting cells. Those B-cells that find matching antigen are stimulated to proliferate, thereby greatly increasing that clonal population. This selection mechanism for clonal expansion continues until the foreign invader has been dealt with. The sequence of events for clonal expansion of B-cells is as follows: (1) encounter with an antigen presenting cell for which the B-cell has affinity, (2) ingestion of the antigen, (3) display of the antigen on a B-cell class II HLA protein, and (4) binding to a primed helper T-cell that also recognizes the antigen and secretes cytokines which stimulate B-cell proliferation.

Many millions of B-cells are manufactured daily by humans. Some B-cells have large nuclei and are large-scale producers of antibodies. Furthermore, other special purpose B-cells are so long lived that they function as memory cells, since they rapidly spring into action, even at a much later time, if their triggering antigen reappears. Still another B-cell subtype expresses lower affinity, but less specific IgM antibodies.

B-cells are distributed in tissues throughout the body, but they have a major presence in the lymph nodes, like T-cells. A small fraction circulate in the vasculature. Activated B-cells are cleared in the liver and spleen and survive only a matter of days, but "naïve" or memory B-cells are protected from this fate. Peripheral B-cells also secrete immunoglobulins such as IgE (which goes mainly to mast cells and basophils) and IgM. The latter is prone to form stable pentameric structures. The binding of IgM to bacterial surfaces causes aggregation to a large polycellular clump which is attacked and destroyed by macrophages.

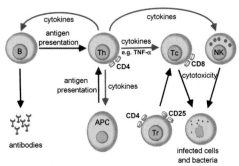

A graphical summary of the interactions and functions of lymphocytes. $T_c$-cells that are self-reactive or no longer needed are disposed of by $T_r$-cells; otherwise $T_c$-cells undergo spontaneous apoptosis after multiple cycles of activation.

## Discrimination Between Self and Non-Self

If a student of biological chemistry were to undertake the theoretical task of designing an immune system that, in principle, can distinguish an individual's own cells (self) from those of invading microbes or viruses (non-self), the result might be along the following lines. Each cell of the person would have to be tagged by a surface molecule that is unique to that individual, in the same way that a combination lock is coded by a unique set of numbers. The molecular tag might be a particular protein or a particular protein that is further garnished by a branching chain of carbohydrate units to ensure even greater diversity. Any immune cell that happened to have a receptor which allows it to bind to the tag, or self-antigen, would then have to be dealt with because such proximity would probably lead to attack and destruction by the immune cell. The individual's cell could deal with the inappropriate recognition and binding by the immune cell in various ways. First, it might simultaneously repel the immune cell and tag it for destruction. Or, it might send a biochemical signal to the cell that renders it incapable of producing any damage. Finally, it might label the immune cell with a marker that would allow regulatory cells to trigger the death of the self-reactive immune cell. This conceptually simple model is not very different from the picture that has emerged after decades of study of this central question in immunology.

We now know that a person's cells do indeed carry surface tags that are unique to the individual and genetically determined. These self- or tolerogenic antigens induce tolerance by the immune system and are essential for distinguishing between self and non-self. Severe failure of this system leads to autoimmune disease. T- and B-lymphocytes and dendritic cells are key players in immunological tolerance.

This discussion will be simplified by focusing on T-cells. Each day about 100 immune-cell precursors enter the thymus and undergo multiple cell divisions to generate about $10^7$-$10^8$ T-cells. However, only about 5% of these T-cells survive the selection process in the thymus which weeds out those T-cells that show an affinity for self-antigens. About 95% of the total production of T-cells either die by apoptosis or by conversion to a harmless (anergic) state. This selection-driven tolerance ("central tolerance") is crucial to the establishment of a properly balanced immune system. If the selection process were to be too severe, a weak immune system incapable of dealing with infection would result. On the other hand, if the selection were to be too lenient, autoimmune disease would become more likely. In general, the immune system seems to err on the side of leniency, because it has developed another level of defense against autoimmunity involving regulatory T-cells ($T_r$). It has been mentioned in the earlier section on T-cells (page 116) that $T_r$-cells can induce apoptosis in $T_c$-cells after an infection has been cleared or inactivate these cells.

The control of self-reactive circulating T-cells by $T_r$-cells occurs in the periphery (and is sometimes called "peripheral tolerance"). The subpopulation of $T_r$-cells carrying both CD4 and CD25 surface proteins is more effective for the elimination of self-reactive T-cells than the $T_r$-cells that carry surface CD4 but not CD25. The molecular mechanisms by which T-cells are downregulated or disposed of involve interleukins as the active agents. T-cell anergy can be induced by inhibition of a tyrosine kinase signaling pathway that downregulates calcium influx into the cell and also by another pathway that blocks IL-2 receptor-mediated activation. There is a whole family of proteins produced by the immune system that suppress cytokine signaling.

Most cancer cells avoid attack by the immune system because they carry self-antigens on their surface.

In summary the immune system has evolved a multilevel quality-control mechanism to minimize damage to host cells by self-reactive T- and B-lymphocytes.

## Immunologic Profile of an Infection

The vast majority of pathogens in the environment are held at bay by the body's physical barriers to infection (e.g., skin, mucus, gastric acidity, and antimicrobial chemicals and enzymes). However, once these outer defenses are breached and an infection sets in, a variety of effective immunological countermeasures emerge. One simplified scenario can be outline for a bacterial infection. The presence of foreign organic matter (e.g., lipopolysaccharide of microbial origin) may provoke attack by components of the innate immune response (e.g., dendritic cells or macrophages), which can lead to activation of the complement system and release of chemotactic agents and cytokines. The activated dendritic cells then generate surface antigen from the target and migrate to a nearby lymph node. There they meet T-cells and B-cells which recognize the antigen as foreign, become activated and mount a response. Immunoglobulins are formed and secreted, and reactive complement fragments and C-reactive protein are generated. Activated monocytes, macrophages, NK cells, $T_c$-cells and B-cells then converge on the invading microbes and clear the infection, at which point the destructive immune cells are caused to die. However, memory B- and T-cells persist and remain in long-term reserve against reinfection.

## Disorders of the Immune System

There are at least one hundred illnesses involving abnormal function of the immune system. These cover a spectrum ranging from systemic immune malfunction (e.g., rheumatoid arthritis) to highly localized organ destruction, such as of pancreas (type I diabetes), adrenals (Addison's disease) or thyroid (Hashimoto's thyroiditis). The susceptibility of an individual may vary with genetic inheritance, gender, history of infection, or age. Causation is almost always complex and multifactorial, but leads pathologically to the harmful action of autoantibodies, autoreactive T-cells or phagocytes on the tissues and organs. Autoimmunity is basically a failure of immune tolerance that can, in principle, stem from:

(1) functional defects in any one of the immune cell types,

(2) the improper secretion of the mediators which coordinate or regulate them,

(3) failure to recognize invaders as foreign,

(4) failure to distinguish self from non-self,

(5) failure to control populations of effective, non-effective or deleterious cells,

(6) failure to generate adequate diversity of immunoglobulins and/or HLA molecules,

(7) purely random events such as the chance similarity of self and non-self antigens.

Autoimmune diseases can be diagnosed in various ways including the nature of the symptoms of the illness or the detection of autoantibodies (often termed "antinuclear antibodies" or ANAs). If high blood levels of ANAs are detected, more precise diagnostic tests are used to determine the type, e.g., rheumatoid arthritis, lupus or Sjögren's syndrome subtype. In addition, elevated levels of C-reactive protein, reactive complement fragments, cytokines or other inflammatory species are indicative. The therapies of autoimmunity are still limited. The most commonly used of these are glucocorticoids such as prednisone, immunosuppressants, or monoclonal antibodies against specific cytokines such as anti-TNF-$\alpha$ or anti-IL-1 antibodies.

## Allergy

Allergy, now termed type 1 hypersensitivity, is characterized by excessive production of IgE in response to harmless common antigens by mast cells and basophils, and their overactivation. As a result, inflammatory mediators (e.g., histamine, $PGD_2$, and leukotrienes) are released causing symptoms ranging from mild to life-threatening (anaphylaxis).

## The Immune System and Aging, Immunosenescence

The intensity and effectiveness of the human immune response to infection increases from birth to a maximum in middle age. In the elderly (>70 years) there is a substantially increased susceptibility to infection, which is especially serious for viral infections such as flu, pneumonia and shingles (*Varicella zoster*). The protective effect of vaccines is reduced in the elderly with regard to the level of the resulting antibodies and the duration of the protective effect. The production of immune-stimulating cytokines (e.g., IL-2) is diminished. In addition, the antibodies generated by elderly individuals exhibit lowered affinity for antigen than those measured for young adults. The levels of antinuclear antibodies are on average considerably higher for individuals between the ages of 70 and 80 than those in their 40s, and the incidence of autoimmune disease is increased. In the elderly much of the thymic tissue is lost by age 70 and the clonal diversity of CD8-bearing $T_c$-cells is diminished. The factors that determine the decline in immune function with age are especially difficult to evaluate because there are so many, and because animal models (e.g., mice) have been unreliable.

1. Abbas, A. K., Lichtman, A. H. Cellular and Molecular Immunology, 5th Edition (**2005**); 2. Cruse, J. M., Lewis, J. R. E. & Editors. Illustrated Dictionary of Immunology (**1995**); 3. Janeway, C. A., Travers, P., Walport, M. & Shlomchik, M. Immunobiology: The Immune System in Health and Disease, 6th Edition (**2004**); 4. Paul, W. E. & Editor. Fundamental Immunology (**1986**); Refs. p. 173

*The medical repair or replacement of damaged or non-functioning organs, now in its infancy, is likely to play a much more important role in the treatment of human illnesses in the coming decades.*

# IMMUNOSUPPRESSIVE AGENTS

# AZATHIOPRINE (IMURAN™)

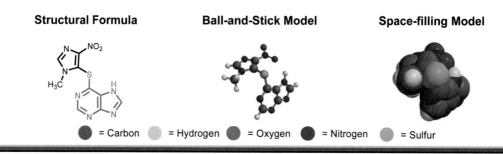

| **Structural Formula** | **Ball-and-Stick Model** | **Space-filling Model** |

● = Carbon  ● = Hydrogen  ● = Oxygen  ● = Nitrogen  ● = Sulfur

*Year of discovery:* Late 1950s (Wellcome Research Laboratories); *Year of introduction:* 1968; *Drug category:* Immunosuppressant; *Main uses:* Prevention of organ transplant rejection and treatment of severe rheumatoid arthritis; *Other names:* Azasan; *Related drugs:* Sirolimus (Rapamune), Mycophenolate mofetyl (Cellcept), FTY720.

Azathioprine is a potent immuno-suppressant used to prevent rejection of organ transplants and to treat patients with severe, erosive rheumatoid arthritis not responsive to conventional treatment. It is also indicated for the treatment of other auto-immune diseases including systemic lupus erythematosus, inflammatory myositis, and inflammatory bowel disease.

Azathioprine was discovered at Wellcome Research Laboratories by George Hitchings and Gertrude Elion in a pioneering study of the effect of DNA base analogs on rapidly dividing cells. In the early 1950s, several of these purine derivatives were tested at the Sloan-Kettering Institute against a wide range of rodent tumor and leukemia cell lines.

**Adenine**          **6-Mercaptopurine**

Of special interest was the adenine analog, 6-mercaptopurine, which in clinical trials could produce complete remission of acute leukemia and substantially increased the life expectancy of terminally ill children. This compound was approved by the FDA for the treatment of leukemia in 1954. Subsequent research to improve on 6-mercaptopurine led to the development of azathioprine. Coincidentally, in 1954 a surgeon, Joseph Murray, performed the first successful organ transplant (kidney) between identical twins and theorized that transplantation between non-identical twins would be possible using immunosuppression. Shortly thereafter, Robert Schwartz and William Dameshek

examined the effect of 6-mercaptopurine on the immune response and demonstrated immunosuppression. Inspired by these findings, a British surgeon, Roy Calne, investigated the effect of 6-mercaptopurine on kidney transplants in dogs and found that the graft survival of the treated animals was longer than that of the controls. Further studies showed that azathioprine was superior to 6-mercaptopurine.[1] Azathioprine was used in combination with prednisone (*see* page 44) for kidney transplants in humans in 1962.

Nobel Prizes in Medicine were awarded to Elion and Hitchings in 1988 and to Murray in 1990 for their research on immunosuppression and organ transplantation.

Azathioprine is a prodrug that is gradually cleaved in the body to 6-mercaptopurine, from which 6-thioguanine nucleotides are bio-synthesized.

**6-Thioguanine**

These compounds interfere with *de novo* DNA biosynthesis and inhibit T-cell proliferation. The major side effect of azathioprine is bone marrow suppression. Although still used extensively after four decades, azathioprine gradually is being displaced by a newer drug, mycophenolate mofetil, which is associated with less bone marrow suppression and fewer incidents of acute rejection (*see* page 123).[2]

1. *Science* **1989**, *244*, 41-47; 2. *Current and Future Immunosuppressive Therapies Following Transplantation* **2001**, 85-110; Refs. p. 173

# MYCOPHENOLATE MOFETIL (CELLCEPT™)

| Structural Formula | Ball-and-Stick Model | Space-filling Model |
|---|---|---|

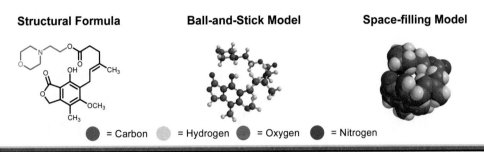

● = Carbon    ○ = Hydrogen    ● = Oxygen    ● = Nitrogen

*Year of discovery:* Late 1980s (Syntex, now Roche); *Year of introduction:* 1995; *Drug category:* Immunosuppressant; *Main uses:* Prevention of organ transplant rejection and treatment of lupus nephropathy; *Related drugs:* Azathioprine (Imuran).

Mycophenolate mofetil is an immuno-suppressant indicated for the prevention of kidney, liver, heart, and lung transplant rejection, as well as for the treatment of inflammation of the kidney caused by syste-mic lupus erythematosus, an autoimmune disease.[1]

The development of mycophenolate mofetil was based on the observation by Anthony Allison and Elsie Eugui that certain inherited enzyme deficiencies in purine biosynthesis leading to RNA and DNA dramatically affect the immune response.

One such enzyme is inosine-5'-mono-phosphate dehydrogenase (IMPDH), which is involved in the synthesis of guanosine nucleo-tides, essential building blocks for RNA and DNA. Allison speculated that inhibition of this enzyme could lead to immunosuppression, because B- and T-lymphocytes are critically dependent on this pathway for proliferation whereas other cells have alternative paths.

A series of known IMPDH inhibitors were tested and mycophenolic acid (MA), a fungal fermentation product, was identified as the most potent immunosuppressant. The mode of binding of this compound to the target enzyme is depicted in the panels below; **Panel A** shows the entire enzyme while **Panel B** gives a close-up view of the drug (shown in red) in the binding site.[2] Detailed studies revealed that in addition to inhibiting B- and T-cell proliferation, MA also suppresses antibody formation by B-lymphocytes and the recruitment of lymphocytes and monocytes into the grafted organ, leading to potent immunosuppression. It was also shown to be more lymphocyte specific than its predecessor, azathioprine (*see* page 122), causing less bone marrow suppression. The

oral bioavailability of mycophenolic acid was greatly improved by conversion to its 2-morpholinoethyl ester, mycophenolate mofetil. Cleavage of the ester moiety (shown in red) in the body leads to the active drug, mycophenolic acid.[3]

Mycophenolate mofetil was introduced in 1995 as CellCept. It is now widely used because of its effectiveness and relative safety, and has annual sales of over $1 billion.

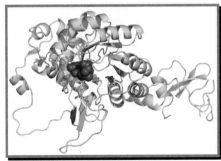

Panel A

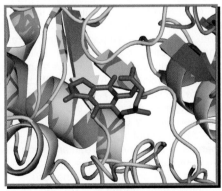

Panel B

1. *BioDrugs* **1999**, *12*, 363-410; 2. *Cell* **1996**, *85*, 921-930 (1JR1); 3. *Transplantation* **2005**, *80*, S181-190; Refs. p. 173

# CYCLOSPORIN (NEORAL™)

| **Structural Formula** | **Ball-and-Stick Model** | **Space-filling Model** |

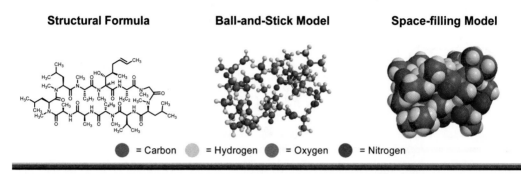

● = Carbon ● = Hydrogen ● = Oxygen ● = Nitrogen

*Year of discovery of immunosuppressant activity:* 1976 (Sandoz, later Novartis); *Year it entered the market:* 1983; *Drug category:* Immunosuppressant; *Main uses:* Prevention of organ transplant rejection, treatment of rheumatoid arthritis and psoriasis; *Other brand names:* Sandimmune, Gengraf, Restasis; *Related drugs:* Tacrolimus (Prograf).

Cyclosporin A has played an important role in the modern era of organ transplantation. It allowed for the first time selective suppression of T-cells without the bone marrow toxicity that is associated with azathioprine therapy (*see* page 122). Cyclosporin A extends kidney graft survival rates and greatly improves heart, liver, lung and even combined heart+lung transplantation. It also has been applied to the treatment of several autoimmune diseases including severe rheumatoid arthritis and psoriasis.[1]

Cyclosporin A is a fungal metabolite that was first isolated in the early 1960s at Sandoz (later Novartis) in an effort to identify compounds with antibiotic activity. However, due to a narrow spectrum of antimicrobial activity, it was never developed as an antibiotic. The reemergence of cyclosporin A came as a result of a screening program to find other biological activities in the crude fungal extract containing it. The extract was found to inhibit lymphocyte proliferation without affecting other somatic cells. The active immunosuppressive ingredient was found to be cyclosporin A. The exact chemical structure of cyclosporin A was first reported in 1976. Animal tests of the drug were conducted by Roy Calne, who participated in the development of azathioprine. His experiments supported the earlier findings that the compound suppresses the immune response more potently than other drugs. The first human trials started in 1976 and the drug was approved for use in the US in 1983. Long-term use of cyclosporin A can lead to nephrotoxicity, susceptibility to infection, hypertension, and hyperlipidemia.[2]

Cyclosporin A acts by forming a complex with a binding protein, cyclophilin, and this complex inhibits the enzyme calcineurin. The three-component complex of cyclosporin A, cyclophilin and calcineurin is depicted in the figure below.[3] Calcineurin dephosphorylates the nuclear factor for activated T-lymphocytes (NFAT), and causes its translocation into the nucleus, where it activates the transcription of several cytokines, including interleukin-2 (IL-2), which stimulate the growth, differentiation, and survival of antigen activated T-cells. Inhibition of this process results in diminished T-cell response to antigen stimulation. Cyclosporin also increases the expression of the cytokine TGF-β, which is a potent inhibitor of T-cell proliferation and of cytotoxic T-lymphocyte formation.

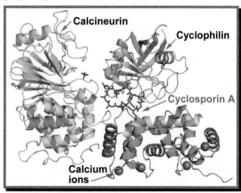

Ternary complex of cyclosporin A, cyclophilin and calcineurin.

1. *Transplant. Proc.* **2004**, *36*, 13S-15S; 2. *Transplant. Proc.* **1999**, *31*, 14S-15S; 3. *Proc. Natl. Acad. Sci. U. S. A.* **2002**, *99*, 13522-13526 (**1MF8**); **Refs. p. 173**

# TACROLIMUS (PROGRAF™)

| Structural Formula | Ball-and-Stick Model | Space-filling Model |
|---|---|---|

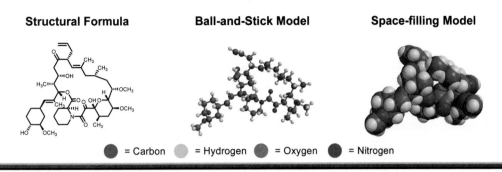

● = Carbon ○ = Hydrogen ● = Oxygen ● = Nitrogen

*Year of discovery of immunosuppressant activity:* 1987 (Fujisawa, later Astellas); *Year of introduction:* 1994; *Drug category:* Immunosuppressant; *Main uses:* Prevention of organ transplant rejection and treatment of eczema; *Other brand names:* FK-506, Fujimycin, Protopic; *Related drugs:* Cyclosporin (Neoral), Sirolimus (Rapamune).

Tacrolimus (FK506) is an immunosuppressant used for the prevention of graft rejection after transplantation and also for the topical treatment of eczema. The advantage of tacrolimus over steroids in dermatological use is that it does not cause thinning of the skin.[1]

The sequence of discovery, mechanism of action, and application of tacrolimus resembles in many ways that for cyclosporin (*see* page 124). Tacrolimus is a microbially produced macrolide that was isolated in 1984 as a result of an antibiotic screening program at Fujisawa Pharmaceuticals in Japan. In 1987, it was discovered that it also exhibits potent immunosuppressive activity. As with cyclosporin, the immunosuppression was due to the inhibition of T-cell activation. Tacrolimus exerts its effect by binding to an intracellular protein, FK506 binding protein-12 (FKBP-12), which further complexes with calcineurin, $Ca^{2+}$ and calmodulin, a small $Ca^{2+}$ binding protein. The ternary complex of tacrolimus, FKBP-12 and calcineurin is depicted in the figure below.[2] Whereas free calcineurin is an active phosphatase that causes T-cell activation, the complex with FK506 and FKBP-12 is inactive.

The first human trials with tacrolimus were conducted at the University of Pittsburgh in 1989. It was approved for human immunosuppression in 1994. Tacrolimus is superior to cyclosporin in improving graft survival and preventing acute rejection after kidney or liver transplantation. Individuals receiving tacrolimus require lower doses of steroids, but are at higher risk of developing diabetes mellitus.[3]

A related macrolide antibiotic, sirolimus (rapamycin) was discovered at Ayers Laboratories in Canada in the early 1970s. Originally, sirolimus was developed as an antifungal agent, but due to its powerful immunosuppressive properties it was advanced as an immunosuppressant by the successor company, Wyeth. Sirolimus was first approved for use in kidney transplantation in 1999, and is sold under the name Rapamune. Sirolimus binds to FKBP-12, but unlike tacrolimus, this complex does not affect calcineurin. Instead, it binds to a kinase enzyme, the mammalian target of rapamycin (mTOR), causing inhibition of cytokine-induced T-cell proliferation.

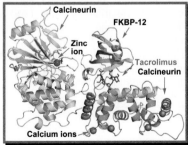

Ternary complex of tacrolimus.

**Sirolimus (Rapamune™)**

1. *Transplant.* **2004**, *77*, S41-S43; 2. *Cell* **1995**, *82*, 507-522 (1TCO); 3. *Principles of Drug Development in Transplant. Autoimmun.* **1996**, 159-163; Refs. p. 173

# FTY720 (FINGOLIMOD)

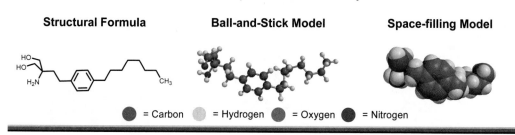

**Structural Formula**        **Ball-and-Stick Model**        **Space-filling Model**

● = Carbon   ● = Hydrogen   ● = Oxygen   ● = Nitrogen

*Year of discovery:* 1995 (Yoshitomi); *Drug category:* Immunosuppressant; *Main uses:* Clinical trials have shown positive signs of benefit for the autoimmune disease multiple sclerosis.

FTY720, also known as fingolimod, was first synthesized at Yoshitomi Pharmaceuticals in Japan in 1995 as a synthetic analog of ISP-1 (myriocin, thermozymocidin), a compound that was isolated from a fungus in a search for novel anticancer agents. ISP-1 did not have antitumor activity, but surprisingly it was a potent immunosuppressant, although too toxic for further development. Extensive studies of ISP-1 derivatives led to simple synthetic analogs that retain the immunosuppressive activity.

**ISP-1**

The potent immunosuppressive activity and the simple and unique structure of FTY720 came to the attention of scientists at Sandoz (later Novartis), and after investigation of its immunosuppressive properties they purchased commercial rights to the drug.[1]

Early studies revealed that FTY720 exerts its biological effect in a significantly different way from the other known immunosuppressants. Whereas earlier drugs such as cyclosporin (*see* page 124) and tacrolimus (*see* page 125) block the activation of B- and T-lymphocytes, FTY720 localizes B- and T-lymphocytes in the lymph nodes, thereby preventing attack on implanted grafts by T-cells. Although B-lymphocytes are retained in the nodes, those B-cells that recognize the circulating virus release an antibody specific for that virus into the circulation and periphery. Bacteria are attacked in the same way. Consequently, the use of FTY720 causes more discriminating and less inflammatory immunosuppression than the older drugs.

The localization of lymphocytes in the lymph nodes is thought to result because FTY720 resembles sphingosine, a natural lipid molecule that participates in protein signaling.

**Sphingosine**

Workers at Sandoz in collaboration with K. Lynch at the University of Virginia discovered that a phosphorylated derivative of FTY720 (FTY720-P) forms in the body and acts on sphingosine-1-phosphate (S1P) receptors. The role of these receptors in immune cell trafficking was elucidated by immunologist J. Cyster at the University of California, San Francisco. A subtype of these receptors, S1P-1 is present in large amounts on lymphocytes. When lymphocytes return to the lymph nodes to multiply, the level of these receptors on the cell membrane is reduced. When they are ready to exit, the level is increased, sensitizing them to S1P molecules and provoking the migration of these cells from the nodes to the circulation. FTY720-P strongly activates the S1P-1 receptors, which in response, pull back inside the cell, thus, the cell is unable to leave the lymph node.[2]

Clinical studies have shown FTY720 to be a promising treatment for multiple sclerosis, an autoimmune disorder affecting the nervous system. In multiple sclerosis, myelin, a fatty layer that insulates nerve fibers and allows conduction of electrical signals (*see* page 221), is attacked and disintegrated by the immune system.

FTY720 may cause a substantial lowering of heart rate (bradycardia), a side effect which demands termination of the treatment.

1. *Transplant. Proc.* **2004**, *36*, 531S-543S; 2. *Transplant. Proc.* **1999**, *31*, 2779-2782; Refs. p. 174

# INFECTIOUS DISEASES

## Malaria

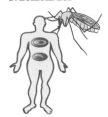

**ARTEMISININ**
(antimalarial)

**Human
Immunodeficiency
Virus**

**BACTERIAL
DRUG EFFLUX
PUMP**

**CIPRO**
(antibacterial)

**LAMISIL**
(antifungal)

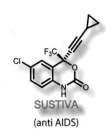

**SUSTIVA**
(anti AIDS)

**IVERMECTIN**
(antiparasitic)

**Penicillin V**

**Imipenem**

# *Chemistry can cure disease.*

**Azithromycin (Zithromax™)**

**Azithromycin**, a macrolide antibiotic, acts by inhibiting protein synthesis by reversible binding to the large 50S subunit of the bacterial ribosome (*see* page 134).

# ANTIBIOTICS

# AMOXICILLIN (AMOXIL™)

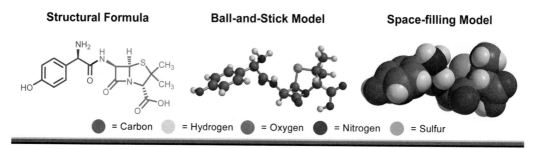

**Structural Formula**     **Ball-and-Stick Model**     **Space-filling Model**

● = Carbon     ○ = Hydrogen     ● = Oxygen     ● = Nitrogen     ○ = Sulfur

*Year of discovery:* 1972; *Year of introduction:* 1979 (GlaxoSmithKline); *Drug category:* β-Lactam/penicillin; *Main uses:* For the treatment of bacterial infections, bronchitis, pneumonia, urinary tract infections, gonorrhea, blood infections, typhoid fever and the eradication of *Helicobacteria pylori* in the gut; *Other brand names:* Isimoxin, Ospamox; *Related drugs:* Augmentin (Amoxicillin + Clavulanic acid), Unasyn (Ampicillin + Sulbactam), Piperacillin (Pipracil).

It was discovered serendipitously by Alexander Fleming in 1928 that a fungus of the *Penicillum* family produces a material that has powerful antimicrobial activity. For that discovery, he received the Nobel Prize in Medicine in 1945 along with H.W. Florey and E.B. Chain. The latter two followed up on Fleming's finding years later and their studies led to the successful application of penicillin to humans in 1943. Because many soldiers in World War II were dying of wound-related infections, a massive wartime program was undertaken by the Allies on the development of Fleming's antibiotic, which became known as penicillin. A consortium of US and UK scientists in universities and in industry worked together under the government wartime research program to determine ways to purify penicillin, determine the chemical structure and find a method for mass production. That effort succeeded, and by 1944 sufficient penicillin G had been made for routine use by the military.

Penicillin belongs to the class of drugs known as β-lactams which all feature a four-membered cyclic amide (shown at left). The penicillin nucleus (called 6-aminopenicillanic acid or 6-APA, shown in red) contains an amino-substituted β-lactam ring sharing a common edge with the five-membered thiazolidine ring. The penicillin nucleus is essential for antibiotic activity, whereas the side chains (shown in blue) determine the pharmacological properties.[1]

The production of structurally diverse natural penicillins became possible when it was discovered that the chemical composition of the fermentation medium influenced the nature of the side chain group. For example

the addition of phenylacetic acid or phenoxyacetic acid to the fermentation medium resulted in the production of penicillin G and penicillin V, respectively.[2]

**β-Lactam**     **6-APA**     **Thiazolidine**

*Chemical Synthesis*

**phenylacetic acid**     **Semisynthetic Penicillins**     **phenoxyacetic acid**
*Penicillum chrysogenium*          *Penicillum chrysogenium*

**Penicillin G**     **Penicillin V**

Later, the mass production of the side chain-free 6-aminopenicillanic acid allowed the assembly of penicillin derivatives by chemical synthesis (semisynthetic penicillin derivatives) which led to compounds with improved potency and bioavailability that are still in use today.[3]

Amoxicillin, discovered in 1972, is an orally active aminopenicillin that is effective against both Gram-positive and Gram-negative bacteria. Penicillins, including amoxicillin, work by inhibiting the growth of the bacterial cell wall that is essential for the survival of microorganisms. Bacterial cell walls consist of peptide-sugar copolymers that are cross-linked by bridges. This essential cross-linking reaction is catalyzed by the enzyme transpeptidase, which is inhibited by penicillins. However, over time bacteria become resistant by expressing enzymes that degrade penicillins. The most common mechanism is the production of β-lactamases that cleave the amide bond in the four-membered ring and render the drug inactive. An X-ray picture of the deactivated amoxicillin (red) bound covalently to Ser64 (cyan) of the enzyme β-lactamase is shown below.[4]

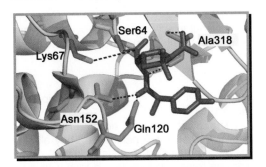

In connection with the search for new antibiotics of microbial origin, a structurally-related β-lactam natural product, clavulanic acid, was discovered. While it has negligible antimicrobial activity, clavulanic acid was found to be an irreversible inhibitor of β-lactamases. Shortly afterwards, Glaxo marketed the combination of amoxicillin and clavulanic acid as Augmentin. This became a major antibiotic because of its effectiveness against β-lactamase-producing resistant strains. Another inhibitor of β-lactamase, sulbactam, was discovered at Pfizer, which introduced a formulation of sulbactam with ampicillin (Unasyn and Ampictam).

**Clavulanic acid**          **Sulbactam**

The synthesis of penicillin derivatives that do not have a sulfur atom in the five-membered ring led to a family of broad-spectrum antibiotics, the carbapenems, that are active against both Gram-positive and Gram-negative bacteria. Carbapenems are not cleaved by several classes of β-lactamases and consequently are useful against many β-lactamase-producing, penicillin-resistant strains. Imipenem, the most widely used carbapenem, was developed at Merck and sold in combination with cilastatin, a dipeptidase inhibitor, since imipenem alone is rapidly deactivated by the dipeptidase enzyme

in the kidney. The imipenem/cilastatin sodium combination is sold as Primaxin™. It is effective against a number of serious Gram-negative infections, especially with those involving cephalosporin-resistant strains.

**Imipenem**          **Cilastatin sodium**

Monocyclic β-lactam antibiotics (monobactams) are stable to many (but not all) β-lactamases. The best known member of this family is aztreonam (Azactam™, Squibb) which is active only against Gram-negative bacteria. Aztreonam is generally well-tolerated and is especially useful in patients who are allergic to penicillin. The electron-withdrawing group SO₃H (green) on the nitrogen atom is important, since it increases the reactivity of the β-lactam ring towards cleavage. In the absence of this group monobactams are not active enough to inhibit the transpeptidase enzyme. It was by screening a library of monobactams that the cholesterol-absorption inhibitor ezetimibe was discovered (*see* page 65).

**Aztreonam (Azactam™)**          **Ezetimibe (Zetia™)**

Penicillins are biosynthesized by an iron-containing enzyme from the tripeptide cysteinyl-valine aminoadipate, as shown in the equation below. This elegant design requires enzymatic catalysis and can not be effected by purely chemical reaction conditions.

**Cysteinyl-valine aminoadipate**

Penicillin
Synthase

1. *Drug Interactions in Infectious Diseases (2nd Edition*, Humana Press, *2005*); 255-287; 2. *Antimicrob. Agents* **2005**, 113-162; 3. *Int. J. Antimicrob. Agents* **2007**, *29*, 3-8; 4. *Chem. Biol.* **2002**, *9*, 971-980 (1LL9); Refs. p. 174

# CEFACLOR (CECLOR™)

| **Structural Formula** | **Ball-and-Stick Model** | **Space-filling Model** |

● = Carbon    ○ = Hydrogen    ● = Oxygen    ● = Chlorine    ● = Nitrogen    ○ = Sulfur

_Year of discovery:_ Late 1960s (Eli Lilly); _Year of introduction:_ 1979; _Drug category:_ β-Lactam/cephalosporin; _Main uses:_ For the treatment of bacterial infections, bronchitis, pneumonia, and urinary tract infections; _Related drugs:_ Cefprozil (Cefzil), Ceftazidime (Fortaz), Cefotaxime (Claforan), Cefepime (Maxipime).

Cephalosporins (Cephs) are chemically related to penicillins since they also feature the four-membered β-lactam ring.[1] The first Ceph, cephalosporin C, was isolated in the 1950s from cultures of the mold _Cephalosporium acremonium_. Cephalosporin C can be converted via hydrolysis to 7-aminocephalosporanic acid (7-ACA) that is used as the starting material for semisynthetic Cephs. A similar precursor, 6-aminopenicillinic acid, is available for the preparation of semisynthetic penicillin derivatives (_see_ page 130).

**Cephalosporin C**        **7-ACA**

Modification of 7-ACA at the 3- and 7-positions gives rise to a great variety of semisynthetic Cephs that are significantly more active than cephalosporin C. Differences in side chain structure at the 7-position directly affect antibacterial activity, whereas modifications made at the 3-position influence the metabolism and the pharmacokinetic properties of the drugs.

Ceph antibiotics are classified by generation (one to four) based on their therapeutic properties including spectrum of activity against Gram-positive and Gram-negative bacteria, stability towards β-lactamases and the method of administration (oral or injectable). With each successive generation of Cephs, activity against Gram-negative bacteria has been improved at the expense of Gram-positive activity. Cefaclor was discovered at Eli Lilly in the late 1960s and is classified as a second-generation Ceph. It was one of the first orally administered Cephs, and it remains a major drug. A fourth-generation cephalosporine, cefepime, with an extended spectrum of

activity, has been available as Maxipime™ since 1994.[2]

The mechanism of action of Cephs is similar to that of the penicillins. They all bind to penicillin-binding proteins, including the enzyme transpeptidase that is responsible for the cross-linking of the bacterial cell wall, which is crucial for the structural strength and shape of the microorganism. Almost all bacteria have a cross-linked cell wall. In the last step of cell wall biosynthesis, the peptidoglycan (glycan) chains are cross-linked by bacterial transpeptidases. The OH group of Ser62 in the active site attacks the carbonyl group (C=O) of the penultimate D-alanine (Ala) residue of one of the glycan chains, and the resulting acyl-enzyme reacts with the amino (NH₂) group of another glycan chain, giving rise to cross-linked chains. Cephalosporins mimic the C=O group of this D-Ala residue, and by irreversibly binding to Ser62 (cyan), they deactivate the enzyme and deny access to the second glycan chain. The X-ray image of a Ceph-transpeptidase complex (below) clarifies this final and crucial step of cross-linking. Here, the peptide side chain (magenta) of the bound cephalosporin (red) substitutes for the peptidoglycan chain and Ser62 has already opened the β-lactam ring.[3]

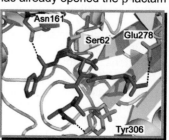

1. _Antimicrob. Agents_ **2005**, 222-268; 2. _Anti-Infec. Agents in Med. Chem._ **2007**, _6_, 71-82; 3. _Proc. Natl. Acad. Sci. U. S. A._ **2001**, _98_, 1427-1431 (**1HVB**); Refs. p. 174

# DOXYCYCLINE (VIBRAMYCIN™)

| Structural Formula | Ball-and-Stick Model | Space-filling Model |
|---|---|---|

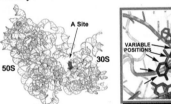

● = Carbon    ○ = Hydrogen    ● = Oxygen    ● = Nitrogen

*Year of discovery:* 1962 (Pfizer); *Year of introduction:* 1967 (as Vibramycin); *Drug category:* Tetracycline/broad-spectrum antibiotic; *Main uses:* For the treatment of respiratory tract, urinary tract and eye infections, anthrax, bubonic plague, syphilis, cholera, acne, gonorrhea, elephantiasis, and malaria; *Other brand names:* Adoxa, Doryx; *Related drugs:* Minocycline (Minocin), Lymecycline (Tetralysal), Demeclocycline (Declomycin).

The tetracycline class of broad-spectrum antibiotics, produced by soil-dwelling microbes of the *Streptomyces* family, are active against both Gram-positive and Gram-negative bacteria. The first members to be used in medicine were chlortetracycline (from Lederle Laboratories) and oxytetracycline (from Pfizer), both introduced in the early 1950s. They were followed by tetracycline, which was produced from chlortetracycline by the chemical replacement of Cl by H. All three became important in the treatment of bacterial infections because they are orally active against a wide variety of bacteria and very well-tolerated. Doxycycline, an improved synthetic tetracycline, was introduced by Pfizer in 1962 and became the most widely used member of the class.

**Chlortetracycline**

**Oxytetracycline**

Doxycycline is rapidly absorbed after oral administration, distributes to all important organs via blood (except the brain) and is well-tolerated by most people. With the exception of doxycycline, tetracyclines are excreted via the kidney. For this reason, doxycycline is the safest tetracycline for persons with impaired kidney function.

The antibiotic activity of tetracyclines is due to the selective inhibition of bacterial protein synthesis. Tetracyclines bind to the 30S subunit of the bacterial ribosome (the site of protein synthesis) and prevent the binding of amino-acyl *t*-RNA to the acceptor site (A-site) on the *m*-RNA-ribosome complex. As a result, protein synthesis is halted since no new amino acid can be delivered to the growing peptide chain. The pictures below shows tetracycline (TC, red) bound in the A-site of the 30S subunit. The close-up view below shows how the nucleoside residues (green) of the 30S subunit interact with the tetracycline molecule.[1]

**TC in the 30S Subunit**    **Close-Up View of TC**

Tetracyclines are produced on a massive scale (over 30,000 tons/year). At one time tetracyclines were widely used to accelerate the growth of chickens and cattle. The emergence of resistant strains is only a matter of time when antibiotics are used on such a vast scale. Unfortunately, the discovery of entirely new antibiotics by microbial screening approaches has passed the point of diminishing returns. Biotarget structure-guided design and chemical modification of known antibiotics provides a way of dealing with resistant strains. Two examples of the application of this strategy to tetracyclines are shown below: (1) tigecycline (Wyeth, semisynthetic) and (2) pentacycline (A.G. Myers, totally synthetic). Both are highly active against resistant strains.

**Tigecycline (Tygacil™)**    **Pentacycline**

1. *EMBO J.* **2001**, *20*, 1829-1839 (1I97); **Refs. p. 174**

# AZITHROMYCIN (ZITHROMAX™)

| Structural Formula | Ball-and-Stick Model | Space-filling Model |
|---|---|---|

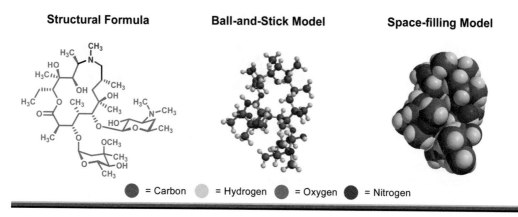

● = Carbon  ○ = Hydrogen  ● = Oxygen  ● = Nitrogen

*Year of discovery:* 1980 (Pliva); *Year of introduction:* 1986 (as Zithromax by Pfizer); *Drug category:* Azalide/macrolide antibiotic; *Main uses:* For treatment of bacterial skin, ear and respiratory infections as well as for STD; *Other brand names:* Zitromax, Vinzam, Zmax, Aztrin, Sumamed; *Related drugs:* Erythromycin (Ilosone, E-Mycin), Clarithromycin (Biaxin).

Azithromycin is a member of a family of antibiotics, the macrolide class, which are large-ring cyclic esters.[1] Erythromycin, the first macrolide antibiotic, was isolated at Eli Lilly in 1949 from a soil-dwelling fungus now called *Saccharopolyspora erythrea*. Erythromycin was marketed by Lilly as Ilosone in 1952 for use against a wide variety of Gram-positive bacterial infections. It is especially useful in individuals who are allergic to penicillin. Erythromycin displays a spectrum of activity similar to that of the penicillins. Because of its instability to gastric acid, erythromycin is administered as enteric coated tablets. Erythromycin distributes readily and widely in the body, except for the brain and spinal fluid. An acid-stable, orally active semisynthetic derivative of erythromycin, clarithromycin (6-O-methylerythromycin), was developed by the Japanese pharmaceutical company Taisho and introduced in the US by Abbott.

Erythromycin (Ilosone™)    Clarithromycin (Biaxin™)

A different semisynthetic analog of erythromycin, azithromycin, was prepared in 1980 at Pliva, and was found to be superior to both erythromycin and clarithromycin. Azithromycin can be given orally or intravenously. It has a long half-life and achieves high concentrations in blood and body tissue.[2] Just four daily doses or a single injection of azithromycin can effectively clear an infection.

Azithromycin, erythromycin, clarithromycin, and all other macrolides act by inhibiting protein synthesis by reversible binding to the large 50S subunit of the bacterial ribosome, the biochemical machine that converts amino acids to proteins (*see* below).[3] Upon binding, azithromycin blocks the exit channel and prevents elongation of the growing polypeptide chain. (Cf. binding of doxycycline to the small 30S subunit, *see* page 133).

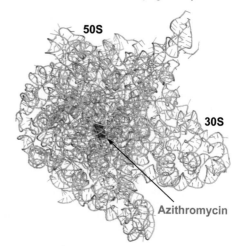

Ribosomal binding site of azithromycin.

1. *Pharmacol. Res.* **2004**, *50*, 211-222; 2. *Exp. Opin. Pharmacother.* **2005**, *6*, 2335-2351; 3. *Structure* **2003**, *11*, 329-338 (1NWY); **Refs. p. 174**

# CIPROFLOXACIN (CIPRO™)

| Structural Formula | Ball-and-Stick Model | Space-filling Model |
|---|---|---|

● = Carbon  ● = Hydrogen  ● = Oxygen  ● = Fluorine  ● = Nitrogen

*Year of discovery:* ca.1986 (Bayer Corporation); *Year of introduction:* 1988; *Drug category:* Fluoroquinolone/broad-spectrum antibiotic; *Main uses:* For the treatment of bacterial infections of the urinary and respiratory tract, sexually-transmitted diseases, anthrax; *Other brand names:* Ciproxin, Ciprobay; *Related drugs:* Norfloxacin (Noroxin), Levofloxacin (Levaquin), Gatifloxacin (Tequin).

Ciprofloxacin, a very widely used broad-spectrum antibiotic, is a non-natural, totally synthetic compound based on the quinolone ring system (shown in red above). The antimicrobial activity of quinolones was discovered more than 50 years ago, as a result of the finding that certain intermediates used for the synthesis of the antimalarial drug resochin were antibacterial. After testing a large number of synthetic analogs, a highly active compound, nalidixic acid, was discovered in 1962. It was found to be orally active and especially effective in the treatment of urinary tract infections.

**Original Lead Compound**

**Nalidixic acid**

**Norfloxacin (Noroxin™)**

**Levofloxacin (Levaquin™)**

Subsequently, a fluorine derivative, norfloxacin, was discovered that had fewer side effects and a much broader spectrum of activity than nalidixic acid. It was introduced into clinical use in the early 1980s. A further modified fluoroquinolone analog of norfloxacin, ciprofloxacin, was discovered in 1986 by Bayer and later marketed as Cipro. Ciprofloxacin proved to be especially effective against Gram-negative bacteria in low doses. The third-generation quinolone, levofloxacin, has also come into widespread use because of greater potency.

Fluoroquinolone antibiotics are orally active, well-tolerated and especially useful for the treatment of infections of the urinary and respiratory tracts, including sexually transmitted diseases. Because they distribute into the cerebrospinal fluid, fluoroquinolones effectively treat meningitis.[1] Ciprofloxacin is successful against many bacteria which are resistant to β-lactams, tetracyclines, and aminoglycosides, as well as for the treatment and prevention of anthrax.

Fluoroquinolones act by preventing the replication and repair of bacterial DNA and RNA. This is achieved by the inhibition of the enzyme DNA gyrase (topoisomerase II), which is responsible for the supercoiling of the DNA double helix. During replication, the double helix of the DNA molecule is unwound to allow the synthesis of RNA chains. However, during this process a number of supercoils are formed ahead of the enzyme RNA polymerase. DNA gyrase relaxes these supercoils and allows the replication to continue. The X-ray image of a nalidixic acid-DNA complex is shown below.[2] By binding to the DNA gyrase complex, quinolone antibiotics disrupt DNA replication.

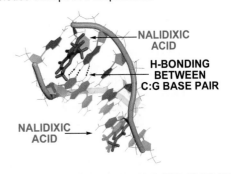

NALIDIXIC ACID

H-BONDING BETWEEN C:G BASE PAIR

NALIDIXIC ACID

1. *Profiles Drug Subst., Excip. Rel. Meth.* **2004**, *31*, 209-214;
2. *Nucleic Acids Res.* **2005**, *33*, 4838 (**2BQ2**); Refs. p. 175

# TRIMETHOPRIM (TRIPRIM™)

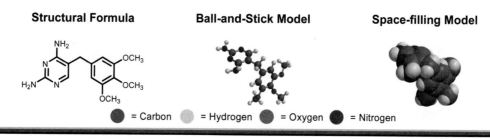

**Structural Formula**  **Ball-and-Stick Model**  **Space-filling Model**

● = Carbon  ● = Hydrogen  ● = Oxygen  ● = Nitrogen

*Year of discovery:* 1956 (G.H. Hitchings); *Year of introduction:* 1973 (in combination with sulfamethoxazole as Bactrim or Cotrimoxazole or Septra); 1980 (as the single drug Triprim); *Drug category:* Diaminopyrimidine/dihydrofolate reductase inhibitor (antifolate); *Main uses:* For the treatment of urinary tract infections, travelers' diarrhea, respiratory and middle ear infections; *Other brand names:* Proloprim, Monotrim.

Tetrahydrofolate, a metabolite of folic acid, is essential for normal cell division in both humans and bacteria. Humans require dietary folic acid because they can not produce it. Most bacteria, however, make their own folic acid from a pteridine derivative (orange) and *para*-aminobenzoic acid (blue). This process is catalyzed by the enzyme dihydropteroate synthase (DHPS). The two-step conversion of folic acid to tetrahydrofolate is catalyzed by dihydrofolate reductase (DHFR). It was recognized in the 1940s that the inhibition of either one (or both) of these enzymes would block bacterial growth and reproduction.

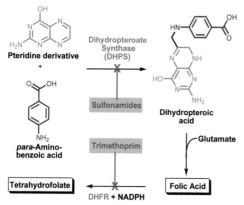

Sulfanilamide and several of its derivatives are wide-spectrum antibiotics, effective against both Gram-positive and Gram-negative bacteria. They bind to the enzyme DHPS, as mimics of *para*-aminobenzoic acid and inhibit folic acid biosynthesis. Unfortunately, sulfonamide-resistant strains emerged fairly soon after the introduction of "sulfa" drugs. For this reason sulfonamides now are rarely used alone. Resistance develops because a minute fraction of bacteria can use

the low levels of folate provided by the host and these bacteria proliferate.

**Sulfanilamide**  **Sulfamethoxazole**

However, the combination of diamino-pyrimidines such as trimethoprim and sulfon-amides is very effective in antibacterial therapy.[1] Trimethoprim is a selective inhibitor of the bacterial enzyme DHFR (cf. metho-trexate, page 46) and prevents the production of tetrahydrofolate from folic acid. It was found that the 20:1 combination of sulfamethoxazole and trimethoprim is synergistic and provides optimal antibiotic therapy.

Trimethoprim binds selectively to *bacterial* DHFR enzyme rather than the *human* enzyme because of subtle differences in the tetra-hydrofolate binding site.[2] The X-ray structure of trimethoprim bound to the bacterial enzyme DHFR has been determined.[3] The picture below shows NADPH (silver) and trimethoprim (red) bound to DHFR (yellow).

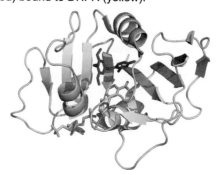

1. *Arch. Intern. Med.* **2003**, *163*, 402-410; 2. *Curr. Med. Chem.* **2006**, *13*, 377-398; 3. *J. Mol. Biol.* **2000**, *295*, 307-323 (**1DG5**); Refs. p. 175

# AMIKACIN (AMIKIN™)

| Structural Formula | Ball-and-Stick Model | Space-filling Model |
|---|---|---|

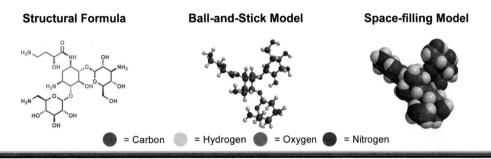

● = Carbon ⬤ = Hydrogen ⬤ = Oxygen ● = Nitrogen

*Year of discovery:* 1972; *Year of introduction:* 1981 (Bristol-Myers Squibb); *Drug category:* Aminoglycoside (amino-sugar) antibiotic; *Main uses:* For the treatment of serious infections by aerobic Gram-negative bacilli that are resistant to other aminoglycosides; *Related drugs:* Gentamicin (Genoptic), Kanamycin (Kantrex).

Aminoglycosides are naturally occurring bactericidal organic molecules in which one or two amino-sugar units (blue) are linked to a central sugar nucleus (red). The first member of the class, streptomycin, was discovered in 1943 by S.A. Waksman and A. Schatz who isolated this compound from the soil microbe *Streptomyces griseus*. Streptomycin effectively attacks aerobic Gram-negative bacteria that are not sensitive to the penicillins. Within a few years, millions of tuberculosis patients had been successfully treated with streptomycin. As a result, the death rate from tuberculosis plummeted. Waksman was awarded the Nobel Prize in Medicine in 1952.

**Streptomycin**   **Tobramycin (Tobi™)**

Unfortunately, streptomycin-resistant bacterial strains appeared soon after its introduction. However, a number of novel and more potent aminoglycosides were discovered as a result of intensive research. Among these aminoglycosides were gentamicin C, kanamycin A, tobramycin, netilmicin and amikacin. Intravenous gentamicin is now mostly used for the treatment of urinary tract infections and pneumonia. Inhaled tobramycin (Tobi™) is effective in treating people with cystic fibrosis, a chronic respiratory disease in which there is a high risk of infection. Amikacin, a semisynthetic derivative of kanamycin A, was designed to be resistant to

the enzymes acetylase, adenylase and phosphorylase that catalyze the inactivation of all previously known aminoglycosides. Amikacin is prepared by attaching the γ-amino-α-hydroxybutyryl group (L-haba group, shown in green above) to kanamycin A.[1] Patients receiving amikacin should be monitored to detect hearing loss or kidney damage.

The mechanism of action of aminoglycosides has not been fully elucidated. However, their primary target is the bacterial small ribosomal unit (30S), the site of protein synthesis. Amikacin binds to the A-site in the 16S rRNA and interferes with the initiation of protein synthesis. This leads either to the accumulation of abnormal initiation complexes or to the misreading of the mRNA template. The latter leads to the incorporation of incorrect amino acids and the formation of proteins that are toxic to the microorganism. The X-ray picture **A** (below) shows two amikacin molecules (red) bound to the A-site.[2] The close-up view **B** shows the L-haba group specifically interacting with two G:C base pairs.

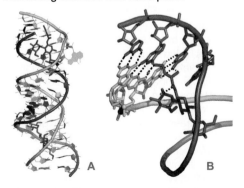

1. *Anti-Infec. Agents in Med. Chem.* **2006**, *5*, 255-271; 2. *Biochimie* **2006**, *88*, 1027-1031 (2G5Q); Refs. p. 175

# VANCOMYCIN (VANCOCIN™)

| Structural Formula | Ball-and-Stick Model | Space-filling Model |
|---|---|---|

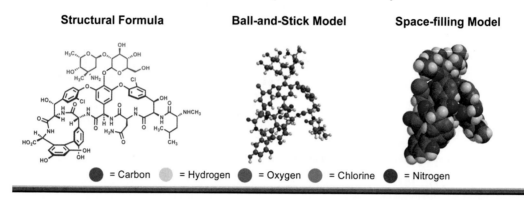

● = Carbon  ○ = Hydrogen  ● = Oxygen  ● = Chlorine  ● = Nitrogen

*Year of discovery:* ca. 1956; *Year of introduction:* 1958 (as Vancocin by Eli Lilly); *Drug category:* Glycopeptide antibiotic; *Main uses:* For the treatment of life-threatening infections by Gram-positive bacteria that are resistant to all other known antibiotics; *Related drugs:* Teicoplanin (Targocid).

Vancomycin, the first member of the glycopeptide antibiotic class, was isolated from the soil microbe *Amycolatopsis orientalis* obtained from the jungles of Borneo. Vancomycin was shown to be bacteriocidal towards Gram-positive (but not Gram-negative) bacteria. It was marketed as Vancocin by Lilly and became widely used as the antibiotic of "last resort" once multidrug-resistant bacterial strains began to emerge in the 1960s and 1970s. Another widely used drug with a similar spectrum of activity is teicoplanin (Targocid™, Sanofi Aventis), a lipoglycopeptide that is only available in Europe.

**Teicoplanin (Targocid™)**

The structures of vancomycin and teicoplanin are closely related and feature heptapeptide scaffolds (blue) linked to sugar and aminosugar moieties (red). Teicoplanin is classified as a lipoglycopeptide because it contains a long hydrophobic acyl chain (green).[1] Vancomycin and teicoplanin are currently reserved for the treatment of life-threatening bacterial infections such as the methicillin-resistant *Staphylococcus aureus* (MRSA). Vancomycin must be administered intravenously, since it is orally inactive. Due to the emergence of vancomycin-resistant, Gram-positive bacterial strains during the past 15 years, restrictions in

the clinical use of vancomycin were introduced by the US Centers for Disease Control.

Glycopeptides halt the development of Gram-positive bacteria by inhibiting the biosynthesis of the peptidoglycan layer of the cell wall. Both vancomycin and teicoplanin bind to late-stage peptidoglycan intermediates containing a D-alanyl-D-alanine terminus. This strong binding blocks further elaboration of the growing glycan chain by transglycosidation or transpeptidation. The D-Ala-D-Ala terminus can form five hydrogen bonds with vancomycin (*see* X-ray image below).[2] In vancomycin-resistant strains the peptidoglycan chains have D-alanyl-D-lactate as terminal units. With this modified terminus only four hydrogen bonds are possible and, as a result, there is a 1000-fold *decrease* in antibiotic potency against the resistant strains.

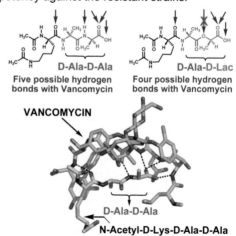

D-Ala-D-Ala
Five possible hydrogen bonds with Vancomycin

D-Ala-D-Lac
Four possible hydrogen bonds with Vancomycin

**VANCOMYCIN**

D-Ala-D-Ala

**N-Acetyl-D-Lys-D-Ala-D-Ala**

1. *Chem. Rev.* **2005**, *105*, 425-448; 2. *Protein Data Bank* **2001**, (http://www.pdb.org/) (1FVM); Refs. p. 175

# LINEZOLID (ZYVOX™)

| Structural Formula | Ball-and-Stick Model | Space-filling Model |
|---|---|---|

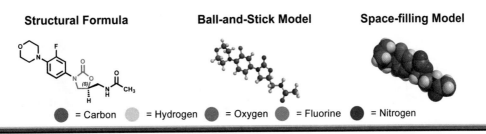

● = Carbon   ● = Hydrogen   ● = Oxygen   ● = Fluorine   ● = Nitrogen

*Year of discovery:* 1993 (by Steven J. Brickner at Upjohn); *Year of introduction:* 2000 (as Zyvox by Pharmacia, now Pfizer); *Drug category:* Oxazolidinone; *Main uses:* For the treatment of serious Gram-positive bacterial infections that are resistant to all other known antibiotics; *Related drugs:* AZD2563 (under development by AstraZeneca).

Linezolid, a completely synthetic antibiotic of the oxazolidinone class, was discovered in 1993 by S.J. Brickner and coworkers at Upjohn. Members of this class of drugs share a common structural element, a five-membered oxazolidinone ring (highlighted in red). The antibiotic properties of oxazolidinones were first recognized in 1984 at DuPont during the investigation of substances (e.g., DuP-721, *see* structure below) for the treatment of plant diseases caused by bacterial and fungal pathogens.[1] Subsequent animal studies, however, revealed serious toxicity associated with many of these compounds. In the late 1980s Brickner initiated a search for oxazolidinones that are both potent and safe for human use. This effort led to the discovery of eperezolid and linezolid. Extensive studies of the relationship between structure, activity and toxicity revealed that linezolid possesses the best pharmacokinetic and safety profile and also high potency against resistant Gram-positive bacteria.[2] It was the first novel antibiotic in 40 years, since all the new antibiotics between 1960 and 2000 were derived simply by chemical modification of members of previously known classes.

The three critical structural features needed for good antibacterial activity (red boxes) are: (1) the (*S*) spatial configuration at C-5; (2) the acetaminomethyl group at C-5 and (3) an aromatic ring attached to the oxazolidinone nitrogen atom. The fluorine and electron-donating nitrogen substituents on the aromatic ring led to improved efficacy and increased safety of the drug, but are not critical for antibiotic activity.

Linezolid acts by the inhibition of bacterial protein synthesis.[3] Specifically it is thought to bind to the 50S ribosomal subunit so as to prevent its joining with the 30S small ribosomal subunit–mRNA complex and N-formylmethionine–tRNA to form the 70S assembly which catalyzes protein synthesis. Because N-formylmethionine (fMet) is the initial subunit of a new peptide chain, linezolid functions to block the very first step in the biosynthesis of bacterial proteins, thereby preventing bacterial growth and replication. An X-ray crystal structure has been obtained by T.A. Steitz showing the binding of linezolid to the 50S subunit of the ribosome.

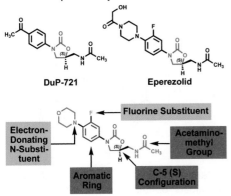

DuP-721          Eperezolid

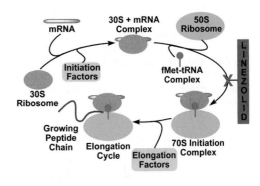

1. *Antimicrob. Agents* **2005**, 604-630; 2. *Angew. Chem., Int. Ed. Engl.* **2003**, *42*, 2010-2023; 3. *Exp. Opin. Pharmacother.* **2005**, *6*, 2315-2326; Refs. p. 175

# ISONIAZID (LANIAZID™)

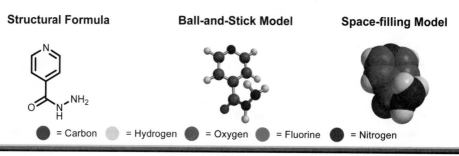

| Structural Formula | Ball-and-Stick Model | Space-filling Model |

● = Carbon  ○ = Hydrogen  ● = Oxygen  ● = Fluorine  ● = Nitrogen

*Year of discovery:* 1912 (by Hans Meyer and Josef Mally of Germany); *Year of introduction:* 1952 (Bayer, Hoffmann-La Roche, and Squibb independently); *Drug category:* Antimycobacterial agent; *Main uses:* For the treatment of tuberculosis as a monotherapy or in combination with other tuberculosis drugs. It is also used to reduce tremors in patients with multiple sclerosis; *Other brand names:* Nydrazid, Rimifon, Ditubin, Dinacrin, INH; *Related drugs:* Rifampin (or Rifampicin), Streptomycin, *para*-aminosalicylic acid.

Tuberculosis, a highly contagious pulmonary infection caused by *Mycobacterium tuberculosis,* has taken countless human lives for thousands of years. Its effects are evident in the pathological signs of tubercular damage found on the spines of ancient Egyptian mummies. Tuberculosis is responsible currently for almost 2 million human fatalities annually. The early antibiotics, "sulfa" drugs and penicillins, are totally ineffective against tuberculosis. The first clinically useful anti-tuberculosis agent was streptomycin. However, a few years after its introduction, streptomycin-resistant strains emerged. Combination therapy of tuberculosis with streptomycin and *para*-aminosalicylic acid, a *para*-aminobenzoic acid mimic that prevents bacterial synthesis of folic acid (*see* trimethoprim and methotrexate, pages 136 and 46), was only moderately more useful.

In 1951, three companies, Bayer, Roche, and Squibb independently introduced the potent antituberculosis agent isonicotinic acid hydrazide (INH or isoniazid), a compound that had been prepared first in 1912. Isoniazid proved to be specific and highly effective against tuberculosis, was well-tolerated and relatively inexpensive. Despite 50 years of use, combination therapy with isoniazid is still a first-line treatment of tuberculosis.

Following the success of isoniazid, the search for more potent structural analogs began. As a result, ethionamide and pyrazinamide were discovered, both of which are still in thera-peutic use (*see* structures on the left).

The initial phase of tuberculosis chemo-therapy destroys rapidly growing bacteria. The subsequent clearance of semidormant bacte-ria requires prolonged therapy, usually six months. The drug regimen must be strictly followed to avoid the emergence of resistant strains. Currently the most widely used combi-nations of antibiotics against tuberculosis employ isoniazid, rifampin, streptomycin, or ethambutol. Ethambutol is highly successful when applied in combination with isoniazid since the two act on different aspects of cell wall biosynthesis.

**Ethambutol**

**Rifampin**

*p*-Aminosalicylic acid

**Ethionamide**

**Pyrazinamide**

Many antibiotics have no effect on tuberculosis bacteria because they can not cross the cell wall. The tubercular lipid bilayer is unusually rigid and impervious to most lipophilic drugs. The mycobacterial cell wall contains three types of covalently bound macromolecules: (1) peptidoglycans, polymers of peptides and carbohydrates; (2) arabinogalactan, a polymeric carbohydrate, and (3) mycolic acids, very waxy solids. Mycolic acids are high molecular weight fatty acids that are found either in the form of esters with arabinose or as free lipids. An α-mycolate, the most abundant type of mycolic acid, is shown below.

**α-Mycolic acid**

The mechanism of action of isoniazid appears to involve an NADH-dependent reductase enzyme which plays a role in the synthesis of the mycolic acids.[1]

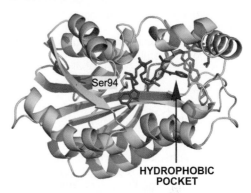

**Isoniazid**   **Isonicotinoyl radical**

**InhA inhibitor**

Isoniazid undergoes oxidation once inside *Mycobacterium tuberculosis* by the catalase-peroxidase enzyme KatG to afford the isonicotinoyl radical, which in the presence of the NADH-dependent reductase enzyme and its cofactor NADH (or NAD$^+$) leads to the deactivation of the reductase enzyme. Specifically, reaction of the isonicotinoyl radical with NAD$^+$ gives rise to a compound (shown above) that serves as the inhibitor of the NADH-dependent reductase enzyme.

The X-ray crystal structure of the NADH-dependent reductase enzyme-inhibitor comp-

lex has been determined.[2] The image below shows the contacts between the inhibitor and the enzyme. The isonicotinoyl moiety is embedded in a hydrophobic pocket, flanked by hydrophobic residues (amino acids Phe149, Gly192, Tyr158, and Trp22 are shown in green). A mutation of the amino acid Ser94 (shown in cyan) to Ala has been linked to isoniazid resistance.

**HYDROPHOBIC POCKET**

The eradication of tuberculosis has proved elusive for reasons other than the emergence of resistant strains.[3] The organism *M. tuberculosis* is able to assume a latent or dormant state in which susceptibility to antibiotics is minimal.[4] In addition, the microorganism is able to evade the immune response, either because of its ability to take up residence in sites that cannot be reached by immune cells or because it can assume a non-immunogenic form. Although rifampin and pyrazinamide have strongly cidal effects on *M. tuberculosis*, the current best treatment with six-month combination therapy using isoniazid, rifampin, pyrazinamide, and ethambutol only has a success rate of 90% in achieving patient sterilization. It has also been a challenge to develop methodology for the screening of new substances to determine their effectiveness against latent *M. tuberculosis* and their efficiency in achieving sterilization in humans. Nonetheless, because of the obvious need for more antitubercular drugs, there are currently at least five new entities in development that target different biochemical mechanisms.

In principle, effective prevention of tuberculosis should be possible using vaccines. Currently, at least one new vaccine is in clinical trials. However, it is too early to tell whether this approach will be successful within the next decade.

1. *Mol. Microbiol.* **2006**, *62*, 1220-1227; 2. *Science* **1998**, *279*, 98-102 (**1ZID**); 3. *Exp. Rev. Anti-Infect. Ther.* **2006**, *4*, 759-766; 4. *Nat. Rev. Microbiol.* **2007**, *5*, 39-47; **Refs. p. 176**

# ANCILLARY ANTIBIOTICS

Chloramphenicol, a remarkably simple halogen-containing antibiotic, was isolated in 1947 from the microbe *Streptomyces venezuelae*.[1] Once the structure of chloramphenicol had been determined, large-scale industrial production by chemical synthesis followed in 1949. Although common in chemical practice, the nitrobenzene (blue) and the dichloroacetic acid (red) units of chloramphenicol are unusual for a natural product.[2]

**Chloramphenicol**

Chloramphenicol is a wide-spectrum antibiotic, active against 95% of Gram-negative bacteria and a large number of Gram-positive aerobic and anaerobic strains. During the 1950s it was successfully used to quell epidemics of typhoid fever and meningitis. However, because of a serious side effect (blood dyscrasia) in a small fraction of those treated, its application as an oral or intravenous antibiotic was discontinued in the US in the late 1970s. Chloramphenicol is extensively used in poorer countries because of its low cost. In the US it is mainly used to treat ocular infections. Chloramphenicol acts by binding to the large 50S ribosomal subunit and blocks bacterial protein synthesis (*see* azithromycin, page 134). It also binds to ribosomes in mammalian mitochondria and harms rapidly dividing blood cell precursors in the bone marrow.

Synthetic 2-nitroimidazole (Azomycin) was prepared in 1955 and later shown to be effective against *Trichomonas vaginalis*, the microbe that causes the vaginal infection trichomoniasis.

**Azomycin**      **Metronidazole (Flagyl™)**

Extensive synthesis and evaluation of other imidazoles led to the identification of metronidazole, a 5-nitroimidazole derivative, as the most potent against *T. vaginalis* both *in vivo* and *in vitro*.[3]

Metronidazole (Flagyl™, Pfizer) is effective against a wide range of parasites and anaerobic bacteria. It is also used to treat acne rosacea, a common facial condition.

Once inside a microorganism, metronidazole accepts an electron to give a very reactive intermediate that attacks DNA and/or proteins and eventually kills the microbe. It has been shown that metronidazole damages the helical DNA structure and thus inhibits its replication.

Cycloserine, a broad-spectrum antibiotic, was first isolated in 1955 from the microbe *Streptomyces orchidaceous*. It is currently used against tuberculosis (Seromycin™, Lilly) in conjunction with drugs such as rifampicin, ethambutol, and streptomycin, but only when other treatments have failed.

**Cycloserine (Seromycin™)**

Cycloserine acts by inhibiting the biosynthesis of peptidoglycans that form the bacterial cell wall.[4]

Bacitracins, cyclopolypeptides isolated from the microbe *Bacillus subtilis* in 1943, are commonly used as topical antibiotics for the treatment of skin and eye infections. The structure of the most abundant, bacitracin A, is shown below.

**Bacitracin A**

Bacitracins are often used topically in combination with other antibiotics such as neomycin B and polymyxin B (Neosporin™). Bacitracins inhibit bacterial growth by blocking the formation of dolichol phosphate, a $C_{55}$-component of membranes that transports intermediates for cell wall biosynthesis.[5]

1. *Adv. Phytochem.* **2003**, 109-184; 2. *Antimicrob. Agents* **2005**, 925-929; 3. *Antimicrob. Agents* **2005**, 930-940; 4. *Antibiotic and Chemotherapy (7th Edition,* Churchill Livingstone, **1997**); 344-345; 5. *Antimicrob. Agents* **2005**, 377-400; Refs. p. 176

# DRUG RESISTANCE

It came as an unpleasant surprise that shortly after the introduction of each of the various classes of antibiotics new pathogenic strains emerged that were no longer susceptible. The emergence of resistant microorganisms is now known to be a predictable consequence of widespread antibiotic use (and misuse). All that is required for this to happen is the survival of genetically modified organisms that can reproduce even in the presence of a particular antibiotic.[1] The modifications can come from (1) spontaneous mutations, (2) overexpression of certain genes and overproduction of their gene products or (3) acquisition of genetic material from other microorganisms, for example by plasmid DNA exchange. These genetic changes can confer resistance in the following ways:

1. *By an enzyme-catalyzed reaction that renders the antibiotic inactive.* There are many examples of this type including hydrolytic cleavage of the β-lactam ring of penicillins and cephalosporins by β-lactamases and acetylation or phosphorylation of aminoglycosides catalyzed by acetyltransferase or phosphotransferase enzymes.

2. *Mutations in the organisms* can result in changes in the biological target that are protective because they prevent the antibiotic from binding to it. An example of this is the development of bacterial resistance to a tetracycline as a result of mutations that reduces the binding volume of the A site in the 30S subunit of the bacterial ribosome so that it no longer accommodates the tetracycline. Another example is the mutation of vancomycin resistant strains that attenuates hydrogen bonding between the developing cell wall and the antibiotic and reduces vancomycin potency by three orders of magnitude (*see* page 138).

3. *Upregulation of drug efflux from the bacterial cell.* Most living cells are endowed with membrane-associated proteins that serve as active transporters of organic substances either into or out of the cell. Some microorganisms have multiple efflux pumps that can handle a variety of molecular types. Usually these pumps are powered by ATP in the sense that the overall hydrolysis of ATP provides the driving force for the conformational changes of the efflux protein that push the ligand molecule out of the cell.

Examples of this process are the efflux pump-mediated removal of tetracyclines or quinolone antibiotics from microorganisms that have become resistant because of an ability to overproduce efflux pumps. Such mutant microbes grow more slowly since there is a major diversion of energy to the efflux process.

4. *Decreased cell-wall permeability by mutation-induced genetic changes.* Such changes contribute to the resistance shown by the microbes responsible for tuberculosis.

5. *Genetic changes that alter the metabolic pathways or needs of the microbe so as to circumvent antibacterial actions.* An early example of this mechanism for resistance was the finding that sulfanilamide-resistant strains had developed the capability of utilizing folic acid at levels provided by the host organism.

6. *Biofilm formation.* Bacteria communicate with one another using small organic molecules as signals, for example derivatives of the amino acid homoserine. These messenger molecules can activate the bacterial colony to produce virulent proteins which are toxic to the host or to secrete carbohydrates that join to form tough, protective films, or biofilms. These films resist penetration by antibiotics and consequently function to protect bacteria. Biofilm formation contributes to persistent infection *in vivo*. It is a major medical problem in device implantation therapy, for instance hip or knee replacements, and is the main reason for failed procedures.

The development of resistance to drugs also accompanies the application of antiviral drugs to viral infections. Viral resistance is generally due to mutations that change the structure of the protein target of the antiviral drug.

Much information on viral resistance to chemotherapy has emerged from the massive recent studies on the treatment of HIV AIDS. Three major classes of drugs have been employed in medicine over the past decade: (1) nucleoside derivatives as inhibitors of the enzyme reverse transcriptase, which catalyzes the formation of DNA from viral genomic RNA; (2) non-nucleoside reverse transcriptase inhibitors and (3) inhibitors of HIV protease which cleaves two large precursor proteins into smaller fragments that are essential for the reproduction of HIV.

Resistance to these agents quickly developed when they were used separately. Combination therapy with three or more agents is currently the standard treatment. However, about half of the patients on such therapy develop resistance to at least one of these drugs.

Resistance develops as a result of the ability of HIV to undergo genomic mutations at a surprisingly rapid rate. Consequently, mutant strains evolve with one, two, or even more mutations that render the drugs ineffective because of changes in the structure of the viral target.

Drug resistance is a major problem in cancer chemotherapy. In cancer, as with infectious diseases, the population of cells exposed to a drug is in the billions and so the combination of high mutation rates and large numbers of cells ensures the evolution of survivors and their proliferation by natural selection. Resistance in cancer cells may also involve overexpression of efflux pump proteins or acquisition of evasive biochemistry (*see* page 184 for an overview of cancer).[2]

## Structure and Function of Efflux Pumps

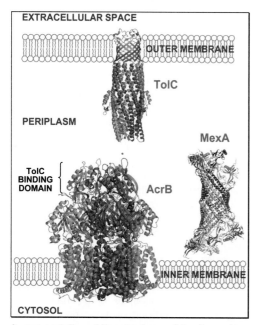

A representation of the structures of the three drug efflux pump components (resting state), proteins TolC, MexA and AcrB. TolC is barrel-shaped; its upper α-barrel domain is embedded in the outer membrane and forms an exit channel to the extracellular space. The lower β-barrel domain extends into the periplasmic space. AcrB is a trimeric protein at the inner membrane that provides entry from the cytosol. MexA is a periplasmatic

adaptor, linking TolC and ArcB when the efflux pump is in its fully assembled functioning form (*see* image below).[3-5]

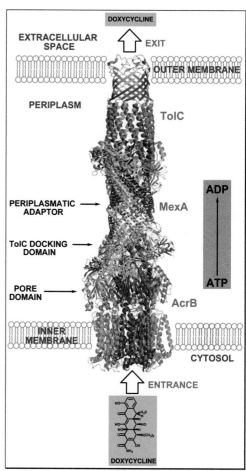

This image is a representation of the fully assembled drug efflux pump, connecting the intracellular with the extracellular space. The MexA periplasmatic adaptor fills the gap between AcrB and TolC so there is no opening of the channel into the periplasm. The substrate enters the AcrB assembly and exits into the extracellular space through the opening of the TolC.

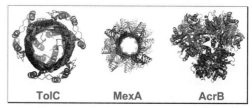

Top-side views of the three drug efflux pump components TolC, MexA, and AcrB. All three proteins have a cylindrical cavity through which the various drug molecules can pass.

1. *Evolution of Microbial Pathogens* **2006**, 221-241; 2. *Nat. Rev. Microbiol.* **2006**, *4*, 629-636; 3. *Science* **2006**, *313*, 1295-1298 (**2GIF**); 4. *J. Biol. Chem.* **2004**, *279*, 25939-25942 (**1VF7**); 5. *J. Mol. Biol.* **2004**, *342*, 697-702 (**1TQQ**); Refs. p. 176

# ANTIVIRAL AGENTS

# ON VIRUSES AND VIRAL DISEASES

Viruses are infectious particles that consist of DNA or RNA genomic material and protein (without and with attached carbohydrates). They are only about one-thousandth the size of a bacterial or mammalian cell and can undergo replication only inside such a cell. Once inside the infected cell, viruses reprogram the cell and use it to replicate themselves. The newly formed viruses then exit the cell and infect other cells (*see* scheme on next page). Some viruses coexist with the host organism and are harmless (equilibrium type), but others cause damage and disease (non-equilibrium type). There are about 2200 recognized viral species; the list continues to grow.

Viruses were discovered in 1892 by the Russian botanist, Dmitri Iwanowski who recognized that tobacco mosaic disease was caused by a particle much smaller than bacteria. Iwanowski's discovery marks the beginning of virology. A virus was identified as the cause of human disease, yellow fever, for the first time by Walter Reed in 1901. The electron microscope, introduced in the 1940s, provided the first images of viral structure. This development, together with advances in biochemistry, played a crucial role in the understanding of how viruses work.

Although viruses exist in all shapes and forms, a key difference between them is in the type of nucleic acid they use to store their genetic information. There are two main categories, RNA and DNA viruses. Greatly simplified schematic representations of an RNA and a DNA virus are shown at right. Viruses lack the cellular machinery to replicate. Most viruses enclose their RNA or DNA in a protein protective layer known as the capsid. The capsid is usually enclosed by a viral envelope, the surface of which is studded with glycoproteins. These glycoproteins are essential for viral entry into a target cell and infection. In addition to the genetic material, viruses contain enzymes (not shown here) that are necessary for viral survival and replication within the host cell (*see* the schematic representation of HIV-1 on page 152).

The current classification of viruses, based on the type of genomic material they carry and their mode of replication, was proposed by David Baltimore, a Nobel Prize-winning biologist. In this system the viruses belong to one of seven groups:

Group I-II: double and single-stranded DNA viruses;

Group III: double-stranded RNA viruses;

Group IV-V: positive and negative-sense single-stranded RNA viruses;

Group VI: Diploid single-stranded RNA viruses that use reverse transcription;

Group VII: Circular double-stranded DNA viruses that use reverse transcription;

Examples of these viruses are: Group I (herpesvirus, smallpox, and chickenpox); Group II (parvovirus); Group III (rotavirus); Group IV (hepatitis A, SARS, hepatitis C, yellow fever); Group V (Ebola, influenza); Group VI (HIV) and Group VII (hepatitis B).

Viruses adapt to changing and challenging environments because they replicate rapidly and mutate. Consequently they evolve at a much faster rate than host cells. For example such mutated viruses are able to find new avenues of entry into host cells.

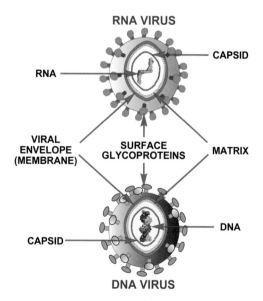

**RNA VIRUS**

RNA — CAPSID

VIRAL ENVELOPE (MEMBRANE) — SURFACE GLYCOPROTEINS — MATRIX

CAPSID — DNA

**DNA VIRUS**

There is an upper limit to the mutation rate of viruses.[1] Beyond that rate, a large number of lethal mutations occur that result in non-infectious virus and eventually extinction, a phenomenon known as "error catastrophe".[2] It is estimated that the average mutation rate of RNA viruses is one mutation per genome per replication.[3] That mutation rate (which is close to the catastrophe limit) reflects the absence of proofreading capability of the viral reverse transcriptase enzyme. In comparison, the mutation rate of DNA viruses is only one mutation per genome per 1000 replications because these viruses utilize the DNA polymerase enzyme of the host cell that does have proofreading capability. The genome of DNA viruses is often more than 10 times larger than the genome of RNA viruses. Consequently, DNA viruses evolve much more slowly than RNA viruses.

The high mutation rates coupled with the use of the host cell's molecular machinery makes viruses more difficult to fight with drugs than bacteria. However, during the past three decades, a great deal of information has been collected regarding the life cycle of various viruses. The life cycle of an RNA virus (retrovirus) is shown below.[4,5] The structural elements of the host cell and the virus are shown in blue and the important stages of the viral life cycle are shown in black. The points where drugs can interfere with these viral life cycle events are highlighted in the yellow boxes. For example, viral entry can be inhibited by the use of CCR5 receptor inhibitors (e.g., maraviroc, page 156) and the translation of viral RNA can be prevented with reverse transcriptase inhibitors (e.g., nevirapine and efavirenz, pages 153 and 154). In addition, the assembly of viral proteins and of new viral RNA into newly formed viruses are effectively inhibited by protease inhibitors such as lopinavir and ritonavir (page 155). Finally, the release of viruses can be halted by the administration of neuramidinase inhibitors (e.g., oseltamivir, page 150).

1. *Trends Microbiol.* **2003** *11*, 543-546.
2. *J. Virol.* **2006** *80*, 20-26.
3. *Proc. Natl. Acad. Sci. U. S. A.* **1999** *96*, 13910-13913.
4. Acheson, N. H. Fundamentals of Molecular Virology (John Wiley & Sons, **2006**).
5. Dimmock, N. J., Easton, A. J. & Leppard, K. Introduction to Modern Virology (Blackwell, **2007**); Refs. p. 176

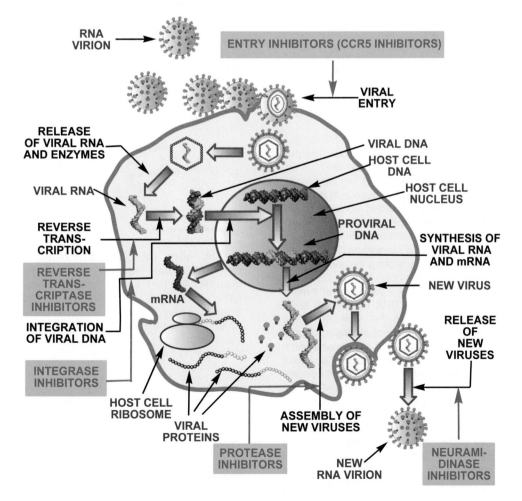

# ACYCLOVIR (ZOVIRAX™)

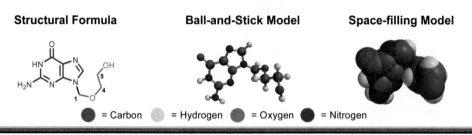

| **Structural Formula** | **Ball-and-Stick Model** | **Space-filling Model** |

⬤ = Carbon    ⬤ = Hydrogen    ⬤ = Oxygen    ⬤ = Nitrogen

*Year of discovery:* 1970s (by Burroughs Wellcome); *Year of introduction:* 1982; *Drug category:* Antiviral; Viral DNA synthesis inhibitor; *Main uses:* For the treatment of herpes infections (e.g., cold sores, genital herpes, chicken pox); *Other brand names:* Zovir; *Related drugs:* Valacyclovir (Valtrex), Penciclovir (Denavir), Ganciclovir, Valganciclovir (Vitrasert).

Acyclovir, an acyclic guanine nucleoside analog, was one of the first highly effective systemic antiviral agents in clinical practice. It followed earlier nucleoside antiviral agents such as idoxuridine and ribavirin (*see* page 149) that were not widely used.

**Idoxuridine**       **Ribavirin**

The success of acyclovir demonstrated convincingly that it is possible to find antiviral agents that are relatively nontoxic and selective for the pathogen.

Acyclovir was approved in 1982 for the treatment of infections by various herpes viruses such as genital herpes, cold sores, shingles, and chicken pox. It is unique among the early antiviral agents because it is highly selective for virus infected cells and is minimally cytotoxic. Acyclovir mimics deoxyguanosine, which is a critical building block for DNA, and differs from deoxyguanosine in the sugar moiety (red) by lacking carbons #2 and #3 (*see* scheme to the right).[1]

Acyclovir is actually a prodrug that inhibits the replication of viral DNA because it is an excellent substrate for the viral thymidine kinase enzyme (only present in virally infected host cells), but not for the mammalian enzyme.[2] In uninfected human cells acyclovir is not phosphorylated. Once inside an infected host cell, acyclovir is first converted to the monophosphate by the viral enzyme and then transformed to acyclovir triphosphate by kinases of the host cell (*see* scheme). Acyclovir triphosphate is a potent competitive inhibitor of viral DNA polymerase and retards the incorporation of the natural substrate, deoxyguanosine triphosphate, into a growing

DNA chain. However, when phosphorylated acyclovir is attached to viral DNA (highlighted in red below), it causes DNA chain termination because no further lengthening is possible; viral replication stops.

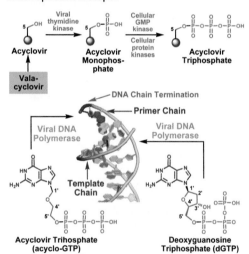

**Acyclovir Trihosphate (acyclo-GTP)**      **Deoxyguanosine Triphosphate (dGTP)**

The valine ester derivative of acyclovir, valacyclovir, has the advantage of greater bioavailability (5x) and is now the preferred form.[3] In addition, there are other effective analogs in clinical use, for example ganciclovir and its valine ester, valganciclovir. These valine esters are hydrolyzed in the body by esterases to the parent drugs.

**Valacyclovir (Valtrex™)**     **Valganciclovir (Vitrasert™)**

1. *Antiviral Res.* **2006**, *71*, 134-140; 2. *Br. J. Pharmacol.* **2006**, *147*, 1-11; 3. *Exp. Rev. Anti-Infect. Ther.* **2006**, *4*, 367-376; **Refs. p. 176**

# RIBAVIRIN (VIRAZOLE™)

| Structural Formula | Ball-and-Stick Model | Space-filling Model |
|---|---|---|

● = Carbon    ● = Hydrogen    ● = Oxygen    ● = Nitrogen

*Year of discovery:* 1970 (ICN); *Year of introduction:* 1980 (for RSV in children); 1998 (in combination with interferon for HCV); *Drug category:* Nucleoside antimetabolite/antiviral; *Main uses:* For the treatment of hepatitis C (HCV) in combination with interferon-α or peginterferon-α; *Other brand names:* Rebetol, Copegus, Cotronak, Ribasphere. Sold as a generic drug since 2005; *Related drugs:* Viramidine, HCV-796 (Wyeth & Viro-Pharma), Peginterferon-α.

Ribavirin, a broad-spectrum antiviral compound discovered in 1970, is currently a first-line therapy against hepatitis C in combination with interferon-α. Transmission of the *Hepatitis C virus* (HCV) requires its introduction into the bloodstream (e.g., by blood transfusion or intravenous drug use). The disease affects nearly 200 million people worldwide, including 4 million in the US. Individuals infected with hepatitis C usually show few if any symptoms within the first six months, making detection of the disease difficult. If left untreated, HCV infection eventually can lead to serious liver damage (cirrhosis) or cancer over a period of years (10-20). Liver failure in HCV infected individuals is the main reason for liver transplants in the US.

Ribavirin was first synthesized and shown to be active against a number of RNA and DNA viruses in the 1970s. It was first approved for the treatment of severe respiratory syncytial virus (RSV) infections in children in 1989. Following the discovery of the hepatitis C virus, ribavirin was tested as monotherapy in HCV infected individuals, but even prolonged administration did not result in the clearance of the virus. However, the combination of ribavirin with interferon-α (INF-α), a natural protein synthesized by immune cells in response to viral infections, led to clearance of HCV infection in many of the patients so treated.[1] The combination of INF-α and ribavirin is strongly synergistic, since either agent alone is relatively ineffective. Recently, peginterferon-α (INF-α covalently coupled to polyethyleneglycol) has been introduced as an improvement over INF-α that requires only weekly dosing because it is inactivated more slowly *in vivo*.[2] Also, a

prodrug for ribavirin, viramidine, is currently undergoing Phase III clinical trials. Viramidine (shown below) is eliminated from the body more rapidly than ribavirin and is expected to have fewer side effects.

**Viramidine**

Although the mechanisms of action of INF-α and ribavirin have not been established, it is possible that ribavirin diminishes the rate of viral replication sufficiently to allow more effective destruction of infected cells by INF-α-activated immune cells.[3]

The INF-α/ribavirin therapy of HCV is successful for about 90% of patients who have been infected for less than six months (acute phase) and successful in about 50% of those patients who have been infected for more than six months. Because the large doses used in ribavirin/INF-α therapy cause serious side effects (e.g., hemolytic anemia and flu-like symptoms), safer therapies are still needed. A potent HCV polymerase inhibitor, HCV-796, is currently in Phase II clinical development for the treatment of HCV in combination with pegINF-α, with and without ribavirin.

**HCV-796**

1. *N. Engl. J. Med.* **1998** *339*, 1485-1492; 2. *N. Engl. J. Med.* **2006** *355*, 2444-2451; 3. *Nature* **2005** *436*, 967-972; Refs. p. 176

# OSELTAMIVIR (TAMIFLU™)

| Structural Formula | Ball-and-Stick Model | Space-filling Model |
|---|---|---|

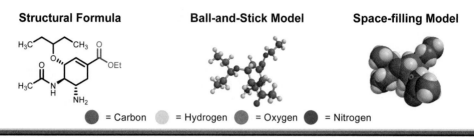

● = Carbon    ● = Hydrogen    ● = Oxygen    ● = Nitrogen

*Year of discovery:* Early 1990s (by Gilead Sciences); *Year of introduction:* 1999 (Roche); *Drug category:* Neuraminidase inhibitor/antiviral; *Main uses:* For the treatment and prevention of influenza A and B viral infections; *Related drugs:* Zanamivir (Relenza), Peramivir.

Oseltamivir, a second-generation oral neuramidinase inhibitor, is widely used for the treatment and prevention of influenza A and B viral infections, including avian flu. Influenza is a serious respiratory illness that affects millions of people annually ("seasonal flu") and results in many deaths as well as an estimated $10 billion in lost productivity. The symptoms of flu and the common cold are similar, but flu symptoms are generally more severe. Older people and those with a weakened immune system are especially at risk to develop complications such as bronchitis and pneumonia which are the main causes of influenza-related deaths. In recent years a more virulent influenza A type virus, the avian flu, has emerged. Certain strains of the avian flu (e.g., H5N1) are able to infect humans and cause high mortality. Since the effective treatment of the various flu infections has traditionally been difficult due to the lack of potent antiviral drugs, the possibility of a catastrophic worldwide avian flu epidemic has arisen, as well as the urgent need for effective antiviral drugs. During the 1990s, the structure of the influenza neuramidinase enzyme that plays a crucial role in viral replication and infectivity was elucidated. In particular, neuramidinase cleaves sialic acid residues that connect the newly formed viruses to the surface of the host cell that it had infected. The synthesis of a large number of structural mimics of sialic acid led to the discovery of potent neuraminidase inhibitors that prevent the flu viruses from being released and limit the spread of the disease.

The first useful neuraminidase inhibitor, zanamivir (Relenza™), was discovered in 1989 using target-guided design. This drug, however, has poor bioavailability, can be administered only by inhalation and as a result it has seen limited clinical use.

**Sialic acid**     **Zanamivir (Relenza™)**

The first orally active neuramidinase inhibitor, oseltamivir, was developed by C.U. Kim at Gilead Sciences in the early 1990s and marketed as Tamiflu by Roche.[1,2] Oseltamivir itself is a prodrug that is activated by cleavage of the ethyl ester moiety (shown in blue above) to the carboxyl group (COOH). Currently oseltamivir is considered the only effective treatment for avian flu. The X-ray structure of oseltamivir carboxylate bound to neuramidinase (shown below) has been determined.[3] Strong binding is the result of hydrogen bonding between Arg224 and Glu276 as well as hydrophobic interactions. However, resistant strains have already been reported. Resistant strains have mutations in certain amino acid residues (e.g., Arg292 to Lys and His274 to Tyr) that significantly reduce the binding affinity of oseltamivir to the enzyme.

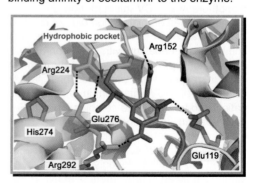

1. *J. Am. Chem. Soc.* **1997**, *119*, 681-690; 2. *Future Virology* **2006**, *1*, 577-586; 3. *Nature* **2006**, *443*, 45-49 (2HT8); Refs. p. 177

# ZIDOVUDINE (RETROVIR™, AZT)

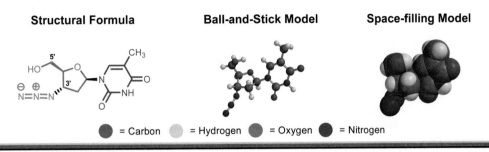

**Structural Formula**     **Ball-and-Stick Model**     **Space-filling Model**

● = Carbon     ○ = Hydrogen     ● = Oxygen     ● = Nitrogen

*Year of discovery:* 1964 (B.A. Karmanos Cancer Inst. and Wayne State Univ. School of Medicine); *Year of introduction:* 1987 (Burroughs-Wellcome, now GSK); *Drug category:* Antiretroviral agent/pyrimidine nucleoside reverse transcriptase inhibitor; *Main uses:* For the treatment of adults and children with HIV infection and for the prevention of mother-to-child transmission of HIV as a mono- or combination therapy (e.g., Combivir and Trizivir); *Other brand names:* Retrovir, Azidothymidine (AZT), Retrovis; *Related drugs:* Stavudine, Zalcitabine, Didanosine, Lamivudine.

Zidovudine, also known as azidothymidine (AZT), was the first antiretroviral drug approved for the treatment of persons infected with the human immunodeficiency virus (HIV). HIV is a retrovirus that attacks and gradually destroys the immune system of the infected individual, leading to the acquired immuno-deficiency syndrome (AIDS). Because of the weakened immune system, other infections result, which culminate in organ failure and death. It has been estimated that there are approximately 40 million people worldwide, 38 million adults and 2 million children, who are infected with HIV. Each year there are approximately 4 million new cases of HIV/AIDS. The annual global death toll is about 3 million and the total number of deaths due to AIDS since 1981 is about 25 million. HIV infection occurs by contact with body fluids (e.g., blood, semen, vaginal fluid, breast milk) that contain the virus. The probability of an HIV infection can be significantly reduced by the screening of blood products used for transfusions, avoiding the use of non-sterile hypodermic needles and adherence to safe sexual practices.

During the early 1980s, when the medical community realized the potential of HIV/AIDS to become a pandemic, research to find effective anti-HIV medications intensified. AZT, first synthesized in 1964 as an anti-cancer agent, was found to be an effective drug against HIV and shown to prolong the life of patients with AIDS.[1] It was first approved in 1987 for the treatment of, and later, for the prevention of HIV infections (e.g., mother-to-child transmission or exposure to the virus).

The genomes of retroviruses, such as HIV, consist of single-stranded RNA which must be converted to double-stranded DNA in order for viral replication to occur, since only double-stranded DNA can be integrated into the chromosomes of the host cell. The RNA to DNA conversion is carried out by the enzyme reverse transcriptase that is present in each HIV virus along with two copies of the retroviral RNA. AZT is converted to the 5'-triphosphate derivative which is a potent and specific inhibitor of the enzyme HIV reverse transcriptase. In addition, when AZT-5'-triphosphate is incorporated into DNA, chain termination results.[2] (*See* also acyclovir on page 148).

Currently AZT is mainly used in combination with other antiretroviral drugs to avoid the emergence of resistant strains.[3] HIV replicates rapidly and lacks the proofreading enzymes that normally correct errors in the copying of DNA. Consequently, resistance toward AZT develops as a result of mutations in the reverse transcriptase. The combination of AZT with lamivudine (Combivir™, GSK) or with abacavir and lamivudine (Trizivir™, GSK),[4] suppresses the replication of HIV by an order of magnitude more than the individual drugs.

**Lamivudine**          **Abacavir**

1. *Proc. Natl. Acad. Sci. U. S. A.* **1985**, *82*, 7096-7100; 2. *Cell. Mol. Life Sci.* **2006**, *63*, 163-186; 3. *Combination Therapy of AIDS* **2004**, 41-51; 4. *Future Virology* **2007**, *2*, 23-30; Refs. p. 177

# ZALCITABINE (HIVID™)

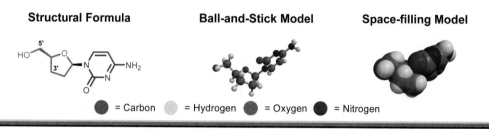

| Structural Formula | Ball-and-Stick Model | Space-filling Model |

● = Carbon  ● = Hydrogen  ● = Oxygen  ● = Nitrogen

*Year of discovery:* 1987 (National Cancer Institute); *Year of introduction:* 1992 (as a monotherapy by Roche and in combination with AZT); *Drug category:* Antiretroviral agent/pyrimidine nucleoside reverse transcriptase inhibitor; *Main uses:* For the treatment of HIV infection and AIDS; *Related drugs:* Zidovudine (AZT), Stavudine (Zerit), Didanosine (Videx), Lamivudine (Epivir).

Zalcitabine, also known as 2',3'-dideoxy-cytidine (ddC), is a synthetic pyrimidine nucleoside HIV reverse transcriptase (RT) inhibitor. Originally discovered at the National Cancer Institute, it is now sold as Hivid by Roche.[1]

Understanding the important stages in the life cycle of HIV has been critical to the development of effective antiretroviral drugs against HIV/AIDS.[2] The structure of HIV, depicted in a simplified form below, provides valuable insights into the key molecular players.

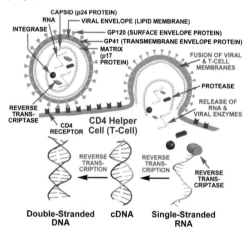

**Double-Stranded DNA**   **cDNA**   **Single-Stranded RNA**

The genetic material, two copies of single-stranded RNA, and essential viral enzymes such as RT, protease, and integrase are surrounded by the capsid, a protective protein layer made up by the aggregation of viral p24 protein building blocks. A second layer, the matrix, which is composed of p17 viral protein subunits, encloses the capsid and ensures the structural integrity of the virus particle. The matrix is surrounded by an outer lipid membrane, the viral envelope. The surface of the HIV virus is covered with two glycoproteins

(gp): (1) gp41 which is embedded in the membrane and (2) gp120 which protrudes from the surface and is anchored to gp41.

HIV attacks, and eventually destroys the CD4 T-helper cells of the immune system (page 116). In order for HIV to infect the T-cell and replicate, its genome (RNA) has to be introduced into the cytoplasm of the T-cell. First, gp120 binds to the CD4 receptor located on the T-cell surface. Upon binding a conformational change takes place, causing the viral envelope to fuse with the T-cell membrane. (*See* more about this process on page 157). The contents of the capsid are then released into the T-cell cytoplasm and three viral enzymes guide the replication cycle of HIV. The viral enzyme reverse transcriptase copies the single-stranded viral RNA to form a complementary DNA chain (cDNA) from which double-stranded DNA is made. This transcription process uses nucleotides from the T-cell cytoplasm (e.g., deoxycytidine).

**Deoxycytidine**

Zalcitabine is a derivative of deoxycytidine derived by the replacement of the hydroxyl group (OH) at the 3'-position by a hydrogen atom. *In vivo*, zalcitabine is converted to the 5'-triphosphate that is a potent inhibitor of the viral enzyme RT. In addition, upon its incorporation into DNA, the absence of a 3' hydroxyl group in the developing molecule prevents attachment of another nucleotide and halts DNA chain growth. Replication of the virus is thus halted. (*See* also acyclovir and AZT on pages 148 and 151).

1. *J. Antimicrob. Chemother.* **1989**, *23*, 29-34. 2. *Drug Des. Rev.-Online* **2004**, *1*, 83-92; Refs. p. 177

# NEVIRAPINE (VIRAMUNE™)

| **Structural Formula** | **Ball-and-Stick Model** | **Space-filling Model** |
|---|---|---|

● = Carbon　　◐ = Hydrogen　　● = Oxygen　　● = Nitrogen

*Year of discovery:* Early 1990s (Rega Institute for Medical Research); *Year of introduction:* 1996 (Boehringer Ingelheim); *Drug category:* Non-nucleoside HIV reverse transcriptase inhibitor; *Main uses:* For the treatment of HIV infections as part of combination therapy with zidovudine and didanosine; Also as a single-dose therapy in HIV-infected pregnant women to prevent mother-to-child transmission of the virus; *Related drugs:* Delavirdine (Rescriptor), Efavirenz (Sustiva).

Nevirapine is an anti-HIV drug that blocks the HIV enzyme reverse transcriptase and thereby prevents viral replication.[1] It was discovered in Belgium in the early 1990s and currently is a first-line treatment for HIV-1 infection in combination with other antiretroviral drugs such as HIV protease inhibitors and nucleoside reverse transcript-tase inhibitors. These drugs are synergistic since they act by different molecular mechanisms. Nevirapine is also used to prevent mother-to-child transmission of HIV. Nevirapine does not cure HIV since it only inhibits the replication of the virus. However, even after the administration of a few doses, a large drop in the levels of circulating HIV is usually observed.

Nevirapine inhibits the enzyme HIV-1 reverse transcriptase (RT), but not the enzyme HIV-2 RT. Unlike nucleoside RT inhibitors such as zidovudine and zalcitabine, nevirapine does not bind at the active site of the enzyme, but instead, at a non-substrate binding site, referred to as the allosteric non-nucleoside reverse transcriptase inhibitor (NNRTI)-binding pocket. The NNRTI-binding pocket is located only 10 Å below the substrate-binding site of the enzyme. The binding of nevirapine results in a change of the catalytically active conformation of the enzyme into a fixed *inactive* conformation. Locked in this inactive conformation, the enzyme can not copy the single-stranded viral RNA to cDNA, the intermediate leading to genomic double-stranded DNA. The figures below show ligand-free HIV-1 RT and also the same enzyme complexed with cDNA. (The template viral RNA strand is shown in cyan;

*see* also scheme on page 152). The structure of HIV-1 RT with nevirapine (shown as red spheres) occupying the NNRTI-binding pocket (transparent mesh representation) is shown in the bottom panel. Part of the tricyclic core of nevirapine binds in a face-to-face attraction with the flat rings of tyrosines 181 and 188.[2] A single mutation of Tyr181 to Cys181 results in the loss of significant binding energy and a 100-fold lower binding affinity.[3] This and the high error rate of RT explain why resistance can develop rapidly even after a few doses of nevirapine monotherapy.

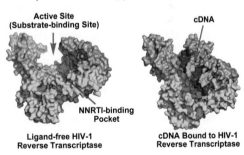

**Active Site (Substrate-binding Site)**　　**cDNA**

**NNRTI-binding Pocket**

**Ligand-free HIV-1 Reverse Transcriptase**　　**cDNA Bound to HIV-1 Reverse Transcriptase**

**Nevirapine**

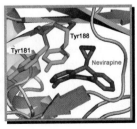

Tyr188
Tyr181
Nevirapine

**Nevirapine in HIV-1 Reverse Transcriptase (1VRT)**

1. *Int. J. Clin. Pract.* **2006**, *61*, 105-118; 2. *Nat. Struct. Biol.* **1995**, *2*, 293-302. (**1VRT**); 3. *J. Mol. Biol.* **2001**, *312*, 795-805 (**1JLB**); Refs. p. 177

# EFAVIRENZ (SUSTIVA™)

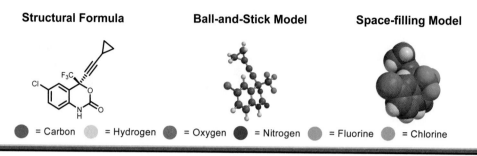

| Structural Formula | Ball-and-Stick Model | Space-filling Model |
| --- | --- | --- |

● = Carbon  ○ = Hydrogen  ● = Oxygen  ● = Nitrogen  ● = Fluorine  ● = Chlorine

*Year of discovery:* Early 1990s (DuPont Merck now Bristol-Myers Squibb); *Year of introduction:* 1998 (Bristol-Myers Squibb/Merck); *Drug category:* Non-nucleoside reverse transcriptase inhibitor; *Main uses:* For the treatment of HIV-1 infections in adults and children as a fixed dose combination therapy (Atripla); *Related drugs:* Nevirapine (Viramune), Delavirdine (Rescriptor).

Efavirenz is a potent second-generation HIV-1 non-nucleoside reverse transcriptase inhibitor (NNRTI). The viral enzyme reverse transcriptase (RT) is responsible for copying single-stranded viral RNA into the double-stranded DNA that is required for the replication of HIV. This RT operates with a high error rate which, when coupled with the rapid rate of HIV replication, leads to high rates of genetic mutation and the rapid appearance of drug resistant strains by natural selection over drug-susceptible strains. These mutant HIV strains are often resistant to other classes of antiretroviral drugs as well, severely limiting treatment options. The first-generation NNRTIs such as nevirapine and delavirdine also produce resistant strains, suffering, for instance, a 100-fold loss in binding affinity to HIV-1 RT as a result of the mutation of a single amino acid residue (Tyr181 to Cys181) in the active site of the viral enzyme.

The rapid emergence of resistant HIV-1 strains spurred the development of second-generation NNRTIs that are more potent and somewhat less sensitive to single point mutations. Efavirenz was discovered in the early 1990s, and in 1998 it was introduced as Sustiva™ by BMS and Merck (as Stocrin™ in Europe). It is now used in combination with other classes of antiretroviral drugs, for example, with lamivudine and tenofovir (or zidovudine) to suppress HIV replication in individuals who have not been previously treated. Efavirenz is also useful for patients who already failed one or two classes of antiretroviral drugs. Efavirenz was shown to be well-tolerated even after long-term treat-ment (10 months or more). Recently, a new

formulation combining fixed doses of efavi-renz, emtricitabine and tenofovir was approved. It is now available under the trade name of Atripla™ (Gilead Science and BMS).[1] The potency of efavirenz allows once-a-day dosing which increases patient compliance and reduces the probability of the emergence of resistant strains. Atripla is the first multi-class antiretroviral drug formulated in a single pill.

The effect of mutations at the NNRTI-binding pocket of the enzyme reverse transcriptase has been extensively studied. The X-ray crystal structures of the wild-type RT and mutated RT, with and without efavirenz, have been elucidated.[2] Efavirenz does not bind to Tyr181 and Tyr188 in contrast to nevirapine and, therefore, the nevirapine-induced mutations to Cys181 or Cys188 do not impact the binding of efavirenz to the enzyme. In fact, two other mutations (Lys103 to Arg and Leu100 to Ile) must occur simultaneously in order to produce high levels of resistance towards efavirenz. (Lys103 is cyan whereas Leu100 is magenta.)

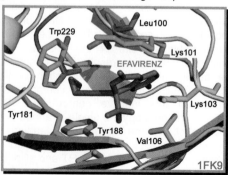

1. *Drugs* **2006**, *66*, 1501-1512; 2. *Structure (London)* **2000**, *8*, 1089-1094. (1FK9); Refs. p. 177

# LOPINAVIR + RITONAVIR (KALETRA™)

| **Structural Formula** | **Ball-and-Stick Model** | **Space-filling Model** |
|---|---|---|

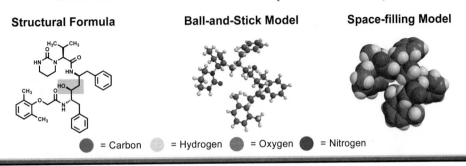

● = Carbon   ● = Hydrogen   ● = Oxygen   ● = Nitrogen

*Year of discovery:* Early 1990s (Abbott Laboratories); *Year of introduction:* 2000 (only in combination with ritonavir as Kaletra); *Drug category:* HIV protease inhibitor; *Main uses:* For the treatment of HIV-infection in combination with other antiretroviral agents; *Related drugs:* Ritonavir (Norvir), Indinavir (Crixivan), Saquinavir (Fortovase & Invirase), Nelfinavir (Viracept), Fosamprenavir (Lexiva).

The anti-HIV drug Kaletra is a combination of two HIV protease inhibitors, lopinavir and ritonavir. The enzyme HIV protease plays an important role in the late phase of the viral life cycle in which the large proteins Gag and Gag-Pol are cleaved to generate smaller structural proteins that associate to form the body of the virus. Cleavage of Gag-Pol also leads to essential viral enzymes, such as reverse transcriptase, protease and integrase. HIV protease must cleave Gag and Gag-Pol at specific peptide bonds to generate functional viral proteins. These proteins are required for the assembly of the new viral particles. Without them, the newly formed viruses can not mature or become infectious. For this reason, targeting the enzyme HIV protease with specific inhibitors has been the focus of much research since the early 1990s.[1]

1989 by A. Wlodawer at NIH.[2] This finding greatly facilitated the rational design of competitive HIV protease inhibitors. The early inhibitors were transition-state mimics which had strong affinity for the enzyme and were stable to it (*see* yellow boxes). The first FDA approved HIV protease inhibitor, saquinavir (Invirase™, Roche), was soon followed by ritonavir (Norvir™, Abbott). Abbott later developed a more potent protease inhibitor, lopinavir, which was found to be even more effective when co-formulated with ritonavir. Ritonavir inhibits the liver enzyme cytochrome P450-3A4 that rapidly metabolizes lopinavir. As a result, when used in combination with ritonavir, lopinavir has an extended half-life and is effective at lower doses. The X-ray crystal structure of lopinavir (red) bound to HIV-1 protease is shown below.[3]

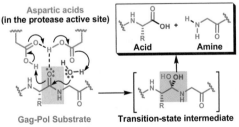

**Aspartic acids** (in the protease active site)

**Acid**   **Amine**

Gag-Pol Substrate   **Transition-state intermediate**

The enzyme HIV protease contains two aspartic acid residues in the active site that play a crucial role in enzymatic activity (*see* above). Specifically, one of the peptide bonds of the substrate (gray box) is attacked by an enzyme activated molecule of water (blue) to form a metastable (transient) intermediate (yellow box), which then undergoes spontaneous peptide bond cleavage to afford acid and amine products. The three-dimensional structure of HIV protease was first determined in

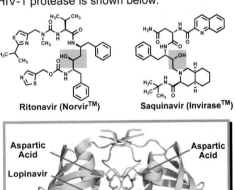

Ritonavir (Norvir™)   Saquinavir (Invirase™)

**Aspartic Acid**   **Aspartic Acid**   **Lopinavir**

1. *Drug Dev. Res.* **2006**, *67*, 501-510; 2. *Science* **1989**, *245*, 616-621; 3. *Bioorg. Med. Chem.* **2002**, *10*, 2803-2806. (1MUI); Refs. p. 178

# UK-427857 (MARAVIROC)

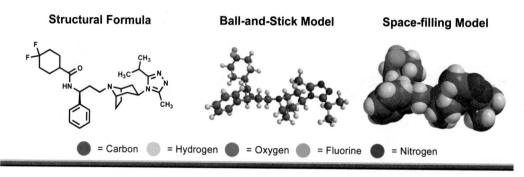

**Structural Formula**     **Ball-and-Stick Model**     **Space-filling Model**

● = Carbon   ● = Hydrogen   ● = Oxygen   ● = Fluorine   ● = Nitrogen

*Year of discovery:* Early 2000s; *Year of introduction:* 2007; *Drug category:* CCR5 receptor antagonist/HIV-entry inhibitor; *Main uses:* For the treatment of HIV/AIDS infected individuals; *Related drugs:* Enfuvirtide (Fuzeon).

Maraviroc is an anti-HIV-1 drug that blocks the attachment of HIV to the CCR5 surface receptors of T-cells, a crucial event in HIV infection.[1]

Chemokine receptors, such as CCR5 and CXCR4, are naturally expressed on the surfaces of immune cells and promote their activation or migration by chemotaxis.[2] These receptors have extracellular, transmembrane and intracellular domains. The former bind natural chemokines and play a role in intracellular activation pathways (*see* more on this in the discussion of the immune system, page 112). The finding that the CCR5 receptor is a key co-receptor for HIV cellular entry for most HIV strains triggered the search for compounds that block this receptor.[3]

Maraviroc was identified at Pfizer after a multistage research effort. Since the structure of CCR5 was unknown, thousands of compounds in the Pfizer compound library were screened to find molecules with an affinity for CCR5. From a number of "hits" that were identified, one molecule, UK-107543, was selected as a lead compound (*see* structure below).

**UK-107543**

During the next phase of drug discovery, approximately 1000 analogs of the lead compound were synthesized and tested for optimum potency and pharmacological and pharmacokinetic properties. Maraviroc was finally selected as a potent inhibitor of CCR5

(0.22 nM) that prevents HIV entry into T-cells and possesses good pharmacological properties.

The mechanism of HIV entry has been studied intensively and is now known in great detail.[2] The surface of HIV is studded with the envelope glycoproteins gp120 and gp41 (*see* below). These glycoproteins have 3 subdomains each. Gp120 protrudes from the surface of the virion, whereas gp41 is embedded in the viral envelope.

**HIV IN 3D REPRESENTATION**

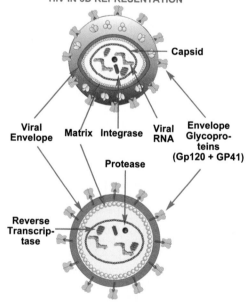

Capsid

Viral Envelope   Matrix   Integrase   Viral RNA   Envelope Glycoproteins (Gp120 + GP41)

Protease

Reverse Transcriptase

**HIV CROSS SECTION**

The genomic material, viral RNA, as well as essential viral enzymes (integrase, protease, and reverse transcriptase) are protected by the capsid.

The crucial stages of viral entry at the molecular level are illustrated below in panels A through D. Initially the gp120 glycoprotein on the surface of HIV-1 particle binds to a CD4 receptor on the surface of a host T-cell. This binding causes the gp120 glycoprotein (light blue) to assume a conformation in which the three subdomains change their shape so as to interact with CCR5, also located on the surface of the T-cells (shown in green).

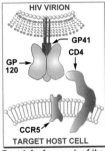

**Panel A:** Approach of the HIV virion to the surface of the target host cell.

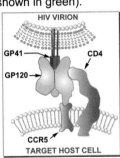

**Panel B:** HIV glycoprotein gp120 binds to the CD4 and CCR5 receptors

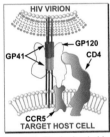

**Panel C:** The hydrophobic fusion domain of gp41 penetrates the membrane of the target host cell.

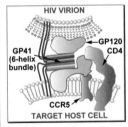

**Panel D:** The hydrophobic and hydrophilic domains of gp41 fold over each other to form a 6-helix bundle hairpin.

The binding to the CCR5 co-receptor brings about a major conformational change in the gp41 glycoprotein that allows the hydrophobic fusion domain (shown in yellow on panel C) to penetrate the membrane of the target host cell. At this point, the gp41 hydrophobic and hydrophilic domains fold upon themselves to form a so-called 6-helix bundle hairpin structure that is stabilized by disulfide bridges (panel D). This hairpin structure pulls the viral membrane into intimate contact with the host cell membrane and induces the fusion of the two membranes. During the fusion process, a pore is formed that is large enough to allow the viral capsid to enter the cytoplasm of the host cell. The capsid is then partially dissolved releasing viral RNA and enzymes (*see* page 152).

Maraviroc is thought to act by locking CCR5 into an *inactive* conformation that is unable to interact with the gp120 envelope glycoprotein as shown in Panel B. This inactive conformation of the maraviroc-CCR5

complex does not lead to cytokine signaling, chemotaxis, or inflammation.

In some individuals (an estimated 10-15% of Europeans) the gene that codes for CCR5 has 32 base pairs missing from the DNA sequence, giving rise to the biosynthesis of dysfunctional CCR5 proteins that are not expressed on the cell surface. This gene variant, referred to as CCR5Δ32, confers almost absolute protection from infection by HIV viruses that use CCR5 for entry (about 90% of all HIV infections). However, when these individuals who are homozygous for the 32 base pair deletion are exposed to HIV viruses that use CXCR4 receptors for entry (X4 viruses), they become infected. On the other hand, individuals who are heterozygous for CCR5Δ32 and not completely immune to HIV infection display much longer disease progression times upon infection. Remarkably, people with CCR5Δ32 mutations are fully immunocompetent, which indicates that the absence of functional CCR5 receptors is not detrimental to health and that CCR5 antagonists such as maraviroc will likely be well-tolerated.

In advance stages of infection, HIV nearly always begins to use the CXCR4 co-receptor for T-cell entry instead of the CCR5 receptor. The cause of the switch in co-receptor affinity (tropism) that results in rapid disease progression has not been demonstrated, but may be due to viral mutation.

An emerging area of research on new anti-AIDS drugs is the inhibition of HIV integrase, an enzyme responsible for the incorporation of viral double-stranded DNA into the DNA of the infected cell. This step is critical in the life cycle of HIV. The integration of viral DNA into cellular DNA enables the HIV to reprogram the infected cell so that it produces multiple copies of HIV. Once integrated into the chromosome of the infected cell, retroviral DNA is copied prior to the division of the infected cell. MK-0518 (below), a compound developed at Merck & Co, is a potent inhibitor of HIV integrase, which is currently in phase III clinical trials. No major adverse effects of MK-0518 have been reported. The drug has been shown to be synergistic with all currently approved anti-HIV drugs.

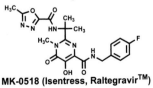

**MK-0518 (Isentress, Raltegravir™)**

1. *Drugs of the Future* **2005**, *30*, 469-477; 2. *JAMA* **2006**, *296*, 815-826; 3. *Drugs* **2005**, *65*, 879-904; Refs. p. 178

*Fungal infection (mycosis) is a persistent medical problem, especially in immunocompromized individuals.*

*Although several types of antifungal therapeutic agents are available, there are still difficult-to-treat conditions.*

# ANTIFUNGAL AGENTS

# AMPHOTERICIN B (FUNGIZONE™)

| Structural Formula | Ball-and-Stick Model | Space-filling Model |
|---|---|---|

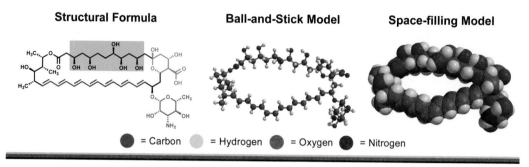

● = Carbon    ○ = Hydrogen    ● = Oxygen    ● = Nitrogen

*Year of discovery:* 1955 (Squibb Institute for Medical Research); *Year of introduction:* 1958; *Drug category:* Polyene macrolide antifungal (antimycotic) agent; *Main uses:* For the intravenous treatment of severe systemic fungal infections; *Related drugs:* Nystatin, Candicidin, Pimaricin, Trichomycin, Abelcet, AmBisome (liposomal fomulation of amphotericin B).

Amphotericin B, a "polyene" macrolide, is a naturally occurring molecule that is used for the treatment of severe systemic fungal infections.[1] Such infections can be life-threatening especially in those whose immune system has been compromised by infection with AIDS, anticancer drugs, or immuno-suppressive therapy. Fungal infections generally do not respond to antibiotics.

Fungi are eukaryotes that have a distinct cell nucleus containing proteins and DNA, as well as various cytoplasmic organelles such as the endoplasmatic reticulum, Golgi apparatus and mitochondria. The structure and biochemical machinery of fungal cells is very similar to that of mammalian cells. However, fungal and mammalian cells differ because fungi have rigid cell walls whereas mammalian cells do not. In addition, fungal membranes contain primarily ergosterol, whereas mammalian membranes contain cholesterol (*see* structures below).

**Cholesterol**      **Ergosterol**

Polyene macrolides, first isolated in the early 1950s from soil microbes, proved to be a highly effective class of antifungal agents.[2] More than 200 antifungal polyene macrolides have been discovered in the interim. Amphotericin B, isolated from *Streptomyces nodosus* in 1955, is a lactone with a 38-membered ring, a heptaene unit (blue), a hemiketal (green) and an aminosugar (red) unit as well as a hydroxyl-rich (polyol) region (highlighted in yellow above). The heptaene

portion of seven double bonds in conjugation is hydrophobic, whereas the rest of the molecule is hydrophilic, rendering the molecule biphilic (amphiphilic). The heptaene part of amphotericin B interacts strongly with hydrophobic ergosterol within the fungal cell membrane. The hydrophilic polyol region of several amphotericin B molecules forms a channel in the membrane that allows the uncontrolled leakage of $Na^+$, $K^+$, and $Ca^{2+}$ ions out of the fungus, eventually leading to cell death (*see* below; only two of 4-6 amphotericins in the aggregate are shown in the picture for clarity). Since amphotericin B has higher affinity for ergosterol than cholesterol, it preferentially binds to fungal cells. However, the dose of amphotericin B must be controlled because that selectivity is limited. There is a need for novel and potent antifungal drugs, since the incidence of severe systemic fungal infections in immunocompro-mized patients has increased and because drug resistance has emerged.

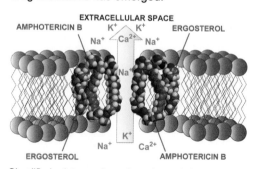

Simplified picture of an ion channel formed from amphotericin B and ergosterol in a fungal membrane.

1. *Int. J. Antimicrob. Agents* **2006**, 27, S12-S16; 2. *Curr. Med. Chem.* **2003**, 10, 211-223; Refs. p. 178

# FLUCONAZOLE (DIFLUCAN™)

| Structural Formula | Ball-and-Stick Model | Space-filling Model |
|---|---|---|

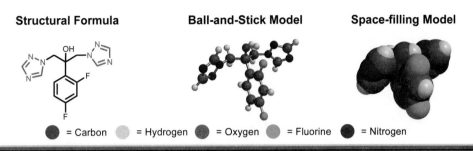

● = Carbon   ○ = Hydrogen   ● = Oxygen   ● = Fluorine   ● = Nitrogen

*Year of discovery:* Mid 1980s (Pfizer); *Year of introduction:* 1990 (both oral and intravenous); *Drug category:* Triazole antifungal drug/Cytochrome P450 14α-demethylase inhibitor; *Main uses:* For the treatment and prevention of superficial and systemic fungal infections; *Related drugs:* Clotrimazole (Lotrimin), Itraconazole (Sporanox), Voriconazole (Vfend), Posaconazole (Noxafil).

Fluconazole, a synthetic bistriazole, is a potent antifungal drug used for fungal infections, both superficial (e.g., candidiasis) and systemic (e.g., cryptococcal meningitis).

Imidazoles, the first orally active, broad-spectrum antifungal agents, were first used as antifungal agents in the early 1970s. Before the discovery of imidazole drugs (e.g., clotrimazole and miconazole, *see* structures below), amphotericin B was the most widely used antifungal agent (*see* page 160). However, the early imidazoles are effective only when used against superficial fungal infections and their oral bioavailability is poor due to their highly lipophilic nature.

Clotrimazole ← Imidazole ring → Miconazole    X, Y = F, Cl, Br

In 1978, Pfizer began a program to find effective, safe and wide-spectrum antifungal agents that were active against both superficial and life-threatening systemic fungal infections and could be administered both orally and intravenously. The most potent antifungal imidazole at that time was ketoconazole, the structure of which was used as the starting point for the design of analogs. A decade and thousands of compounds later, several highly active bis-1,2,4-triazole tertiary alcohols were identified. These compounds all had halogen substituents in the 2 and 4 positions of the phenyl ring (shown in green above). The most potent of the series, the 2,4-difluorophenyl derivative, fluconazole, was safe, water soluble and exceptionally active in humans.[1] The corresponding dichloro derivative (X=Y=Cl) is less active and teratogenic.

Fluconazole acts by specifically inhibiting the cytochrome P450 fungal enzyme C-14α-demethylase. Oxidation of the methyl group at C-14 is one of the 20 enzymatic steps required for the biosynthetic conversion of lanosterol to ergosterol. The 14α methyl group is one of the three methyl groups (highlighted in yellow boxes) that must be removed for the lanosterol to ergosterol transformation. Since ergosterol is a critical component of the fungal cell membrane, inhibition of ergosterol biosynthesis prevents fungal cell replication.

Lanosterol    ~ 20 steps    Ergosterol

Fluconazole binds tightly to the iron (Fe) center of the fungal cytochrome P450 enzyme (*see* X-ray below).[2] One of the nitrogens of fluconazole coordinates to the Fe ion. Unfortunately, triazole-resistant fungal strains have already emerged, necessitating the development of new agents and the use of antifungal combination therapy.[3] A second-generation triazole, voriconazole (Vfend™, Pfizer), which is both potent and active against resistant strains, was approved for human use in 2002.

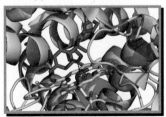

1. *Contemp. Org. Synth.* **1996**, *3*, 125-132; 2. *Proc. Natl. Acad. Sci. U. S. A.* **2001**, *98*, 3068-3073. (**1EA1**); 3. *Drugs* **2005**, *65*, 1461-1480; Refs. p. 178

# CASPOFUNGIN (CANCIDAS™)

| **Structural Formula** | **Ball-and-Stick Model** | **Space-filling Model** |
| --- | --- | --- |

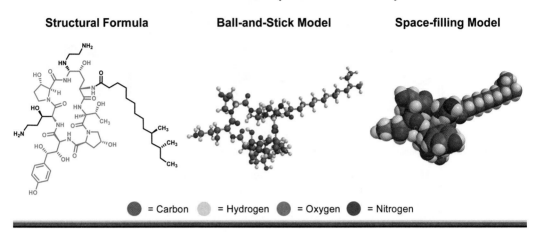

● = Carbon   ● = Hydrogen   ● = Oxygen   ● = Nitrogen

*Year of discovery:* 1989 (Merck); *Year of introduction:* 2001; *Drug category:* Echinocandin/lipopolypeptide antifungal agent/1,3-β-D-glucan synthesis inhibitor; *Main uses:* For the treatment of severe systemic fungal infections (e.g., invasive aspergillosis) especially in those who fail or can not tolerate amphotericin B and/or itaconazole treatment; *Related drugs:* Micafungin (Mycamine), Anidulafungin (Eraxis).

Caspofungin, a semisynthetic cyclic lipid-bearing polypeptide, is an intravenously administered antifungal agent in the echinocandin class of natural products.

Echinocandins were first isolated in 1974 from the fermentation broth of *Aspergillus rugulovavus* and were found to have potent antifungal activity. The predominating compound echinocandin B, was enzymatically degraded to a cyclic hexapeptide core (shown in red) and used to prepare a number of semisynthetic analogs. Caspofungin was the first echinocandin to gain FDA approval in 2001 for the treatment of systemic fungal infections. Invasive fungal infection is a growing problem in health care. It was the 10th leading cause of death in hospitals in 1980 and became the 7th most common cause by the late 1990s.[1] Subsequently, micafungin sodium (Mycamine™, Astellas Pharma) and anidulafungin (Eraxis™, Pfizer) were introduced.[2]

**Micafungin Sodium**
**(Mycamine™)**

**Anidulafungin**
**(Eraxis™)**

Because of the large size of these echinocandins (molecular weights >1200),

and their low oral bioavailability, they are administered intravenously. The three echinocandins have half-lives of 8-13 h and are metabolized mainly in the liver by hydrolysis and N-acetylation to metabolites that do not have antifungal activity. Echinocandins are poor substrates for oxidation by liver cytochrome P450 enzymes (unlike triazoles; *see* page 161), and so few interactions with other drugs have been observed.[3]

Echinocandins are potent inhibitors of the fungal enzyme 1,3-β-D-glucan synthase. Glucans are structurally diverse polymeric sugar molecules that form the rigid cell wall of fungi in combination with protein crosslinks. The function of the fungal cell wall is to protect the cell from changes in osmotic pressure and harmful environmental impacts, but also to allow the influx of nutrients. In contrast, mammalian cells do not have cell walls. Thus, termination of 1,3-β-D-glucan production leads to the disruption of fungal physiology and eventually results in fungal cell death.

Echinocandins are nontoxic at therapeutic doses and generally well-tolerated, but must be used with caution for those with liver abnormalities.

1. *Am. J. Health. Syst. Pharm.* **2006**, *63*, 1693-1703; 2. *Lancet* **2003**, *362*, 1142-1151; 3. *Eur. J. Clin. Microbiol. Infect. Dis.* **2004**, *23*, 805-812; Refs. p. 178

# TERBINAFINE (LAMISIL™)

| Structural Formula | Ball-and-Stick Model | Space-filling Model |
|---|---|---|

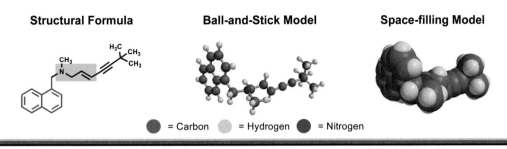

● = Carbon ● = Hydrogen ● = Nitrogen

*Year of discovery:* 1984 (Sandoz); *Year of introduction:* 1996 (Novartis); *Drug category:* Allylamine antifungal agent/squalene epoxidase inhibitor; *Main uses:* For the treatment of fungal infections of the skin (dermatomycosis) and nails (both oral and topical); *Related drugs:* Naftifine, Butenafine.

Terbinafine, a totally synthetic allylamine, is a potent antifungal agent used mainly for the treatment of fungal skin and nail infections.[1] Risk factors for fungal infection include age (>70 years), type 2 diabetes, HIV/AIDS, and immunosuppressive therapy (e.g., for organ transplant and cancer). About half of all nail disorders can be attributed to fungal infection.

The first antifungal allylamine, naftifine, was discovered serendipitously at Sandoz-Werner in 1974 during an investigation of synthetic compounds for activity in the central nervous system. After successful clinical trials, naftifine entered the market as a topical antifungal drug (Exoderil) in 1985. Subsequently, hundreds of analogs were prepared and the antifungal activity was linked to two structural elements: (1) the tertiary allylamine moiety (highlighted in yellow boxes) and (2) the 1-substituted naphthalene ring (shown in blue). Structure-activity correlations led to the development of the more active butenafine (Lotrimin) and eventually terbinafine (Lamisil).

**Naftifine (Naftin™)**    **Butenafine (Lotrimin™)**

Terbinafine is currently the most potent antifungal medicine in the allylamine class. It is active against a wide range of dermatophytes, yeasts and molds. Unlike earlier members of this class, terbinafine can be administered both topically (in the form of a cream) and orally. Terbinafine is a first-line treatment for fungal skin and nail infections, given its lack of toxicity, high potency and the rapidity with which it cures. Terbinafine accumulates in the

skin and around nails, and its concentration remains high even after the cessation of therapy. Consequently, it provides extended protection against fungal relapses.[2]

Terbinafine and all allylamines act by inhibiting the enzyme squalene epoxidase. This enzyme is responsible for the conversion of squalene to 2,3-oxidosqualene, the precursor of all sterols by way of lanosterol (*see* scheme below). Fungi convert lanosterol into ergosterol which is a critical component of fungal cell membranes (*see* page 161). Fungal cell death results not only from the lack of ergosterol but also from the accumulation of squalene.[3]

**Squalene**

Squalene Epoxidase → **Terbinafine**

**2,3-Oxidosqualene**

Lanosterol Cyclase →

**Lanosterol**

**Ergosterol**

To date, terbinafine-resistant fungi have not been reported. However, allylamine-resistance may emerge with more extensive use of this agent. A combination of terbinafine and fluconazole has been effective against triazole-resistant strains.

1. *J. Drug Eval.* **2004**, *2*, 133-155; 2. *Exp. Opin. Pharmacother.* **2005**, *6*, 609-624; 3. *Rev. Contemp. Pharm.* **1997**, *8*, 275-287; Refs. p. 178

*An estimated 2 billion humans are infected with at least one parasite.*

*Malaria causes about 2.5 million deaths annually.*

# ANTIMALARIAL
# AND
# ANTIPARASITIC
# AGENTS

# PARASITIC DISEASES: A FOCUS ON MALARIA

*Parasites* — cellular organisms that can only exist in or on a host — affect almost all forms of life, including animals and humans. An estimated 2 billion humans are infected with at least one parasite. Parasites are classified as endoparasites or ectoparasites depending on whether they live in or on a host. Human parasites generally fall into two main categories: (1) protozoa (one-celled or eukaryotic organisms) and (2) multicellular organisms such as worms (helminths) and insects.

In this section we shall consider only diseases that are caused by endoparasites. Infections by protozoa cause diseases such as malaria, leishmaniasis, sleeping sickness, and chronic diarrhea. Infection by parasitic worms (e.g., roundworms or nematodes, flukes or trematodes, tapeworms or cestodes) results in other diseases, for instance river blindness. Virtually all parts of the human body are vulnerable to parasites. Leishmaniasis affects the skin and spleen; onchocerciasis (river blindness) affects the skin and eyes; giardiasis and crytosporidiosis affect the gut; and malaria affects the blood and central nervous system.

Of all parasitic diseases, malaria is the deadliest since it results in 2.5 million deaths annually. Malaria is caused by the protozoan *Plasmodium* that is spread from human to human by mosquito bite. *Plasmodium* consists of nine species, four of which cause malaria in humans. The life cycle of the most infectious species, *Plasmodium falciparum* is shown below. Human and mosquito hosts are both essential for survival of the parasite. The life cycle starts with an asexual phase and progresses to the later phases that are sexual in nature. The sporozoite forms of the parasite enter the human host through the saliva of the mosquito when it penetrates the skin and withdraws blood. Sporozoites reach the liver in 30 minutes through the blood and infect liver cells. The hepatic stage of the infection lasts for 6-15 days, during which time thousands of daughter parasites, merozoites, are formed. Next, the infected liver cells burst and release thousands of merozoites which go on to infect erythrocytes - red blood cells - and multiply further (erythrocytic stage). The erythrocytes burst and newly formed merozoites are released, causing the generation of cytokines which then lead to fever and chills, the usual symptoms of malaria. Some erythrocytic merozoites enter a sexual phase in which male and female gametocytes are formed. These can be transformed into male and female gametes *only* in the mosquito gut. There, through many steps, new sporozoites are formed and await injection into another human host to continue the life cycle. The drugs that act on the various stages of the cycle are highlighted in yellow boxes in the figure below.[1-3]

1. *Curr. Med. Chem.* **2007**, *14*, 289-314; 2. *Med. Res. Rev.* **2007**, *27*, 65-107; 3. *Mini-Rev. Med. Chem.* **2006**, *6*, 177-202; **Refs. p. 179**

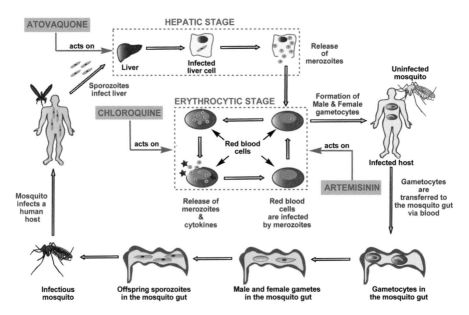

# CHLOROQUINE (ARALEN™)

| Structural Formula | Ball-and-Stick Model | Space-filling Model |
| --- | --- | --- |

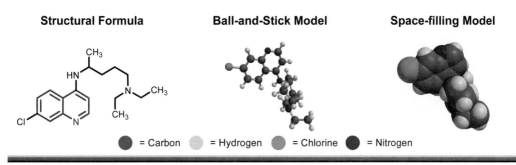

● = Carbon  ● = Hydrogen  ● = Chlorine  ● = Nitrogen

*Year of discovery:* 1934 (Germany) and during 1940s (USA); *Year of introduction:* 1945; *Drug category:* Quinoline antimalarial agent; *Main uses:* For the prophylaxis and treatment of malaria attacks by *Plasmodium vivax, P. ovale,* and *P. malariae.* Also against chloroquine-sensitive strains of *P. falciparum*; *Related drugs:* Quinine, Quinidine, Mefloquine (Lariam), Primaquine.

Chloroquine, a synthetic quinoline derivative first prepared in 1934, is still an important agent for the prevention and treatment of malaria either as monotherapy or in combination with other antimalarial drugs. (The quinoline ring is shown in blue in the structures herein.)

Quinine, a naturally occurring quinoline alkaloid (*see* below) produced by the cinchona tree, has been used as an oral medicine for the prevention of malaria for almost 400 years. Quinine and its stereoisomer quinidine are administered intravenously when used for acute malarial infection. During an intense program of research to eradicate malaria in the mid 20th century, thousands of compounds were synthesized and tested for efficacy in animal models. Pamaquine, an 8-amino substituted quinoline derivative, was identified as a potent antimalarial agent.

**Quinine**

**Quinidine**

**Pamaquine**

**Sontoquine**

Quinine affects the malaria parasite at the stage of its life cycle when it dwells in the bloodstream, whereas pamaquine acts on the parasite in the liver. Consequently, the combination of pamaquine with quinine effectively clears an infection from both the liver and the blood. Pamaquine was followed by second-generation antimalarial agents, sontoquine and chloroquine. Chloroquine was found to be more potent and less toxic than sontoquine and became the most important antimalarial treatment during the 20th century (Aralen™, Sanofi Aventis). It is now a generic drug and is formulated as the phosphate salt, which can be administered both orally and intravenously.

While in the circulation, malaria parasites attack, infect and destroy red blood cells (erythrocytes) and use the proteins and other components therefrom to reproduce.[1] The removal of protein from the oxygen-carrying hemoglobin of erythrocytes leaves behind the oxygen-binding subunit, a heme-iron complex, which is toxic to the parasites. In the absence of chloroquine, this toxin precipitates and is rendered nontoxic to the parasite. Chloroquine interferes with this vital detoxification process.[2] As a consequence, the heme-iron complex remains in solution where it combines with oxygen to form reactive oxygen species that are deadly to the parasite, which is unusually sensitive to oxidative damage. Chloroquine may also act by interfering with enzymes other than the parasite heme polymerase, or even directly by binding to DNA.[3]

Although chloroquine-resistant strains have emerged, chloroquine is still active against the eythrocytic forms of *Plasmodium vivax, Plasmodium malariae* and some strains of *Plasmodium falciparum*. Despite its reduced efficacy, chloroquine is widely used in African countries because of its lower cost. Several new antimalarial drugs are in development.[4]

1. *Tropical Medicine* **2004**, *4*, 97-128; 2. *Coord. Chem. Rev.* **1999**, *190-192*, 493-517; 3. *Mini-Rev. Med. Chem.* **2006**, *6*, 177-202; 4. *Med. Res. Rev.* **2007**, *27*, 65-107; Refs. p. 179

# ARTEMETHER + LUMEFANTRINE (COARTEM™)

| **Structural Formula** | **Ball-and-Stick Model** | **Space-filling Model** |

● = Carbon ○ = Hydrogen ● = Oxygen

*Year of discovery:* 1971 (Shanghai Institute of Materia Medica); *Year of introduction:* 1998; *Drug category:* Trioxane antimalarial; *Main uses:* For the treatment of uncomplicated and severe malaria (*Plasmodium falciparum*) in mono and combination therapies, especially in cases of resistant strains; *Related drugs:* Dihydroartemisinin, Artemether, Arteether, Artesunate, Mefloquine + Artesunate (Artequin), Amodiaquine + Artesunate (ASAQ, Coarsucam).

Artemisinin, a complex natural product, and its derivatives such as dihydroartemisinin, artemether, arteether and sodium artesunate are potent, fast-acting and widely used antimalarial drugs (*see* structures below).

**Artemisinin**

**Dihydroartemisinin**

**Arteether**

**Sodium artesunate**

Artemisinin was first isolated in 1971 from the Chinese medicinal herb *Artemisia annua*, in which it is the main constituent.[1] The herb has been used in China to reduce fever for over a millenium. Once the potent antimalarial activity of artemisinin had been established, several analogs were prepared to improve bioavailability and efficacy, including sodium artesunate which serves as a water-soluble prodrug.

The optimum therapy for malaria depends on several factors including severity of illness, patient age and history and likelihood of resistance to a first-line drug. The recommended standard treatment of uncomplicated malaria involves the oral administration of artemisinin derivatives in combination with longer-acting antimalarial drugs such as lumefantrine and mefloquine.

**Lumefantrine**

**Mefloquine**

The combination of lumefantrine and artemether is highly effective and well-tolerated as an oral antimalarial drug (Coartem™, Novartis) with cure rates above 95%. Artemisinin is produced in China on a large scale by extraction of the dry *Artemisia annua* plant. Increased worldwide demand (>120 million doses) for Coartem led to the establishment of plantations and extraction facilities in Africa as well.

Artemisinins have high efficacy against *Plasmodium falciparum* in the blood stages of the disease. Their mode of action has not been proven, but the endoperoxide (O-O) subunit of the trioxane ring is critical for antimalarial activity. The O-O bond is easily broken by transfer of an electron from $Fe^{2+}$ species and a number of reactive intermediates such as oxygen- and carbon-centered radicals are generated that oxidatively damage the parasite.[2] It may also be relevant that artemisinins strongly inhibit the $Ca^{2+}$-ATPase in the endoplasmatic reticulum of the parasite that is needed for the maintenance of optimal calcium levels.[3]

1. *Med. Chem. Bioact. Nat. Prod.* **2006**, 183-256; 2. *Acc. Chem. Res.* **2004**, *37*, 397-404; 3. *Int. J. Parasitol.* **2006**, *36*, 1427-1441; Refs. p. 179

# ATOVAQUONE + PROGUANIL (MALARONE™)

| Structural Formula | Ball-and-Stick Model | Space-filling Model |
|---|---|---|

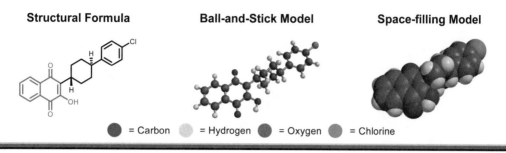

● = Carbon    ● = Hydrogen    ● = Oxygen    ● = Chlorine

*Year of discovery:* 1945 (proguanil), 1980s (atovaquone); *Year of introduction:* 2000 (GlaxoSmith-Kline); *Drug category:* Antiparasitic/hydroxynaphthoquinone; *Main uses:* For the prevention of malaria in travelers and for the treatment of mild-to-moderate cases involving drug-resistant strains of *P. falciparum;* Also as a monotherapy for the treatment of infections caused by the fungus *P. carinii* and the protozoan *T. gondii* especially in immunocompromized patients. *Related drugs:* Pyrimethamine (Daraprim), Chlorproguanil (Lapudrine), Chloroquine (Aralen), Mefloquine (Lariam).

Atovaquone is used in combination with proguanil (Malarone™, GSK) for the prevention and treatment of malaria. The critical part ("pharmacophore") for activity is the naphthoquinone subunit (shown in red above). The combination is often recommended as prophylactic for those traveling to countries where malaria due to *P. falciparum* is common.[1]

A fixed-dose oral combination with proguanil is used because resistance develops quickly with atovaquone alone. Proguanil, a biguanide prodrug, is metabolized to cycloguanil (*see* below) which is a potent and selective inhibitor of the plasmodial enzyme dihydrofolate reductase. Standard treatment with Malarone should begin 2 days before possible exposure and should be continued for a week after final exposure. Malarone is well-tolerated and adverse effects are rare compared to older drugs.

**Mefloquine (Lariam)**

Atovaquone was first developed during the 1980s at Glaxo as a broad-spectrum anti-parasitic agent (marketed as Mepron™) for the treatment of pneumonia caused by *Pneumocystis carinii.* In the 1990s atovaquone was demonstrated to have potent antimalarial activity and to be more effective against the erythrocytic stages of *P. falciparum* than any other known drug.[2] Atovaquone is also effective against the liver stages of the parasitic life cycle, allowing it to be used as a preventive medicine (casual prophylaxis).

Atovaquone functions as a highly lipophilic analog of ubiquinone (coenzyme Q), which is involved in the transport of electrons in parasitic mitochondria. Atovaquone interrupts electron-transfer from dehydrogenase enzymes to cytochromes via ubiquinone and kills the parasite.

**Proguanil**     →  *Metabolism*  →     **Cycloguanil**

Two other commonly used preventive drugs, mefloquine (Lariam™, Roche) and chloroquine (Aralen™, Sanofi Aventis; *see* page 167), often cause side effects in the gastrointestinal and central nervous system. In addition, these older drugs do not always prevent the disease in non-immune individuals and they are ineffective against multidrug-resistant strains.

**Ubiquinone (Coenzyme Q)**

1. *Drugs* **2003**, *63*, 597-623. 2. *Antimicrob. Agents Chemother.* **2002**, *46*, 1163-1173; Refs. p. 179

# MILTEFOSINE (IMPAVIDO™)

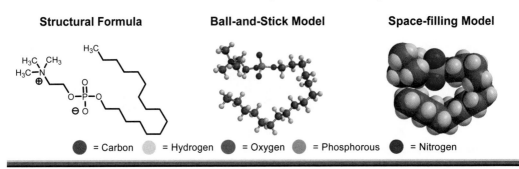

| Structural Formula | Ball-and-Stick Model | Space-filling Model |

● = Carbon   ○ = Hydrogen   ● = Oxygen   ● = Phosphorous   ● = Nitrogen

*Year of discovery:* 1990 (H. Eibl, Germany); *Year of introduction:* 2002 (Zentaris Gmbh); *Drug category:* Antiprotozoal drug/lysophospholipid analogue; *Main uses:* For the treatment of visceral and cutaneous leishmaniasis in immunocompetent patients in India but also for immunocompromized patients in Germany; *Related drugs:* Pentavalent antimony (Sodium Stibogluconate, SSG), Liposomal amphotericin B.

Miltefosine is an oral drug for the treatment of visceral and cutaneous leishmaniasis, a parasitic infection which causes serious illness in many tropical and poorer regions of the world.[1, 2]

Leishmaniasis is caused by the *Leishmania* parasite and it is spread by the bite of the female sandfly *Lutzomyia longipalpis*. There are two main forms of the disease: (1) infection of internal organs (visceral) and (2) infection of the skin (cutaneous). Visceral leishmaniasis affects 500,000 people annually and is usually fatal if left untreated. Each year approximately 1.5 million children and adults develop the two forms of the disease and 70,000 die, mainly from the visceral form.

The disease spreads in a fashion similar to malaria. Female sandflies acquire the amastigote form of the parasite by ingesting blood from infected humans or small mammals. The amastigotes are transformed in the gut of the sandfly to promastigotes, which the sandfly injects into the bloodstream of a human host along with an immunosuppressant. The promastigotes are ingested by white blood cells and revert to amastigotes. In the cutaneous form of the disease the $T_h1$ cells activate phagocytes to destroy the parasite and clear the infection. However, in visceral leishmaniasis, macrophage activation does not occur, and the infection continues.

Phosphatidylcholine is the most abundant phospholipid in animals and plants and the key building block of the lipid bilayer of cell membranes. In the 1980s the anticancer activity of metabolically stable phosphatidylcholine derivatives was discovered and seve-

ral potent compounds such as edelfosine, ilmofosine and miltefosine were prepared.

**Phosphatidylcholine**  **Lysophosphatidylcholine**

**Edelfosine**  **Ilmofosine**

Because these molecules interfere with phospholipid metabolism in cancer cells, phosphatidylcholine analogs were screened against *Leishmania*. This study showed that miltefosine is efficacious against different stages of the parasitic life cycle. It is believed that in addition to inhibiting parasitic phospholipid metabolism, miltefosine may also interfere with membrane signal transduction.

Miltefosine has been available since 2002 as an oral formulation (Impavido). Standard oral treatment of 100 mg per day (2.5 mg/kg/day) for 28 days results in a cure rate of 97% for both forms of leishmaniasis. The use of miltefosine has several advantages over other agents (amphotericin B and sodium stibogluconate) currently used for leishmaniasis: (1) oral vs. parenteral administration; (2) fewer side effects and (3) significantly lower cost.

1. *Mini-Rev. Med. Chem.* **2006**, *6*, 145-151; 2. *Lancet* **2005**, *366*, 1561-1577; **Refs. p. 179**

# NITAZOXANIDE (ALINIA™)

| **Structural Formula** | **Ball-and-Stick Model** | **Space-filling Model** |
| --- | --- | --- |

● = Carbon    ● = Hydrogen    ● = Oxygen    ● = Nitrogen    ● = Sulfur

*Year of discovery:* 1975 (Romark Laboratories); *Year of introduction:* 1996 (in Latin America), 2002 (in US); *Drug category:* Thiazolide antiparasitic agent; *Main uses:* For the treatment of both intestinal protozoal and helminthic infections (e.g., *Giardia intestinalis, Cryptosporidium parvum, Entamoeba histolytica*) that cause diarrhea; *Related drugs:* Tizoxanide, Metronidazole (Flagyl), Tinidazole.

Nitazoxanide, an antiparasitic agent of the thiazolide class, is the first approved drug for the treatment of *Cryptosporidium parvum*-induced diarrhea and the first new drug in 40 years for the treatment of intestinal infections caused by *Giardia intestinalis*.[1] It contains both nitrothiazole (red) and salicylic acid (green) subunits.

Intestinal parasitic infections are the leading cause of death worldwide (ca. 3 million each year), especially in third world countries where sanitation is lacking. Protozoal parasites can be transmitted to uninfected humans by the ingestion of contaminated food and water or by contact with insects, small mammals or other humans. Immunocompromized individuals, the elderly and very young are at increased risk of infection. Two of these diseases, cryptosporidiosis and giardiasis, are the most common protozoal infections in the US. Giardiasis, caused by the protozoan *Giardia intestinalis*, is spread via the fecal-oral route and as a result, children in day care facilities, hospitalized individuals and male homosexuals are at highest risk. Cryptosporidiosis, which causes diarrhea in both animals and humans, is spread by the ingestion of fecally contaminated water and affects travelers, children at day care centers, and those with a weakened immune system. In the developing world, cryptosporidiosis is the main cause of persistent diarrhea and malnutrition in children.

Nitazoxanide, first prepared in 1975, was found to be effective against intestinal tapeworms in animals during the 1980s. Subsequent testing demonstrated that nitazoxanide is a potent inhibitor of over 200 Gram-positive and Gram-negative anaerobic and Gram-positive aerobic microbes.

Nitazoxanide was approved for use in Latin American countries in 1996. In 2002, it entered the US market as an oral suspension (Alinia™, Romark Laboratories) for the treatment of cryptosporidiosis in both adults and children. The current three-day treatment with nitazoxanide (500 mg every 12 hours for adults, 100-200 mg every 12 hours for children) significantly reduces diarrhea and results in the clearance of the microbe. Side effects are rare and usually mild. It has also been shown that treatment for three days with nitazoxanide can substantially reduce the duration of severely dehydrating, rotavirus-caused diarrhea in children.[2]

Nitazoxanide acts by inhibiting the parasitic enzyme pyruvate-ferredoxin oxidoreductase that plays a crucial role in anaerobic energy metabolism. After absorption, nitazoxanide is rapidly converted to the active form of the antibiotic, tizoxanide, by hydrolytic cleavage of the acetyl group ($CH_3CO$; shown in blue above).[3]

**Tizoxanide**

Other available antiprotozoal medications (e.g., metronidazole; *see* page 142) are somewhat toxic at therapeutic doses and may also be ineffective due to the emergence of resistant strains of the parasite.

1. *Clin. Infect. Dis.* **2005**, *40*, 1173-1180; 2. *Lancet* **2006**, *368*, 124-129; 3. *Int. J. Clin. Pharmacol. Ther.* **2000**, *38*, 387-394; Refs. p. 179

# IVERMECTIN (STROMECTOL™)

| Structural Formula | Ball-and-Stick Model | Space-filling Model |
|---|---|---|

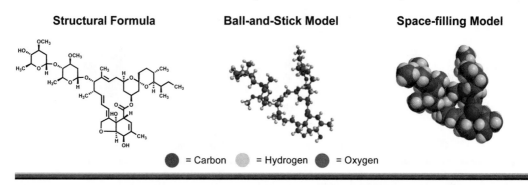

● = Carbon　　● = Hydrogen　　● = Oxygen

*Year of discovery:* Late 1970s (Merck); *Year of introduction:* 1981 (for animal use), 1987 (for human use); *Drug category:* Broad-spectrum antiparasitic, especially effective against nematodes and ectoparasites; *Main uses:* For the treatment of onchocerciasis (river blindness) and strongyloidiasis. Also used in combination with abendazole and diethylcarbamazine against lymphatic filariasis. *Related drugs:* Moxidectin, Milbemycin oxime, Doramectin, Selamectin, Abamectin, Eprinomectin.

Ivermectin, a macrocyclic ester (macrolactone) derived semisynthetically from avermectins, is currently the most effective broad-spectrum, antiparasitic drug for the treatment of human (or animal) infections. Ivermectin is used to prevent the tropical disease river blindness (onchocerciasis).

Onchocerciasis is a serious global health problem which results from infestation by the nematode *Onchocerca volvulus.*[1] The parasite is spread by the bite of the blackfly *Simulium spp.*, which injects immature larval forms of the worms into a human host. Maturation of the larvae usually takes one year. Adult female worms live for up to 15 years and daily produce approximately 1000 larvae (microfilariae), which migrate through the skin causing rash and itching. Microfilariae can migrate over a period of months to the eye where they cause visual impairment and eventually blindness. More than 18 million people become infected annually in Africa and Latin America, and approximately 270,000 lose their eyesight.

In the 1970s, a number of microorganisms were screened for antiparasitic activity by a joint venture of Merck and the Kitasato Institute in Japan.[2,3] One of the microbes, *Streptomyces avermectinius* demonstrated potent activity against a wide range of nematodes and insects and had low mammalian toxicity. Eight structurally similar macrocyclic lactones, the avermectins were responsible for this remarkable activity. The two most potent compounds, avermectin $B_{1a}$ and $B_{1b}$ were 25 times more potent than the most active previously known antiparasitic agents.

Avermectin $B_{1a}$ (R = $CH_2CH_3$)

Avermectin $B_{1b}$ (R = $CH_3$)

The addition of two hydrogens to the C(22)-C(23) double bond (highlighted in a yellow box) yielded dihydro compounds (a mixture of 80% $B_{1a}$ and 20% $B_{1b}$), which had improved oral potency and minimal mammalian toxicity. The mixture, now called ivermectin, was introduced into veterinary practice in 1981. Ivermectin now has annual sales over $1 billion. Based on its efficacy in animals, ivermectin was also screened for activity against human parasites. Human clinical trials, conducted in Africa, revealed that a single annual oral dose of 0.2 mg per kg eradicated microfilarial worms from the eyes and skin after 4 weeks. Ivermectin has been widely used in Africa (Mectizan™) since 1988 for the treatment of river blindness; up to three doses per year of 0.15 mg per kg is recommended. Adverse effects are rare and mild.

Ivermectin acts by opening invertebrate-specific glutamate-gated chloride ion channels found in the nerve and muscle cells of the parasite. This leads to paralysis and death of microfilariae. Although ivermectin does not kill adult worms, it suppresses the production of microfilariae and reduces the worm's lifespan.

1. *Clin. Dermatol.* **2006**, *24*, 176-180; 2. *Nat. Rev. Microbiol.* **2004**, *2*, 984-989. 3. *Comp. Mol. Ins. Sci.* **2005**, *5*, 25-52; Refs. p. 180

# REFERENCES FOR PART IV.

## A Brief Survey of the Immune System (page 112)[1, 2, 4, 5]

## X-Ray Structures (2FL5[6], 1IAK[3])

1. Paul, W. E. & Editor. Fundamental Immunology (**1986**).

2. Cruse, J. M., Lewis, J. R. E. & Editors. Illustrated Dictionary of Immunology (**1995**).

3. Frenont, D.H., Monnaie, D., Nelson, C.A., Hendrickson, W.A., Unanue, E.R. Crystal structure of I-Ak in complex with a dominant epitope of lysozyme. *Immunity* **1998**, *8*, 305-317.

4. Janeway, C. A., Travers, P., Walport, M. & Shlomchik, M. Immunobiology: The Immune System in Health and Disease, 6th Edition (**2004**).

5. Abbas, A. K. & Lichtman, A. H. Cellular and Molecular Immunology, 5th Edition (**2005**).

6. Zhu, X., Wentworth, P., Jr., Kyle, R. A., Lerner, R. A. & Wilson, I. A. Cofactor-containing antibodies: crystal structure of the original yellow antibody. *Proc. Natl. Acad. Sci. U. S. A.* **2006**, *103*, 3581-3585.

## Azathioprine (Imuran, page 122)[1-5]

1. Elion, G. B. The purine path to chemotherapy *Science* **1989**, *244*, 41-47.

2. Tolkoff-Rubin, N. E.; Rubin, R. H. The purine antagonists: azathioprine and mycophenolate mofetil *Ther. Immunol.* **1996**, 44-56.

3. Anstey, A.; Lear, J. T. Azathioprine: clinical pharmacology and current indications in autoimmune disorders *BioDrugs* **1998**, *9*, 33-47.

4. Pirsch, J. D.; David-Neto, E. Mycophenolate mofetil and azathioprine *Current and Future Immunosuppressive Therapies Following Transplantation* **2001**, 85-110.

5. Maltzman, J. S.; Koretzky, G. A. Azathioprine: old drug, new actions *J. Clin. Invest.* **2003**, *111*, 1122-1124.

## Mycophenolate Mofetil (Cellcept, page 123)[1,3-6]

## X-Ray Structure (1JR1)[2]

1. Lipsky, J. J. Mycophenolate mofetil *Lancet* **1996**, *348*, 1357-1359.

2. Sintchak, M. D.; Fleming, M. A.; Futer, O.; Raybuck, S. A.; Chambers, S. P.; Caron, P. R.; Murcko Mark, A.; Wilson, K. P. Structure and mechanism of inosine monophosphate dehydrogenase in complex with the immunosuppressant mycophenolic acid *Cell* **1996**, *85*, 921-930.

3. Bardsley-Elliot, A.; Noble, S.; Foster, R. H. Mycophenolate mofetil: a review of its use in the management of solid organ transplantation *BioDrugs* **1999**, *12*, 363-410.

4. Legendre, C.; Thervet, E. Inosine monophosphate dehydrogenase inhibition: mycophenolate mofetil *Modern Immunosuppressives* **2001**, 77-96.

5. Quiroz, Y.; Herrera-Acosta, J.; Johnson, R. J.; Rodriguez-Iturbe, B. Mycophenolate mofetil treatment in conditions different from organ transplantation *Transplant. Proc.* **2002**, *34*, 2523-2526.

6. Allison Anthony, C.; Eugui Elsie, M. Mechanisms of action of mycophenolate mofetil in preventing acute and chronic allograft rejection *Transplant.* **2005**, *80*, S181-190.

## Cyclosporin (Neoral, page 124)[1-3,5,6]

## X-Ray Structure (1MF8)[4]

1. Kahan, B. D. Cyclosporine: a revolution in transplantation *Transplant. Proc.* **1999**, *31*, 14S-15S.

2. Matsuda, S.; Koyasu, S. Mechanisms of action of cyclosporine *Immunopharmacology* **2000**, *47*, 119-125.

3. Bach, J.-F. Lessons for transplantation of cyclosporine experience in the treatment of autoimmune diseases *Modern Immunosuppressives* **2001**, 201-213.

4. Jin, L.; Harrison, S. C. Crystal structure of human calcineurin complexed with cyclosporin A and human cyclophilin *Proc. Natl. Acad. Sci. U. S. A.* **2002**, *99*, 13522-13526.

5. Calne, R. Cyclosporine as a milestone in immunosuppression *Transplant. Proc.* **2004**, *36*, 13S-15S.

6. Cortesini, R. Cyclosporine-lessons from the first 20 years *Transplant. Proc.* **2004**, *36*, 158S-162S.

## Tacrolimus (Prograf, page 125)[2-6]

## X-Ray Structure (1TCO)[1]

1. Griffith, J. P.; Kim, J. L.; Kim, E. E.; Sintchak, M. D.; Thomson, J. A.; Fitzgibbon, M. J.; Fleming, M. A.; Caron, P. R.; Hsiao, K.; Navia, M. A. X-ray structure of calcineurin inhibited by the immunophilin-immunosuppressant FKBP12-FK506 complex *Cell* **1995**, *82*, 507-522.

2. Carroll, P. B.; Thomson, A. W.; McCauley, J.; Abu-Elmagd, K.; Rilo, H. R.; Irish, W.; McMichael, J.; Van Thiel, D. H.; Starzl, T. E. Tacrolimus and

therapy of human autoimmune disorders *Principles of Drug Development in Transplant. Autoimmun.* **1996**, 211-220.

3. Dumont, F. J. Mechanisms of action of cyclosporine A and tacrolimus (FK506): immunosuppression through immunophilin-dependent inhibition of calcineurin function *Principles of Drug Development in Transplant. Autoimmun.* **1996**, 133-155.

4. Goto, T. Discovery, immunopharmacology and rationale for the development of tacrolimus: a novel immunosuppressant of microbial origin *Principles of Drug Development in Transplant. Autoimmun.* **1996**, 159-163.

5. Bunnapradist, S.; Danovitch, G. Tacrolimus *Modern Immunosuppressives* **2001**, 29-41.

6. Fung, J. J. Tacrolimus and transplantation: a decade in review *Transplant.* **2004**, *77*, S41-S43.

**FTY720 (page 126)[1-5]**

1. Suzuki, S. FTY720: mechanisms of action and its effect on organ transplantation *Transplant. Proc.* **1999**, *31*, 2779-2782.

2. Napoli, K. L. The FTY720 story *Ther. Drug Monit.* **2000**, *22*, 47-51.

3. Brinkmann, V.; Chen, S.; Feng, L.; Pinschewer, D.; Nikolova, Z.; Hof, R. FTY720 alters lymphocyte homing and protects allografts without inducing general immunosuppression *Transplant. Proc.* **2001**, *33*, 530-531.

4. Brinkmann, V.; Pinschewer, D. D.; Feng, L.; Chen, S. FTY720: altered lymphocyte traffic results in allograft protection *Transplant.* **2001**, *72*, 764-769.

5. Kahan, B. D. FTY720: from bench to bedside *Transplant. Proc.* **2004**, *36*, 531S-543S.

**Amoxicillin (Amoxil, page 130)[1, 3-5]**
**X-Ray Structure (1LL9)[2]**

1. Acuna, C. & Rabasseda, X. Amoxicillin-sulbactam: A clinical and therapeutic review. *Drugs of Today* **2001**, *37*, 193-210.

2. Trehan, I., Morandi, F., Blaszczak, L. C. & Shoichet, B. K. Using Steric Hindrance to Design New Inhibitors of Class C β-Lactamases. *Chem. Biol.* **2002**, *9*, 971-980.

3. Bryskier, A. Penicillins. *Antimicrob. Agents* **2005**, 113-162.

4. Neuhauser, M. M. & Danziger, L. H. β-Lactam antibiotics. *Drug Interactions in Infectious Diseases (2nd Edition*, Humana Press **2005**); 255-287.

5. Rolinson, G. N. & Geddes, A. M. The 50th anniversary of the discovery of 6-amino-penicillanic acid (6-APA). *Int. J. Antimicrob. Agents* **2007**, *29*, 3-8.

**Cefaclor (Ceclor, page 132)[2-5]**
**X-Ray Structure (1HVB)[1]**

1. Lee, W. et al. A 1.2-.ANG. snapshot of the final step of bacterial cell wall biosynthesis. *Proc. Natl. Acad. Sci. U. S. A.* **2001**, *98*, 1427-1431.

2. Bryskier, A. & Lebel, M. Oral cephalosporins. *Antimicrob. Agents* **2005**, 222-268.

3. Sonawane, V. C. Enzymatic modifications of cephalosporins by cephalosporin acylase and other enzymes. *Crit. Rev. Biotechnol.* **2006**, *26*, 95-120.

4. Tenover, F. C. Mechanisms of antimicrobial resistance in bacteria. *Am. J. Med.* **2006**, *119*, S3-S10.

5. Mammeri, H. & Nordmann, P. Extended-spectrum cephalosporinases in Enterobacteria-ceae. *Anti-Infec. Agents in Med. Chem.* **2007**, *6*, 71-82.

**Doxycycline (Vibramycin, page 133)[2-5]**
**X-Ray Structure (1I97)[1]**

1. Pioletti, M. et al. Crystal structures of complexes of the small ribosomal subunit with tetracycline, edeine and IF3. *EMBO J.* **2001**, *20*, 1829-1839.

2. Bryskier, A. Tetracyclines. *Antimicrob. Agents* **2005**, 642-651.

3. Charest, M. G., Lerner, C. D., Brubaker, J. D., Siegel, D. R. & Myers, A. G. A Convergent Enantioselective Route to Structurally Diverse 6-Deoxytetracycline Antibiotics. *Science* **2005**, *308*, 395-398.

4. Shlaes, D. M. An update on tetracyclines. *Curr. Opin. Invest. Drugs* **2006**, 7, 167-171.

5. Townsend, M. L., Pound, M. W. & Drew, R. H. Tigecycline: a new glycylcycline antimicrobial. *Int. J. Clin. Pract.* **2006**, *60*, 1662-1672.

**Azithromycin (Zithromax, page 134)[2-5]**
**X-Ray Structure (1NWY)[1]**

1. Schlunzen, F. et al. Structural Basis for the Antibiotic Activity of Ketolides and Azalides. *Structure* **2003**, *11*, 329-338.

2. Abu-Gharbieh, E., Vasina, V., Poluzzi, E. & De Ponti, F. Antibacterial macrolides: a drug class with a complex pharmacological profile. *Pharmacol. Res.* **2004**, *50*, 211-222.

3. Jain, R. & Danziger, L. H. The macrolide antibiotics: A pharmacokinetic and pharmacodynamic overview. *Curr. Pharm. Des.* **2004**, *10*, 3045-3053.

4. Blasi, F. et al. Azithromycin and lower respiratory tract infections. *Exp. Opin. Pharmacother.* **2005**, *6*, 2335-2351.

5. Poehlsgaard, J. & Douthwaite, S. The bacterial ribosome as a target for antibiotics. *Nat. Rev. Microbiol.* **2005**, *3*, 870-881.

## Ciprofloxacin (Cipro, page 135)[1-3, 5]
### X-Ray Structure (2BQ2)[4]

1. Al-Omar, M. A. Ciprofloxacin: drug metabolism and pharmacokinetic profile. *Profiles Drug Subst., Excip. Rel. Meth.* **2004**, *31*, 209-214.

2. Bryskier, A. & Lowther, J. Fluroquinolones and tuberculosis: A review. *Antimicrob. Agents* **2005**, 1124-1145.

3. Greenfield, R. A., Carabin, H., Drevets, D. A. & Gilmore, M. S. Anthrax. *Biodefense* **2005**, 165-216.

4. Siegmund, K. et al. Molecular details of quinolone-DNA interactions: solution structure of an unusually stable DNA duplex with covalently linked nalidixic acid residues and non-covalent complexes derived from it. *Nucleic Acids Res.* **2005**, *33*, 4838-4848.

5. Prats, G., Rossi, V., Salvatori, E. & Mirelis, B. Prulifloxacin: a new antibacterial fluoroquinolone. *Exp. Rev. Anti-Infect. Ther.* **2006**, *4*, 27-41.

## Trimethoprim (Triprim, page 136)[2-5]
### X-Ray Structure (1DG5)[1]

1. Li, R. et al. Three-dimensional Structure of M. tuberculosis Dihydrofolate Reductase Reveals Opportunities for the Design of Novel Tuberculosis Drugs. *J. Mol. Biol.* **2000**, *295*, 307-323.

2. Masters, P. A., O'Bryan, T. A., Zurlo, J., Miller, D. Q. & Joshi, N. Trimethoprim-sulfamethoxazole revisited. *Arch. Intern. Med.* **2003**, *163*, 402-410.

3. Grim, S. A., Rapp, R. P., Martin, C. A. & Evans, M. E. Trimethoprim-sulfamethoxazole as a viable treatment option for infections caused by methicillin-resistant Staphylococcus aureus. *Pharmacotherapy* **2005**, *25*, 253-264.

4. Chan, D. C. M. & Anderson, A. C. Towards species-specific antifolates. *Curr. Med. Chem.* **2006**, *13*, 377-398.

5. Hawser, S., Lociuro, S. & Islam, K. Dihydrofolate reductase inhibitors as antibacterial agents. *Biochem. Pharmacol.* **2006**, *71*, 941-948.

## Amikacin (Amikin, page 137)[1-3, 5]
### X-Ray Structure (2G5Q)[4]

1. Cheer, S. M., Waugh, J. & Noble, S. Inhaled tobramycin (TOBI): a review of its use in the management of Pseudomonas aeruginosa infections in patients with cystic fibrosis. *Drugs* **2003**, *63*, 2501-2520.

2. Magnet, S. & Blanchard, J. S. Molecular Insights into Aminoglycoside Action and Resistance. *Chem. Rev.* **2005**, *105*, 477-497.

3. Davies, J. E. Aminoglycosides: ancient and modern. *J. Antibiot.* **2006**, *59*, 529-532.

4. Kondo, J., Francois, B., Russell, R. J. M., Murray, J. B. & Westhof, E. Crystal structure of the bacterial ribosomal decoding site complexed with amikacin containing the γ-amino-α-hydroxybutyryl (haba) group. *Biochimie* **2006**, *88*, 1027-1031.

5. Li, J. & Chang, C.-W. T. Recent developments in the synthesis of novel aminoglycoside antibiotics. *Anti-Infec. Agents in Med. Chem.* **2006**, *5*, 255-271.

## Vancomycin (Vancocin, page 138)[2-5]
### X-Ray Structure (1FVM)[1]

1. Nitanai, Y., Kakoi, K. & Aoki, K. Complex of Vancomycin with Dl-Acetyl-Lys-D-Ala-D-Ala. *Protein Data Bank* **2001**, (*http://www.pdb.org/*)

2. Kahne, D., Leimkuhler, C., Lu, W. & Walsh, C. Glycopeptide and Lipoglycopeptide Antibiotics. *Chem. Rev.* **2005**, *105*, 425-448.

3. Stevens, D. L. The role of vancomycin in the treatment paradigm. *Clin. Infect. Dis.* **2006**, *42*, S51-S57.

4. Suessmuth, R. D. The chemistry and biology of vancomycin and other glycopeptide antibiotic derivatives. *Med. Chem. Bioact. Nat. Prod.* **2006**, 35-72.

5. Wright, G. D. The antibiotic resistome: the nexus of chemical and genetic diversity. *Nat. Rev. Microbiol.* **2007**, *5*, 175-186.

## Linezolid (Zyvox, page 139)[1-5]

1. Barbachyn, M. R. & Ford, C. W. Oxazolidinone structure-activity relationships leading to linezolid. *Angew. Chem., Int. Ed. Engl.* **2003**, *42*, 2010-2023.

2. Bryskier, A. Oxazolidinones. *Antimicrob. Agents* **2005**, 604-630.

3. Wilcox, M. H. Update on linezolid: the first oxazolidinone antibiotic. *Exp. Opin. Pharmacother.* **2005**, *6*, 2315-2326.

4. Brickner, S. J. Case Histories: Zyvox™ (Linezolid) in *Comprehensive Medicinal Chemistry II* (ed. Triggle, J. B. T. a. D. J.), Vol. *8*, 157-172 (Elsevier-Oxford, **2007**).

5. Zappia, G. et al. The contribution of oxazolidinone frame to the biological activity of pharmaceutical drugs and natural products. *Mini-Rev. Med. Chem.* **2007**, *7*, 389-409.

## Isoniazid (Laniazid, page 140)[2-6]
## X-Ray Structure (1ZID)[1]

1. Rozwarski, D. A., Grant, G. A., Barton, D. H. R., Jacobs, W. R., Jr. & Sacchettini, J. C. Modification of the NADH of the isoniazid target (InhA) from Mycobacterium tuberculosis. *Science* **1998**, *279*, 98-102.

2. Bryskier, A. & Grosset, J. Antituberculosis agents. *Antimicrob. Agents* **2005**, 1088-1123.

3. Liu, J. & Ren, H. P. Tuberculosis: current treatment and new drug development. *Anti-Infec. Agents in Med. Chem.* **2006**, *5*, 331-344.

4. Timmins, G. S. & Deretic, V. Mechanisms of action of isoniazid. *Mol. Microbiol.* **2006**, *62*, 1220-1227.

5. van Helden, P. D. et al. Antimicrobial resistance in tuberculosis: an international perspective. *Exp. Rev. Anti-Infect. Ther.* **2006**, *4*, 759-766.

6. Russell David, G. Who puts the tubercle in tuberculosis? *Nat. Rev. Microbiol.* **2007**, *5*, 39-47.

## Ancillary Antibiotics (page 142)[1-6]

1. O'Grady, F. Cycloserine, amino acid analogs and oligopeptides. *Antibiotic and Chemotherapy (7th Edition)* **1997**, 344-345.

2. Miyachi, Y. Potential antioxidant mechanism of action for metronidazole: implications for rosacea management. *Adv. Ther.* **2001**, *18*, 237-243.

3. Rezanka, T., Jachymova, J. & Spizek, J. Halogenated natural products in Streptomycetes. *Adv. Phytochem.* **2003**, 109-184.

4. Bryskier, A. & Dini, C. Peptidoglycan synthesis inhibitors. *Antimicrob. Agents* **2005**, 377-400.

5. Dubreuil, L. 5-nitroimidazoles. *Antimicrob. Agents* **2005**, 930-940.

6. Fisch, A. & Bryskier, A. Phenicols. *Antimicrob. Agents* **2005**, 925-929.

## Drug Resistance (page 143)[3, 4]
## X-Ray Structures (1VF7, 2GIF, 1TQQ)[1, 2, 5]

1. Akama, H. et al. Crystal Structure of the Membrane Fusion Protein, MexA, of the Multidrug Transporter in Pseudomonas aeruginosa. *J. Biol. Chem.* **2004**, *279*, 25939-25942.

2. Higgins, M. K. et al. Structure of the Ligand-blocked Periplasmic Entrance of the Bacterial Multidrug Efflux Protein TolC. *J. Mol. Biol.* **2004**, *342*, 697-702.

3. Piddock, L. J. V. Multidrug-resistance efflux pumps: Not just for resistance. *Nat. Rev. Microbiol.* **2006**, *4*, 629-636.

4. Rowe-Magnus, D. & Mazel, D. The evolution of antibiotic resistance. *Evolution of Microbial Pathogens* **2006**, 221-241.

5. Seeger, M. A. et al. Structural asymmetry of AcrB trimer suggests a peristaltic pump mechanism. *Science* **2006**, *313*, 1295-1298.

## On Viruses and Viral Diseases (page 146)[1-6]

1. Drake, J. W. & Holland, J. J. Mutation rates among RNA viruses. *Proc. Natl. Acad. Sci. U. S. A.* **1999**, *96*, 13910-13913.

2. Acheson, N. H. Fundamentals of Molecular Virology (John Wiley & Sons, **2006**).

3. Dimmock, N. J., Easton, A. J. & Leppard, K. Introduction to Modern Virology (Blackwell, **2007**).

4. Holmes, E. C. Error thresholds and the constraints to RNA virus evolution. *Trends Microbiol.* **2003**, *11*, 543-546.

5. Baltimore, D. Viruses, Viruses, Viruses in *Engineering & Science*, **2004**, (*http://pr.caltech.edu/periodicals/EandS/articles/LXVII1/Baltimore%20Feature.pdf*)

6. Summers, J. & Litwin, S. Examining the theory of error catastrophe. *J. Virol.* **2006**, *80*, 20-26.

## Acyclovir (Zovirax, page 148)[1-5]

1. Brantley, J. S., Hicks, L., Sra, K. & Tyring, S. K. Valacyclovir for the treatment of genital herpes. *Exp. Rev. Anti-Infect. Ther.* **2006**, *4*, 367-376.

2. Christopher, M. E. & Wong, J. P. Recent developments in delivery of nucleic acid-based antiviral agents. *Curr. Pharm. Des.* **2006**, *12*, 1995-2006.

3. De Clercq, E. & Field, H. J. Antiviral prodrugs - the development of successful prodrug strategies for antiviral chemotherapy. *Br. J. Pharmacol.* **2006**, *147*, 1-11.

4. Golankiewicz, B. & Ostrowski, T. Tricyclic nucleoside analogues as antiherpes agents. *Antiviral Res.* **2006**, *71*, 134-140.

5. Gupta, R. & Wald, A. Genital herpes: antiviral therapy for symptom relief and prevention of transmission. *Exp. Opin. Pharmacother.* **2006**, *7*, 665-675.

## Ribavirin (Virazole, page 149)[1-5]

1. McHutchison, J. G. et al. Interferon alfa-2b alone or in combination with ribavirin as initial treatment for chronic hepatitis C. *N. Engl. J. Med.* **1998**, *339*, 1485-1492.

2. Feld, J. J. & Hoofnagle, J. H. Mechanism of action of interferon and ribavirin in treatment of hepatitis C. *Nature* **2005**, *436*, 967-972.

3. Hoofnagle, J. H. & Seeff, L. B. Peginterferon and ribavirin for chronic hepatitis C. *N. Engl. J. Med.* **2006**, *355*, 2444-2451.

4. Toniutto, P., Fabris, C. & Pirisi, M. Antiviral treatment of hepatitis C. *Exp. Opin. Pharmacother.* **2006**, *7*, 2025-2035.

5. Schiff, E. R. Emerging strategies for pegylated interferon combination therapy. *Nat. Clin. Prac. Gastro. Hepat.* **2007**, *4*, S17-S21.

## Oseltamivir (Tamiflu, page 150)[1-3, 5]
## X-Ray Structure (2HT8)[4]

1. Kim, C. U. et al. Influenza Neuraminidase Inhibitors Possessing a Novel Hydrophobic Interaction in the Enzyme Active Site: Design, Synthesis, and Structural Analysis of Carbocyclic Sialic Acid Analogs with Potent Anti-Influenza Activity. *J. Am. Chem. Soc.* **1997**, *119*, 681-690.

2. Colman, P. Anti-influenza drugs from neuraminidase inhibitors. *Struct.-Based Drug. Disc.* **2006**, 193-218.

3. Laver, G. Antiviral drugs for influenza: Tamiflu past, present and future. *Future Virology* **2006**, *1*, 577-586.

4. Russell, R. J. et al. The structure of H5N1 avian influenza neuraminidase suggests new opportunities for drug design. *Nature* **2006**, *443*, 45-49.

5. Singer, A. C., Nunn, M. A., Gould, E. A. & Johnson, A. C. Potential risks associated with the proposed widespread use of tamiflu. *Environ. Health Perspect.* **2007**, *115*, 102-106.

## Zidovudine (Retrovir, AZT, page 151)[1-5]

1. Mitsuya, H. et al. 3'-Azido-3'-deoxythymidine (BW A509U): an antiviral agent that inhibits the infectivity and cytopathic effect of human T-lymphotropic virus type III/lymphadenopathy-associated virus in vitro. *Proc. Natl. Acad. Sci. U. S. A.* **1985**, *82*, 7096-7100.

2. Tremblay, C. L. & Hirsch, M. S. The basic principles for combination therapy. *Combination Therapy of AIDS* **2004**, 41-51.

3. McIntyre, J. Prevention of mother-to-child transmission of HIV: treatment options. *Exp. Rev. Anti-Infect. Ther.* **2005**, *3*, 971-980.

4. Vivet-Boudou, V., Didierjean, J., Isel, C. & Marquet, R. Nucleoside and nucleotide inhibitors of HIV-1 replication. *Cell. Mol. Life Sci.* **2006**, *63*, 163-186.

5. Milinkovic, A. & Mallolas, J. Fixed-dose combination of abacavir, lamivudine and zidovudine for HIV therapy. *Future Virology* **2007**, *2*, 23-30.

## Zalcitabine (Hivid, page 152)[1-5]

1. Jeffries, D. J. The antiviral activity of dideoxycytidine. *J. Antimicrob. Chemother.* **1989**, *23*, 29-34.

2. Alfano, M. & Poli, G. The HIV life cycle: multiple targets for antiretroviral agents. *Drug. Des. Rev.-Online* **2004**, *1*, 83-92.

3. De Clercq, E. Emerging anti-HIV drugs. *Exp. Opin. Emerg. Drugs* **2005**, *10*, 241-274.

4. Menendez-Arias, L., Matamoros, T., Deval, J. & Canard, B. Molecular mechanisms of resistance to nucleoside analogue inhibitors of human immunodeficiency virus reverse transcriptase. *Drug. Des. Rev.-Online* **2005**, *2*, 101-113.

5. Uckun, F. M. & D'Cruz, O. J. Therapeutic innovations against HIV. *Exp. Opin. Ther. Pat.* **2006**, *16*, 265-293.

## Nevirapine (Viramun, page 153)[4, 5]
## X-Ray Structures (1JLB, 3HVT, 1VRT)[1-3]

1. Wang, J. et al. Structural basis of asymmetry in the human immunodeficiency virus type 1 reverse transcriptase heterodimer. *Proc. Natl. Acad. Sci. U. S. A.* **1994**, *91*, 7242-7246.

2. Ren, J. et al. High resolution structures of HIV-1 RT from four RT-inhibitor complexes. *Nat. Struct. Biol.* **1995**, *2*, 293-302.

3. Ren, J. et al. Structural Mechanisms of Drug Resistance for Mutations at Codons 181 and 188 in HIV-1 Reverse Transcriptase and the Improved Resilience of Second Generation Non-nucleoside Inhibitors. *J. Mol. Biol.* **2001**, *312*, 795-805.

4. Grozinger, K., Proudfoot, J. & Hargrave, K. Discovery and development of nevirapine. *Drug Disc. Dev.* **2006**, *1*, 353-363.

5. Waters, L., John, L. & Nelson, M. Non-nucleoside reverse transcriptase inhibitors: a review. *Int. J. Clin. Pract.* **2006**, *61*, 105-118.

## Efavirenz (Sustiva, page 154)[2-4]
## X-Ray Structure (1FK9)[1]

1. Ren, J. et al. Structural Basis for the Resilience of Efavirenz (DMP-266) to Drug Resistance Mutations in HIV-1 Reverse Transcriptase. *Structure (London)* **2000**, *8*, 1089-1094.

2. Frampton, J. E. & Croom, K. F. Efavirenz/emtricitabine/tenofovir disoproxil fumarate: triple combination tablet. *Drugs* **2006**, *66*, 1501-1512.

3. Arendt, G., de Nocker, D., von Giesen, H.-J. & Nolting, T. Neuropsychiatric side effects of efavirenz therapy. *Exp. Opin. Drug. Safety* **2007**, *6*, 147-154.

4. Vrouenraets, S. M. E., Wit, F. W. N. M., van Tongeren, J. & Lange, J. M. A. Efavirenz: a review. *Exp. Opin. Pharmacother.* **2007**, *8*, 851-871.

## Lopinavir + Ritonavir (Kaletra, page 155)[1, 3-5]
## X-Ray Structure (1MUI)[2]

1. Wlodawer, A. et al. Conserved folding in retroviral proteases: crystal structure of a synthetic HIV-1 protease. *Science* **1989**, *245*, 616-621.

2. Stoll, V. et al. X-ray crystallographic structure of ABT-378 (Lopinavir) bound to HIV-1 protease. *Bioorg. Med. Chem.* **2002**, *10*, 2803-2806.

3. Byrd, C. M. & Hruby, D. E. Viral proteinases: targets of opportunity. *Drug Dev. Res.* **2006**, *67*, 501-510.

4. Mastrolorenzo, A., Rusconi, S., Scozzafava, A. & Supuran, C. T. Inhibitors of HIV-1 protease: 10 years after. *Exp. Opin. Ther. Pat.* **2006**, *16*, 1067-1091.

5. Tan, D. & Walmsley, S. Lopinavir plus ritonavir: a novel protease inhibitor combination for HIV infections. *Exp. Rev. Anti-Infect. Ther.* **2007**, *5*, 13-28.

## UK427857 (Maraviroc, page 156)[1-5]

1. Castagna, A., Biswas, P., Beretta, A. & Lazzarin, A. The appealing story of HIV entry inhibitors: from discovery of biological mechanisms to drug development. *Drugs* **2005**, *65*, 879-904.

2. Ginesta, J. B., Castaner, J., Bozzo, J. & Bayes, M. Maraviroc. *Drugs of the Future* **2005**, *30*, 469-477.

3. Lederman, M. M., Penn-Nicholson, A., Cho, M. & Mosier, D. Biology of CCR5 and its role in HIV infection and treatment. *JAMA* **2006**, *296*, 815-826.

4. Overton, E. T. & Powderly, W. G. Future of maraviroc and other CCR5 antagonists. *Future Virology* **2006**, *1*, 605-613.

5. Citterio, P. & Rusconi, S. Novel inhibitors of the early steps of the HIV-1 life cycle. *Exp. Opin. Invest. Drugs* **2007**, *16*, 11-23.

## Amphotericin (Fungizone, page 160)[1-4]

1. Zotchev, S. G. Polyene Macrolide Antibiotics and their Applications in Human Therapy. *Curr. Med. Chem.* **2003**, *10*, 211-223.

2. Lemke, A., Kiderlen, A. F. & Kayser, O. Amphotericin B. *Appl. Microbiol. Biotechnol.* **2005**, *68*, 151-162.

3. Kleinberg, M. What is the current and future status of conventional amphotericin B? *Int. J. Antimicrob. Agents* **2006**, *27*, S12-S16.

4. Sundriyal, S., Sharma, R. K. & Jain, R. Current advances in antifungal targets and drug development. *Curr. Med. Chem.* **2006**, *13*, 1321-1335.

## Fluconazole (Diflucan, page 161)[1, 3-5]
## X-Ray Strcuture (1EA1)[2]

1. Richardson, K. The discovery of fluconazole. *Contemp. Org. Synth.* **1996**, *3*, 125-132.

2. Podust, L. M., Poulos, T. L. & Waterman, M. R. Crystal structure of cytochrome P450 14α-sterol demethylase (CYP51) from Mycobacterium tuberculosis in complex with azole inhibitors. *Proc. Natl. Acad. Sci. U. S. A.* **2001**, *98*, 3068-3073.

3. Baddley, J. W. & Pappas, P. G. Antifungal combination therapy. Clinical potential. *Drugs* **2005**, *65*, 1461-1480.

4. Grillot, R. & Lebeau, B. Systematic antifungal agents. *Antimicrob. Agents* **2005**, 1260-1287.

5. Kofla, G. & Ruhnke, M. Voriconazole: review of a broad spectrum triazole antifungal agent. *Exp. Opin. Pharmacother.* **2005**, *6*, 1215-1229.

## Caspofungin (Cancidas, page 162)[1-5]

1. Denning, D. W. Echinocandin antifungal drugs. *Lancet* **2003**, *362*, 1142-1151.

2. Theuretzbacher, U. Pharmacokinetics/pharmacodynamics of echinocandins. *Eur. J. Clin. Microbiol. Infect. Dis.* **2004**, *23*, 805-812.

3. Morris, M. I. & Villmann, M. Echinocandins in the management of invasive fungal infections, part 1. *Am. J. Health. Syst. Pharm.* **2006**, *63*, 1693-1703.

4. Morrison, V. A. Echinocandin antifungals: review and update. *Exp. Rev. Anti-Infect. Ther.* **2006**, *4*, 325-342.

5. Turner, M. S., Drew, R. H. & Perfect, J. R. Emerging echinocandins for treatment of invasive fungal infections. *Exp. Opin. Emerg. Drugs* **2006**, *11*, 231-250.

## Terbinafine (Lamisil, page 163)[1-4]

1. Ryder, N. S. & Favre, B. Antifungal activity and mechanism of action of terbinafine. *Rev. Contemp. Pharm.* **1997**, *8*, 275-287.

2. Polak, A. Antifungal therapy - state of the art at the beginning of the 21st century. *Antifungal Agents* **2003**, 59-190.

3. Humphreys, F. Terbinafine. *J. Drug Eval.* **2004**, *2*, 133-155.

4.  Baran, R., Gupta, A. K. & Pierard, G. E. Pharmacotherapy of onychomycosis. *Exp. Opin. Pharmacother.* **2005**, *6*, 609-624.

## Parasitic Diseases: A Focus on Malaria (page 166)[1-3]

1.  Yeh, I. & Altman, R. B. Drug targets for Plasmodium falciparum: a post-genomic review/survey. *Mini-Rev. Med. Chem.* **2006**, *6*, 177-202.

2.  Linares, G. E. G. & Rodriguez, J. B. Current status and progresses made in malaria chemotherapy. *Curr. Med. Chem.* **2007**, *14*, 289-314.

3.  Vangapandu, S. et al. Recent advances in antimalarial drug development. *Med. Res. Rev.* **2007**, *27*, 65-107.

## Chloroquine (Aralen, page 167)[1-5]

1.  Egan, T. J. & Marques, H. M. The role of heme in the activity of chloroquine and related antimalarial drugs. *Coord. Chem. Rev.* **1999**, *190-192*, 493-517.

2.  Abdalla, S. H. Iron and folate in malaria. *Tropical Medicine* **2004**, *4*, 97-128.

3.  Yeh, I. & Altman, R. B. Drug targets for Plasmodium falciparum: a post-genomic review/survey. *Mini-Rev. Med. Chem.* **2006**, *6*, 177-202.

4.  Linares, G. E. G. & Rodriguez, J. B. Current status and progresses made in malaria chemotherapy. *Curr. Med. Chem.* **2007**, *14*, 289-314.

5.  Vangapandu, S. et al. Recent advances in antimalarial drug development. *Med. Res. Rev.* **2007**, *27*, 65-107.

## Artemether + Lumefantrine (CoArtem, page 168)[1-5]

1.  Posner, G. H. & O'Neill, P. M. Knowledge of the Proposed Chemical Mechanism of Action and Cytochrome P450 Metabolism of Antimalarial Trioxanes Like Artemisinin Allows Rational Design of New Antimalarial Peroxides. *Acc. Chem. Res.* **2004**, *37*, 397-404.

2.  Golenser, J., Waknine, J. H., Krugliak, M., Hunt, N. H. & Grau, G. E. Current perspectives on the mechanism of action of artemisinins. *Int. J. Parasitol.* **2006**, *36*, 1427-1441.

3.  Kim, B. J. & Sasaki, T. Recent progress in the synthesis of artemisinin and its derivatives. *Org. Prep. Proced. Int.* **2006**, *38*, 1-80.

4.  Li, Y., Huang, H. & Wu, Y.-L. Qinghaosu (Artemisinin)- a fantastic antimalarial drug from a traditional chinese medicine. *Med. Chem. Bioact. Nat. Prod.* **2006**, 183-256.

5.  Liu, C., Zhao, Y. & Wang, Y. Artemisinin: current state and perspectives for biotechnological production of an antimalarial drug. *Appl. Microbiol. Biotechnol.* **2006**, *72*, 11-20.

## Atovaquone + Proguanil (Malarone, page 169)[1-4]

1.  Baggish, A. L. & Hill, D. R. Antiparasitic agent atovaquone. *Antimicrob. Agents Chemother.* **2002**, *46*, 1163-1173.

2.  McKeage, K. & Scott, L. J. Atovaquone/proguanil: A review of its use for the prophylaxis of plasmodium falciparum malaria. *Drugs* **2003**, *63*, 597-623.

3.  Krungkrai, J. The multiple roles of the mitochondrion of the malarial parasite. *Parasitology* **2004**, *129*, 511-524.

4.  Fotie, J. Quinones and malaria. *Anti-Infec. Agents in Med. Chem.* **2006**, *5*, 357-366.

## Miltefosine (Impavido, page 170)[1-5]

1.  Davis, A. J. & Kedzierski, L. Recent advances in antileishmanial drug development. *Curr. Opin. Invest. Drugs* **2005**, *6*, 163-169.

2.  Murray, H. W., Berman, J. D., Davies, C. R. & Saravia, N. G. Advances in leishmaniasis. *Lancet* **2005**, *366*, 1561-1577.

3.  Berman, J. D. Development of Miltefosine for the Leishmaniases. *Mini-Rev. Med. Chem.* **2006**, *6*, 145-151.

4.  den Boer, M. & Davidson, R. N. Treatment options for visceral leishmaniasis. *Exp. Rev. Anti-Infect. Ther.* **2006**, *4*, 187-197.

5.  Soto, J. & Soto, P. Miltefosine: oral treatment of leishmaniasis. *Exp. Rev. Anti-Infect. Ther.* **2006**, *4*, 177-185.

## Nitazoxanide (Alinia, page 171)[1-4]

1.  Broekhuysen, J., Stockis, A., Lins, R. L., De Graeve, J. & Rossignol, J. F. Nitazoxanide: pharmacokinetics and metabolism in man. *Int. J. Clin. Pharmacol. Ther.* **2000**, *38*, 387-394.

2.  Fox, L. M. & Saravolatz, L. D. Nitazoxanide: a new thiazolide antiparasitic agent. *Clin. Infect. Dis.* **2005**, *40*, 1173-1180.

3.  Farthing, M. J. G. Treatment options for the eradication of intestinal protozoa. *Nat. Clin. Prac. Gastro. Hepat.* **2006**, *3*, 436-445.

4.  Rossignol, J.-F., Abu-Zekry, M., Hussein, A. & Santoro, M. G. Effect of nitazoxanide for treatment of severe rotavirus diarrhoea:

randomised double-blind placebo-controlled trial. *Lancet* **2006**, *368*, 124-129.

## Ivermectin (Stromectol, page 172)[1-5]

1.  Omura, S. & Crump, A. Timeline: Tropical infectious diseases: The life and times of ivermectin - a success story. *Nat. Rev. Microbiol.* **2004**, *2*, 984-989.

2.  Geary, T. G. Ivermectin 20 years on: maturation of a wonder drug. *Trends in Parasitology* **2005**, *21*, 530-532.

3.  Rugg, D., Buckingham, S. D., Sattelle, D. B. & Jansson, R. K. The insecticidal macrocyclic lactones. *Comp. Mol. Ins. Sci.* **2005**, *5*, 25-52.

4.  Wolstenholme, A. J. & Rogers, A. T. Glutamate-gated chloride channels and the mode of action of the avermectin/milbemycin anthelmintics. *Parasitology* **2005**, *131*, S85-S95.

5.  Enk Claes, D. Onchocerciasis--river blindness. *Clin. Dermatol.* **2006**, *24*, 176-180.

# PART V.

*There are more than one hundred types of cancer. Cancer is the second leading cause of death after cardiovascular disease. Approximately 560,000 people die of cancer in the US annually.*

*The concurrent use of multiple therapeutic agents that target the abnormal signaling pathways in a particular patient will lead to a much improved molecular medicine against cancer.*

# MALIGNANT DISEASE

DNA Double Helix

Cisplatin

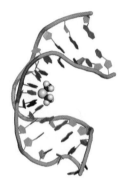

Cisplatin Bound to the
DNA Double Helix

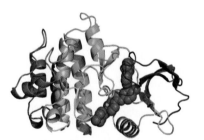

Binding of Gleevec (magenta) to the
Tyrosine Kinase Domain of ABL

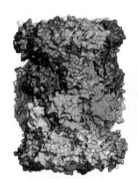

Proteasome

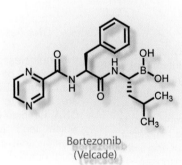

Bortezomib
(Velcade)

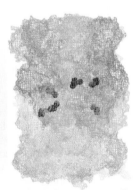

Binding of Six Bortezomib
Molecules to a Proteasome

# AN OVERVIEW OF CANCER

## Introduction

Cancer[1], the unregulated, rapid and pathological proliferation of abnormal cells, is a leading cause of human death (after cardiovascular diseases). This generic term encompasses countless disease states that can involve any human organ, body component and location. Most often affected are lung, intestine, breast, prostate, pancreas, ovary, liver, bladder, blood cells, and bone.

Any individual cancer may fall along a spectrum with regard to the rate of growth, stage of advancement and lethality. At the benign stage the tumor is contained and relatively harmless, but still a threat since it can progress to a malignant form in which proliferation is unconstrained. A localized malignancy can become invasive by forcing its way into surrounding tissues. Invasive tumors inexorably spread via the circulation to other parts of the body. This last stage of cancer – metastatic – is generally fatal.

During each step of this progression cancer cells acquire an increasing number of molecular abnormalities in the critical machinery for cellular regulation. Indeed, biochemical disregulation of the constraints on cell growth and proliferation is the common denominator of all cancers. Especially important is the fact that mutations which increase tumor cell proliferation and survivability are favored by natural selection. Finally, accumulated disregulation may lead to accelerated mutation and runaway disease.

## Causes of Cancer

There are more than 100 types of cancer that are clinically distinctive, and innumerable subsets of each type with regard to the underlying cellular and molecular abnormalities. In addition there are many kinds of small neoplasms that remain benign or are eliminated without becoming malignant. Genetic inheritance can influence the incidence of cancer and predisposes to certain types of cancer, e.g., breast, colon or prostate cancer. Certain viral infections are also known to be a cause of cancer. For instance, there is a strong association of RNA and DNA viral inflection with hepatic cancer (hepatitis B virus), cervical cancer (papilloma virus) and T-cell leukemia (HTLV-1). Bacterial infection by *H. pylori* predisposes to stomach cancer, an indication that persistent microbial infection is also a risk factor.

Many chemicals are strongly carcinogenic, the classic examples being electrophilic organic compounds that can alkylate DNA, e.g., epoxides of polycyclic aromatic (coal tar) hydrocarbons. Naturally produced compounds can also be highly carcinogenic (e.g., the aflatoxins which are formed by microbial action on crops such as peanuts). There are very sensitive tests of tumorgenicity of chemicals, including the Ames test for bacterial mutagenicity.

Most cancers arise from chance events involving changes in cellular DNA (somatic, i.e., non-inherited change). These events include:

(1) collateral damage to DNA caused by reactive chemical fragments produced in the body as a consequence of ordinary metabolism or inflammation;
(2) damage caused by the cells and products of the immune system;
(3) ultraviolet or other radiation;
(4) abnormal products of diseased cells.

Cellular mutations can also occur as a result of errors in DNA copying prior to cell division. Fortunately, such mutations are uncommon for at least three reasons:

(1) the fidelity of DNA replication as measured for DNA polymerase is high (a rough estimate is 1 error per $10^9$ bases);
(2) there is a proofreading mechanism and there are several DNA repair enzymes that detect and correct mismatches in DNA base pairing.

Nonetheless, over a period of time, given the $6 \times 10^9$ bases in the human genome and the billions of cells in the human body, there are bound to be many cells with mutated DNA. Of these, the most harmful will be those which lead to defects in the regulatory proteins that either limit cell proliferation, act as tumor suppressors, or repair mismatched DNA. The products of these harmful mutated genes are oncogenic proteins which cause cell proliferation.

Age is an important risk factor for cancer, which is not surprising since the cumulative burden of DNA mutations of varying origin can only increase with time. However, another contributor to the increase in risk may be the weakening of the immune response beyond middle age. This possibility is supported by the observation that immunosuppression, due to either immunosuppressant medications or infection by HIV, favors tumor cell proliferation and survival.

## Molecular Mechanisms of Cellular Disregulation in Cancer

### I. Disregulation of the Cell Cycle by Abnormal Signaling

During the normal lifetime of a cell it passes through a number of stages, starting with a quiescent or resting state ($G_0$) and progressing to a mitotic phase (M) which eventually leads to cell division. The various states are generally designated as $G_0$, $G_1$, S, $G_2$, and M. There is a "restriction" point (R) in the $G_1$ stage that absolutely determines whether a cell remains in $G_0$ early $G_1$ or goes to the S phase where DNA replication occurs, leading inexorably to the M (mitotic) phase and cell division. Proteins of the pRb group act as the R-point gatekeepers. These R-point proteins allow passage if, and only if, they are heavily phosphorylated (at serine and threonine hydroxyls). The level of phosphorylation, which is so critical to cell cycle progression, is determined by signals emanating outside the cell from the attachment of signaling molecules to cell surface receptors. These signals activate a cascade of tyrosine kinase signaling that results in hyperphosphorylation of the crucial R-point proteins.

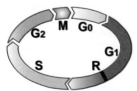

Stages of the cell cycle.

Certain RNA and DNA viruses produce cancer in mammals by transfecting tumor genes - oncogenes - into host DNA. One class of oncogenes produces proteins that are oncogenic because they propagate signaling cascades which defy normal regulation. The Ras oncoprotein, for example, is a mutant tyrosine kinase that is always switched on because it lacks a GTPase domain (*see also* page 195).

The Ras oncoprotein lies at a strategic point of a signaling manifold that affects the transcription of many genes. Signal propagation through Ras leads to:

(1) hyperphosphorylation of the R point protein that opens the gateway to cell division and proliferation;
(2) increased protein synthesis and
(3) inhibition of apoptosis (biochemically programmed cell death of abnormal cells).

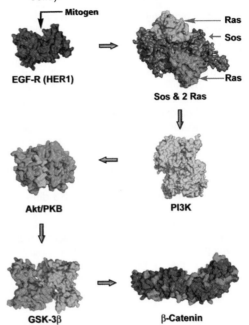

Partial summary of kinases along the HER1-Ras signaling pathway.

Tumor cells often overexpress surface receptors whose ligands (mitogens) promote cell proliferation, or mutated forms of these receptors that respond promiscuously to multiple ligands in addition to the normal mitogen. The epidermal growth factor receptor (EGFR, or HER1 in humans) is prominent in human cancer, either as overexpressed or mutated protein. It is a tyrosine kinase that initiates signaling via autophosphorylation to a form which associates with the proteins Grb2 and Sos. The association of Grb2 and Sos with the Ras protein results in the conversion of a GDP molecule on Ras to GTP. Ras is activated as a kinase in this way. Oncogenic Ras remains in the active state because its mutations disable the GTP → GDP off-switch. Disregulated Ras then continuously activates three proteins on different (and proliferative) signaling pathways, Raf kinase, $PI_3$ kinase and Ral-GEF. The $PI_3$ branch includes the kinases Akt and GSK-3$\beta$ which lead to phosphorylation of the protein $\beta$-catenin. Phosphorylated $\beta$-catenin migrates to the nucleus and activates gene transcription leading to cell proliferation. The Akt pathway results in inhibition of apoptosis.

There are numerous other oncoproteins and multiple signaling pathways that lead to excessive cell proliferation and tumor formation. Although the molecules are all different, the events that occur on these pathways are analogous to those for Ras.

Progression of a cell through the cycle is regulated by many different proteins acting at numerous "check points" on the basis of "instructions" from other proteins. Input of information from these proteins orchestrates the process. Much of the information transfer occurs by kinase-mediated signaling which can become disregulated in tumor cells so as to favor their excessive proliferation.

The power of oncoproteins to cause disregulation is a consequence of their involvement in signaling processes that are highly amplified and that, in the end, lead to the activation of genes favoring cell proliferation.

The human genome codes for 518 different kinases which evolved over about a billion years from an ancestral protein of an early life form. Much effort is now being directed at the development of potent and selective inhibitors of one or just a few of these enzymes. A phylogenetic diagram of the tyrosine kinases in the "kinome" is shown below.[2]

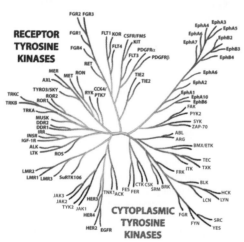

## II. Evasion of Tumor Suppression Mechanisms and Apoptotic Cell Death

Cells are protected against disregulation by the protein p53 which is a transcription factor that can activate genes to produce proteins which either halt cell cycle progression or induce apoptotic cell death. Those properties of p53 and its capacity to sense abnormal cell function make it a very effective tumor suppressor. Cancer cells avoid this obstacle to their propagation in various ways, including:

(1)  production of p53 mutants which are not functional;
(2)  promoting the removal of p53 by overexpression of the enzyme Mdm2 which labels p53 with a ubiquitin marker that leads to protein degradation;
(3)  interfering with the production of p53;
(4)  interfering with the pro-apoptotic pathway;

### Other Mechanisms for Tumor Survival

Normal human cells can reproduce only a limited number of times (about 60 divisions) because the DNA at the ends of chromosomes are capped by a DNA section called a telomere which is shortened each time a cell divides. Tumor cells generally acquire the capacity to regenerate telomeres and thus divide indefinitely.

Tumor cells also generally must evade detection by the immune system. In addition, in order to grow, a massive solid tumor must stimulate the development of blood vessels that can distribute oxygen and nutrients to its cells. Finally, in order to spread, a tumor must express enzymes that allow the breakdown and invasion of surrounding tissue.

### Strategies for the Development of Tailor-Made Anticancer Therapies

Tumors arise over a period of time as a result of the accumulation of mutations that favor cell proliferation and survival. The discovery over the last three decades of many significant biochemical signaling pathways which lead to tumor growth has revealed several promising new avenues for the development of therapeutic agents. It is likely that additional targets of opportunity will emerge in the coming years. Even now, the outlines of future anticancer therapies can be perceived. The starting points will be the development of better biomarkers for cancer and methods for early diagnosis and practical methods for defining the critical mutated genes and proteins in an *individual cancer patient*. The ability to pinpoint the defects underlying a particular person's form of cancer and the concurrent use of multiple therapeutic agents that target the abnormal signaling pathways, will lead to a much improved form of molecular medicine against cancer.

1. Weinberg, R. A. The Biology of Cancer (Garland Science, **2007**); 2. Courtesy of Cell Signaling Technologies Inc. (http://www.cellsignal.com/); **Refs. p. 200**

# CAPECITABINE (XELODA™)

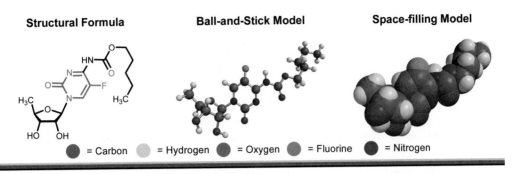

**Structural Formula**     **Ball-and-Stick Model**     **Space-filling Model**

● = Carbon   ○ = Hydrogen   ● = Oxygen   ● = Fluorine   ● = Nitrogen

*Year of discovery:* 1994; *Year of introduction:* 1998 (Hoffmann-La Roche); *Drug category:* Antimetabolite; *Main uses:* Treatment of metastatic breast cancer and colon cancer; *Related drugs:* Tegafur, Floxuridine (Fudr), Fluorouracil (Carac, Efudex, Fluoroplex).

Capecitabine, a cytotoxic inhibitor of cell replication, is used for palliative treatment of breast and colon cancer. It acts as a mimic of uracil, a building block for RNA and DNA synthesis. Although it is toxic to normal host cells, it is much more deadly to tumor cells because of their faster rates of metabolism and cell division.[1]

Capecitabine is transformed after being absorbed in the body to 5-fluorouracil, an anticancer agent that had been developed in the 1950s. Although 5-fluorouracil itself is a potent antiproliferative agent, it is used mainly for intravenous infusion therapy because of low oral bioavailability.

**Uracil**     **5-Fluorouracil**

In the body 5-fluorouracil is converted to fluorodeoxyuridine monophosphate, which binds to the enzyme thymidylate synthase, the normal substrate for which is uridine monophosphate. Thymidylate synthase methylates deoxyuridine monophosphate to form thymidylate, the obligatory precursor to thymidine triphosphate, an essential building block for DNA. Inhibition of thymidylate synthesis blocks DNA production and halts cell division. 5-Fluorouracil is also incorporated into RNA where it obstructs RNA processing and protein synthesis.

Capecitabine, marketed under the name Xeloda by Roche, is used in monotherapy and in combination with docetaxel for invasive breast cancer.[2] The annual sales of capecitabine exceeded $600 million in 2005.

**5-Fluorouracil**     **Fluorodeoxyuridine monophosphate**

Thymidylate synthase

**Deoxyuridine monophosphate**     **Thymidine monophosphate (thymidylate)**

**DNA**

Cytidine mimics such as gemcitabine (Gemzar™, Eli Lilly) also interfere with DNA and RNA biosynthesis. Gemcitabine is efficacious in ameliorating metastatic pancreatic cancer, non-small cell lung cancer and cancers of the ovary, bladder, head and neck. Cladribine (Leustatin™, Ortho Biotech), a purine nucleotide mimic, is used to treat active hairy cell leukemia.

**Gemcitabine (Gemzar™)**     **Cladribine (Leustatin™)**

1. *Clin. Ther.* **2005**, *27*, 23-44; 2. *Anticancer Res.* **2002**, *22*, 3589-3596; Refs. p. 200

# CARBOPLATIN (PARAPLATIN™)

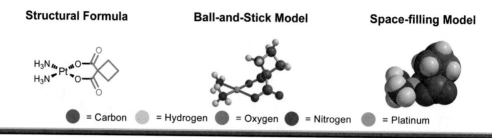

| Structural Formula | Ball-and-Stick Model | Space-filling Model |

● = Carbon   ○ = Hydrogen   ● = Oxygen   ● = Nitrogen   ● = Platinum

*Year of discovery:* 1973; *Year of introduction:* 1989 (Bristol-Myers Squibb); *Drug category:* Platinum-based anticancer agent; *Main uses:* Treatment of advanced-stage ovarian cancer; *Related drugs:* Cisplatin (Platinol), Oxaliplatin (Eloxatin).

Carboplatin, a platinum-based chemotherapeutic agent, is used for the primary treatment of advanced ovarian cancer (in combination with cyclophosphamide) and for palliative treatment of ovarian cancer recurrent after prior chemotherapy.

The development of platinum coordination complexes as anticancer drugs began with the serendipitous discovery in the 1960s that some platinum complexes inhibit bacterial cell division. Subsequent experiments revealed that they also inhibit tumor cell proliferation. The highest antitumor activity was exhibited by *cis*-diamminedichloroplatinum(II), known as cisplatin. Cisplatin, approved in 1978 and introduced by Bristol-Myers Squibb under the name Platinol, has been used against several types of cancers including metastatic ovarian cancer, metastatic testicular cancer and advanced bladder cancer.

**Cisplatin (Platinol™)**      **Oxaliplatin (Eloxatin™)**

The serious side effects associated with the use of cisplatin, which include kidney damage, nerve damage, hearing loss, nausea, and vomiting, spurred the development of improved versions. Carboplatin, available since 1989 under the name Paraplatin, exhibits reduced toxicity and good potency against some cancers that are not susceptible to cisplatin, for instance leukemia and lung cancer. The combination of bleomycin, etoposide, and platins (BEP therapy) can be curative for early bladder cancer.[1]

Oxaliplatin (Eloxatin), another important platinum-based anticancer drug has been approved for the treatment of advanced colorectal cancer in combination with fluorouracil (in 2004).[2] The combined annual sales of platinum based chemotherapeutics exceed $1 billion.

Platins are administered intravenously and diffuse to the tumor cells, where they undergo hydrolysis (loss of chloride or the carboxylate groups) leading to a platinum cation. The positively charged platinum coordinates to the nitrogen atoms of purine bases of DNA (adenine and guanine) forming complexes that distort the DNA helix in such a way that the DNA duplication is hindered (*see* figure below).[3] The inability of the cells to repair the damaged DNA initiates biochemically programmed cell death (apoptosis), which is the final step in cytotoxicity.

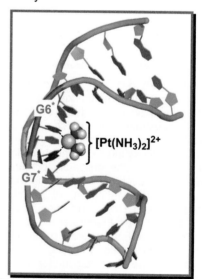

Coordination of platinum-based drugs to DNA.

1. *Nat. Rev. Drug Discov.* **2005**, *4*, 307-320; . *Cancer Invest.* **2001**, *19*, 756-760; *Biochemistry*, **1998**, *37*, 9230-9239 (1A84); **Refs. p. 200**

# VINBLASTINE (VELBAN™)

| Structural Formula | Ball-and-Stick Model | Space-filling Model |
|---|---|---|

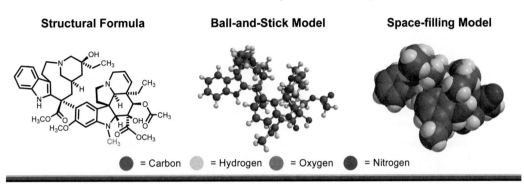

● = Carbon ● = Hydrogen ● = Oxygen ● = Nitrogen

*Year of discovery:* Late 1950s; *Year of introduction:* 1965 (Eli Lilly); *Drug category:* Vinca alkaloid, an inhibitor of microtubule formation; *Main uses:* Treatment of Hodgkin's lymphoma, testicular cancer, Kaposi's sarcoma and other types of cancer; *Related drugs:* Vincristine (Oncovin), Vinorelbine (Navelbine).

Vinblastine, a natural product in the vinca alkaloid family, is used for several types of cancer. It can be a curative therapy for metastatic testicular cancer (in combination with cisplatin and bleomycin). It also is a constituent of the ABVD chemotherapy (dacarbazine, bleomycin, vinblastine, and doxorubicin) against Hodgkin's lymphoma. Other cancers treated with vinblastine include Kaposi's sarcoma, small-cell lung cancer and cancers of the placenta and breast.

Vinblastine, along with a related vinca alkaloid used in chemotherapy, vincristine (Oncovin), was first isolated in the late 1950s from the periwinkle plant of Madagascar. Because extracts of this plant had been found to decrease the number of circulating white blood cells, scientists at Eli Lilly probed the anticancer properties of the extract and showed that the active components, vinblastine and vincristine were effective against leukemia and lymphoma cell lines. A few years later these agents were approved for clinical use in the US. Although the structures of vinblastine and vincristine differ in only one small part (the $CH_3$ shown in red), they are not interchangeable medically. The most important current application of vincristine is the treatment of childhood leukemia in combination with corticosteroids. These chemotherapeutics are both administered intravenously.[1]

Vinblastine and vincristine act to inhibit microtubule formation by binding to tubulin, a structural protein, which is the basic building block of microtubules (*see* figure below), and preventing its aggregation.

**Vincristine (Oncovin™)**

Microtubules are hollow cylindrical protein structures that participate in many important cellular processes. They serve as structural components of cells, and also play a role in intracellular transport processes. Microtubules form the mitotic spindle, which is responsible for the segregation of chromosomes during cell division. Inhibition of microtubule formation arrests cell division at the metaphase stage of the cell cycle.[2]

**Tubulin subunits**

3D Image of an intact microtubule obtained by cryo-electron microscopy by Ken Downing

1. *Anticancer Agents from Natural Products* **2005**, 123-135;
2. *Nat. Rev. Cancer* **2004**, *4*, 253-265; **Refs. p. 200**

# PACLITAXEL (TAXOL™)

| Structural Formula | Ball-and-Stick Model | Space-filling Model |
|---|---|---|

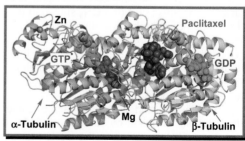

● = Carbon  ○ = Hydrogen  ● = Oxygen  ● = Nitrogen

*Year of discovery:* Late 1960s (National Cancer Institute and Research Triangle Institute); *Year of introduction:* 1992 (Bristol-Myers Squibb); *Drug category:* Microtubule-stabilizing agent; *Main uses:* Treatment of ovarian, breast and non-small cell lung cancer, as well as Kaposi's sarcoma; *Other brand names:* Abraxane; *Related drugs:* Docetaxel (Taxotere).

Paclitaxel, a potent cytotoxic natural product, is mainly used for the treatment of metastatic ovarian and breast cancer. It is also a palliative medicine for AIDS-related Kaposi's sarcoma and non-small cell lung cancer (in combination with cisplatin) in patients who are not candidates for curative surgery or radiation therapy.

Paclitaxel was discovered as a result of a screening program initiated by the National Cancer Institute in the 1960s to identify cytotoxic natural products from plant extracts. Since extracts from the bark of the Pacific yew tree (*Taxus brevifolia*) showed potent cytotoxic activity, the active component, paclitaxel, was isolated and defined structurally by Monroe E. Wall and coworkers at the Research Triangle Institute in the late 1960s. Early clinical development of the drug was sponsored by the National Cancer Institute. In an effort to obtain adequate supplies of paclitaxel and complete Phase III clinical trials, the National Cancer Institute announced a competition for further development of the drug. Bristol-Myers Squibb was selected as the developer in 1991 and paclitaxel was approved and commercialized under the name Taxol in 1992.[1]

Because the supply of paclitaxel from the bark of the slow growing Pacific yew tree was insufficient and controversial, Bristol-Myers Squibb developed a semisynthetic route for the production of the drug that started with 10-deacetylbaccatin, which was obtained from the abundantly available needles of the Himalayan yew tree (*Taxus baccata*). The side chain (red) was attached in a few chemical steps. Currently, paclitaxel is also produced in cell culture using fermentation techniques.

Docetaxel (Taxotere), a related taxane developed by Rhône-Poulenc (later part of Sanofi Aventis), was subsequently approved for the treatment of breast cancer and non-small cell lung cancer. Total sales of paclitaxel and docetaxel are about $4 billion per annum.

**Docetaxel (Taxotere™)**

Paclitaxel acts by stabilizing microtubule structures, a novel mode of action that was elucidated by Susan B. Horwitz at Albert Einstein College of Medicine. Unlike vinblastine, (*see* page 189) which acts by inhibiting microtubule formation, paclitaxel stimulates the formation of these structures. It binds to the β-tubulin subunits (*see* below) leading to microtubules that are hyper-stabilized and unable to disassemble.[2] Since microtubules are essential for the mitotic process, the stabilization of microtubules leads to arrested cell division.[3]

1. *J. Nat. Prod.* **2004**, *67*, 129-135; 2. *J. Mol. Biol.* **2001**, *313*, 1045-1057 (1TUB); 3. *Cancer (New York)* **2000**, *88*, 2619-2628; **Refs. p. 200**

# CYCLOPHOSPHAMIDE (CYTOXAN™)

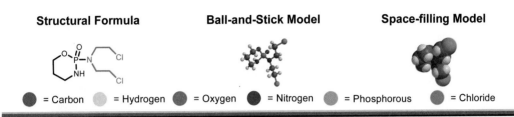

| Structural Formula | Ball-and-Stick Model | Space-filling Model |

● = Carbon    ● = Hydrogen    ● = Oxygen    ● = Nitrogen    ● = Phosphorous    ● = Chloride

*Year of discovery:* 1956; *Year of introduction:* Late 1950s (Asta-Werke AG); *Drug category:* Oxazaphosphorine alkylating agent; *Other brand names:* Neosar; *Main uses:* Treatment of several types of cancer; *Related drugs:* Mechlorethamine (Mustargen), Ifosfamide (Ifex).

Cyclophosphamide is a cytostatic molecule that has been used to slow the progression of several types of cancer, including lymphomas and leukemias, multiple myeloma, neuro- and retinoblastoma and cancer of the breast and ovary. It is usually applied in combination with other anticancer drugs such as methotrexate or fluorouracil. Because of its potent immunosuppressant properties, it has also been employed in the treatment of autoimmune diseases and in the prevention of organ rejection after transplantation.

Cyclophosphamide belongs to the family of "nitrogen mustards", which were the first chemotherapeutics for cancer treatment. Nitrogen mustards were originally developed as lethal chemical warfare agents. Research on the pharmacological action of these substances conducted by Louis Goodman and Alfred Gilman at Yale University in the early 1940s revealed that they are highly toxic to rapidly dividing cells including blood cell precursors in the bone marrow, lymphoid cells, and epithelial cells of the gastrointestinal track. Because of these antiproliferative activities, the effect of these agents on cancer cells was examined and their ability to slow tumor growth was demonstrated. One of the first nitrogen mustards was mechlorethamine (Mustargen), which was primarily used against Hodgkin's lymphoma as a part of a combination regimen.

Cyclophosphamide was developed in the late 1950s at Asta-Werke in Germany in a program to identify more selective nitrogen mustard type antitumor agents.[1] It is a prodrug that is metabolized to a series of compounds that eventually lead to the active species by the pathway shown below.

Cyclophosphamide    →(liver)→    4-Hydroxy-cyclophosphamide    ⇌

Aldophophamide    →    Phosphoramide mustard (ACTIVE METABOLITE)    +    Acrolein

A related drug, ifosfamide (Ifex), also developed at Asta-Werke in the 1960s, is used for the treatment of testicular cancer and several types of sarcomas. Ifosfamide is also a prodrug, but its activation in the body occurs more slowly than the activation of cyclophosphamide accounting for the higher required dosage and different therapeutic spectrum.

Mechlorethamine (Mustargen™)      Ifosfamide (Ifex™)

Nitrogen mustards are very reactive species that form strong C-N covalent bonds with the nitrogen bases of DNA leading to a non-replicating form of DNA and death of the malignant cells.

Mechlorethamine

DNA alkylation

Further activation and alkylation

ALKYLATED DNA        INTERSTRANDED DNA

1. *Cancer* **1996**, *78*, 542-547; **Refs. p. 201**

# TAMOXIFEN (NOLVADEX™)

| Structural Formula | Ball-and-Stick Model | Space-filling Model |
|---|---|---|

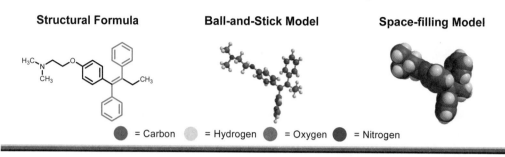

● = Carbon　　● = Hydrogen　　● = Oxygen　　● = Nitrogen

*Year of discovery:* 1962; *Year of introduction:* 1977 (ICI Pharmaceuticals, later part of AstraZeneca)
*Drug category:* Selective estrogen receptor modulator (SERM); *Other brand names:* Soltamox; *Main uses:* Treatment of breast cancer; *Related drugs:* Toremifene (Fareston).

Breast cancer strikes one in every eight women. In 2006, over 200,000 new cases were diagnosed in the US, and more than 40,000 women died, making it the second leading cause of cancer-related deaths among women (after lung cancer). Tamoxifen, a selective estrogen receptor modulator, has been a widely prescribed medication for the treatment of both early and advanced breast cancer in women of all ages. It has also been approved for preventive treatment, since it reduces the risk of breast cancer.

The estrogenic activity of substances that incorporate the diphenylethylene moiety (shown in red in tamoxifen structure) was recognized in the 1930s by James Dodds at the Middlesex Hospital in London. Tamoxifen, first synthesized at ICI Pharmaceuticals in 1962, was originally developed as a possible oral contraceptive, but unexpectedly, it was found to exert the opposite effect and actually induced ovulation in women. It was also discovered serendipitously that it inhibits the proliferation of breast cancer cells.[1]

Tamoxifen behaves as an estrogen receptor antagonist in breast tissue since it blocks the binding of estrogen to one of the two types of estrogen receptors (α). The binding of 4-hydroxytamoxifen, an active metabolite of tamoxifen, to the α-estrogen receptor is depicted below.[2] Tamoxifen exhibits some beneficial estrogenic activity in bone marrow, modestly reducing bone resorption. It acts as an estrogen agonist in uterine tissue, leading to an increased risk of uterine cancer, a concerning aspect of tamoxifen therapy. Its therapeutic profile differs from that of raloxifene (Evista), a synthetic estrogen used for the treatment of osteoporosis, which blocks the β-estrogen receptor more strongly than the α-receptor (*see* page 99).

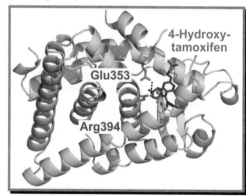

Selective estrogen receptor down-regulators, (SERD), another class of drugs utilized in the treatment of breast cancer, act as pure estrogen antagonists in all tissue, thereby avoiding the undesired estrogenic side effects of selective estrogen receptor modulators. A representative example of this group is fulvestrant (Faslodex, AstraZeneca).

The newest class of breast cancer drugs, aromatase inhibitors, (AI) block the biosynthesis of estrogen in the body. They bind to the aromatase enzyme, which is responsible for the conversion of androstenedione and testosterone to estrogens. This class includes anastrozole (Arimidex™, AstraZeneca), Letrozole (Femara™, Novartis) and exemestane (Aromasin™, Pfizer).

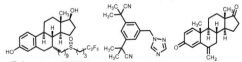

**Fulvestrant (SERD)　Anastrozole (AI)　Exemestane (AI)**

1. *Nat. Rev. Drug Discov.* **2003**, *2*, 205-213; 2. *Cell* **1998**, *95*, 927-937 (**3ERT**); **Refs. p. 201**

# IRINOTECAN (CAMPTOSAR™)

| Structural Formula | Ball-and-Stick Model | Space-filling Model |
| --- | --- | --- |

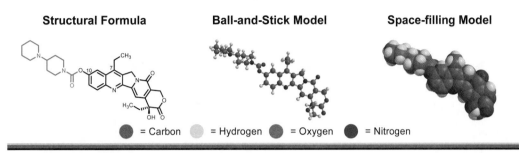

● = Carbon    ● = Hydrogen    ● = Oxygen    ● = Nitrogen

*Year of discovery:* 1985 (Yakult Institute); *Year of introduction:* 1994; *Drug category:* Topoisomerase I inhibitor; *Main uses:* Treatment of colorectal cancer; *Related drugs:* Topotecan (Hycamtin).

Colorectal cancer, the third most common type of cancer in both men and women, affected over 140,000 people in the US in 2006 and accounted for over 55,000 deaths. Early stage colorectal cancer is curable by surgery. Surgical resection followed by chemotherapy with irinotecan, 5-fluorouracil (*see* page 187) and leucovorin (Folfiri) is the treatment of choice for advanced colorectal cancer.

Irinotecan is structurally related to camptothecin, a cytotoxic and antiproliferative natural product that was isolated from the extracts of the Chinese tree *Camptotheca acuminate* in the 1960s by Monroe Wall at the Research Triangle Institute. Clinical trials of camptothecin were conducted in the early 1970s, but were discontinued because of drug toxicity. Further development of camptothecin analogs resumed in the mid 1980s, after researchers at Johns Hopkins University and Smith Kline and French Laboratories (later GlaxoSmithKline) elucidated the mechanism of action of camptothecin.[1]

Irinotecan, a water soluble camptothecin derivative, is a prodrug that is converted in the body to the active agent, 7-ethyl-10-hydroxycamptothecin by cleavage of the side chain (shown in red in structure above). Irinotecan was developed at the Yakult Institute for Microbiological Research in Japan and first approved for use in Japan in 1994. In the US, it was introduced in 1996 under the trade name Camptosar by Pharmacia and Upjohn (later Pfizer).

Topotecan, another semisynthetic water-soluble camptothecin analog, was developed by the Smith Kline and French Laboratories. It was approved for the treatment of ovarian cancer in 1996 and for small-cell lung cancer in 1998.

**Camptothecin**          **Topotecan (Hycamtin™)**

Camptothecin and its analogs act by inhibiting the enzyme topoisomerase I, which plays an important role in DNA replication. DNA normally exists in a supercoiled form, which unwinds during replication, allowing selected regions of the DNA duplex to separate and serve as templates for duplication. Topoisomerase I facilitates the transient cleavage of one strand of DNA, allowing rotation of the broken strand around the intact strand, thereby relieving torsional strain that develops in front of the replication fork. Once the torsional stain is relieved, topoisomerase I reseals the cleavage and dissociates from DNA. Camptothecin analogs stabilize the DNA-topoisomerase I cleavage complex and inhibit the religation step.[2] The collision of the cleavage complex with the replication fork leads to irreversible double-strand breaks that ultimately cause cell death.

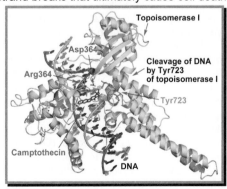

1. *Nat. Rev. Cancer* **2006**, *6*, 789-802; 2. *J. Med. Chem.* **2005**, *48*, 2336-2345 (**1T8I**); Refs. p. 201

# BLEOMYCIN (BLENOXANE™)

| Structural Formula | Ball-and-Stick Model | Space-filling Model |
|---|---|---|

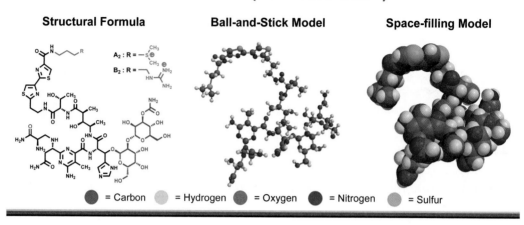

● = Carbon    ● = Hydrogen    ● = Oxygen    ● = Nitrogen    ● = Sulfur

*Year of discovery:* 1962 (Nippon Kayaku); *Year of introduction:* 1969; *Drug category:* Anticancer antibiotic; *Main uses:* Treatment of testicular cancer, head and neck cancer, Hodgkin's and non-Hodgkin's lymphoma; *Related drugs:* Actinomycin D, Daunorubicin, Etoposide, Teniposide.

Bleomycins are a group of microbially produced and structurally related glycopeptide antibiotics that exhibit potent antitumor activity. Members of this family share a common core structure and differ only in one side chain (shown in red in structure above) and in the sugar moieties (shown in blue). The clinically administered drug consists of a mixture of bleomycin $A_2$ (~60%), bleomycin $B_2$ (~30%) and a few other minor components (~10%). Bleomycins are used in combination with other antiproliferative drugs for the treatment of several types of cancer including head, neck, and testicular cancer as well as certain types of lymphoma.

Bleomycins were first isolated from the microbe *Streptomyces verticillus* in 1962 by Hamao Umezawa and colleagues in Japan. They were introduced as intravenously administered anticancer drugs in Japan by Nippon Kayaku in 1969. They were approved for use in the United States in 1973 and then marketed by Bristol Laboratories (later Bristol-Myers Squibb) under the name Blenoxane.

The cytotoxicity of bleomycins is due to their ability to damage the deoxyribose units in DNA, leading to breakage of one or both strands.[1] Cells affected by bleomycins show chromosome aberrations such as breaks, gaps, deletions, and translocations. In their active form, bleomycins complex with an iron ion ($Fe^{2+}$), an oxygen molecule ($O_2$) and a one-electron reductant to form a highly reactive oxygen radical. A model of this complex, in which the reactive $Fe^{2+}$ ion is substituted with the more stable $Co^{3+}$ is shown in the panels below. Panel A depicts a bleomycin complex with intact DNA;[2] panel B shows a complex containing a DNA lesion.[3]

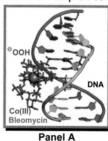
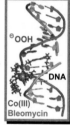

Panel A    Panel B

The first step in the sequence of transformations induced by the active bleomycin complex is hydrogen abstraction from the C4' position of deoxyribose. The resulting radical intermediate then undergoes a series of chemical reactions that eventually lead to cleavage of the DNA.[1]

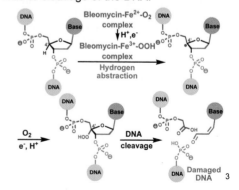

1. *Nat. Rev. Cancer* **2005**, *5*, 102-112; 2. *J. Inorg. Biochem.* **2002**, *91*, 259-268 (**1MXK**); 3. *Biochemistry (Mosc)*. **2001**, *40*, 5894-5905 (**1GJ2**); Refs. p. 201

# IMATINIB (GLEEVEC™)

| Structural Formula | Ball-and-Stick Model | Space-filling Model |

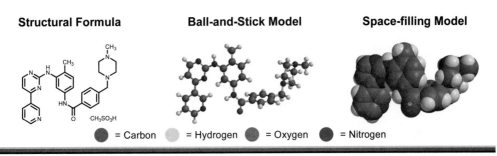

● = Carbon    ○ = Hydrogen    ● = Oxygen    ● = Nitrogen

*Year of discovery:* 1992; *Year of introduction:* 2001 (Novartis); *Drug category:* Protein kinase inhibitor; *Main uses:* Treatment of chronic myelogenous leukemia and gastrointestinal stromal tumors; *Other brand names:* Glivec; *Related drugs:* Gefitinib (Iressa), Erlotinib (Tarceva).

Imatinib, a non-cytotoxic orally-active anticancer agent used for the treatment of chronic myeloid leukemia, gastrointestinal stromal tumors, and a few other malignancies, targets specific enzymes that are required for tumor growth, but are not essential for normal cells. Because of its higher selectivity and decreased toxicity, it represents a major advance over traditional anticancer agents, which act by interfering with DNA synthesis and cell division and also affect healthy cells.

Chronic myeloid leukemia, a relatively rare, but serious form of leukemia, is caused by a single genetic abnormality that results from swapped segments of DNA between chromosomes 9 and 22. This discovery was one of the first linkages between a specific genetic change and human cancer. The fusion of the BCR-ABL genes results in the expression of an enlarged tyrosine kinase (one protein), which is stuck in the activated form and which cannot be regulated. Tyrosine kinases belong to the family of enzymes that transfer a phosphate group from ATP to other proteins. They play a major role in the regulation of cellular processes such as cell division. Constant tyrosine kinase signaling leads to uncontrolled cell proliferation and tumor formation. Due to the involvement of the BCR-ABL tyrosine kinase in the development of chronic myelogenous leukemia, it was proposed by Brian J. Druker to be an attractive target for therapeutic intervention.[1]

Imatinib, a potent and selective inhibitor of BCR-ABL, was identified at the Swiss pharmaceutical company Ciba-Geigy (later Novartis) in 1992 using structure-guided design and chemical synthesis. Although imatinib can be curative for leukemia due to the BCR-ABL mutation, it can also be rendered inactive because of tumor cell mutation of a single amino acid in the kinase. Later studies revealed that it also inhibits other kinases, including c-Kit, which is implicated in gastrointestinal stromal tumors.[2] Imatinib mesylate was approved for the treatment of chronic myeloid leukemia in 2001 and for gastrointestinal stromal tumors in 2002.[3] Sales of imatinib were $1.6 billion in 2004.

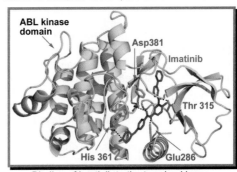

Binding of imatinib to the tyrosine kinase domain of ABL.

Gefitinib (Iressa, AstraZeneca), an inhibitor of the tyrosine kinase domain of the epidermal growth factor receptor (EGFR), has been used for the treatment of non-small cell lung cancer since 2003.

Erlotinib (Tarceva™, Pfizer/OSI/Genentech), an inhibitor of the HER1/EGFR tyrosine kinase, was introduced for the treatment of locally advanced or metastatic non-small cell lung cancer in 2004.

**Gefitinib (Iressa™)**      **Erlotinib (Tarceva™)**

1. *Nat. Rev. Cancer* **2005**, *5*, 172-183; 2. *Cancer Res.* **2002**, *62*, 4236-4243 (**1IEP**); 3. *Nat. Rev. Drug Discov.* **2002**, *1*, 493-502; **Refs. p. 201**

# SUNITINIB (SUTENT™)

| Structural Formula | Ball-and-Stick Model | Space-filling Model |
|---|---|---|

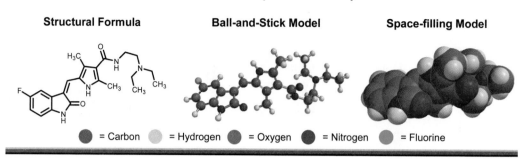

⬤ = Carbon    ⬤ = Hydrogen    ⬤ = Oxygen    ⬤ = Nitrogen    ⬤ = Fluorine

*Year of discovery:* Late 1990s; *Year of introduction:* 2006 (Sugen, later Pfizer); *Drug category:* Selective inhibitor of multiple receptor tyrosine kinases; *Main uses:* Treatment of gastrointestinal stromal tumor and renal cell carcinoma; *Related drugs:* Bevacizumab (Avastin), Sorafenib (Nexavar), Imatinib (Gleevec).

Sunitinib, a selective inhibitor of several receptor tyrosine kinases implicated in tumor growth and pathogenic angiogenesis, has been introduced as therapy for renal cell carcinoma. This common and aggressive type of kidney cancer affects almost 40,000 people and causes more than 10,000 deaths in the US per annum. Interferon-α, a macro-molecular therapeutic agent, produces only modest benefit when used in renal cancer.

Sunitinib is also used for gastrointestinal stromal tumors in patients experiencing disease progression during imatinib therapy (page 195). Gastrointestinal stromal tumors develop in the stromal tissue of the digestive system (most often in the stomach or small intestine). Although these tumors are uncommon (about 1,500 new cases in the US annually), they progress rapidly if left untreated.

Sunitinib was developed at Sugen (later part of Pfizer) and introduced as its maleate salt under the trade name Sutent in 2006.

Patients with renal cell carcinoma often have a genetic disorder, in which the von Hippel-Lindau gene, responsible for encoding a tumor suppressor protein involved in the regulation of the vascular endothelial growth factor (VEGF) and platelet-derived growth factor (PDGF), is mutated. The binding of these growth factors to receptor tyrosine kinases stimulates tumor angiogenesis. Angiogenesis, the generation of new capillaries from pre-existing blood vessels, plays an important role in the progression of cancer. Neovascularization is crucial for tumor growth, which is dependant on a supply of $O_2$

and nutrients, and for tumor metastasis, which requires the development of new blood vessels, since malignant cells must enter the circulation in order to spread. The inhibition of VEGF and PDGF receptors leads to impaired tumor angiogenesis, and is efficacious for cancer therapy.

A large proportion of patients with gastrointestinal stromal tumors have mutations in the c-Kit gene. This genetic defect results in the expression of an overactivated c-Kit protein leading to high cell division rates. Sunitinib blocks the action of c-Kit by binding to the target tyrosine kinase. Sunitinib also inhibits the platelet-derived growth factor receptor PDGFR-α, which is overexpressed in many gastrointestinal tumors.[1]

Sorafenib (Nexavar™, Bayer), another PDGF and VEGF receptor kinase inhibitor, is also available for the treatment of renal cell carcinoma.

**Sorafenib (Nexavar™)**

Bevacizumab (Avastin™, Genentech), a monoclonal antibody that blocks the action of VEGF, is an effective treatment for metastatic colorectal cancer in combination with various chemotherapeutic agents. Bevacizumab was introduced in 2004 as the first selective macromolecular angiogenesis inhibitor.

1. *Drugs* **2006**, *66*, 2255-2266; Refs. p. 202

# BORTEZOMIB (VELCADE™)

| **Structural Formula** | **Ball-and-Stick Model** | **Space-filling Model** |
|---|---|---|

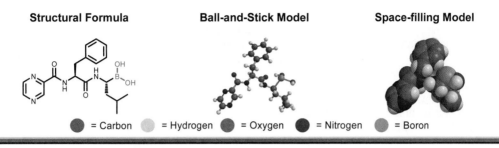

● = Carbon    ● = Hydrogen    ● = Oxygen    ● = Nitrogen    ● = Boron

*Year of discovery:* 1996; *Year of introduction:* 2003 (Millennium Pharmaceuticals); *Drug category:* Proteasome inhibitor; *Main uses:* Treatment of multiple myeloma and mantle cell lymphoma.

Bortezomib, the first member of a new class of drugs called proteasome inhibitors, is mainly used for the treatment of multiple myeloma. In 2006, over 16,000 people in the US were diagnosed with this disease, which is generally fatal within a year if not treated. Multiple myeloma is a malignancy of bone marrow plasma cells (myeloma cells) that leads to disrupted bone marrow function, reduction of red blood cell levels (and also white blood cells), suppression of immune function, and severe bone erosion. Bortezomib is also employed for mantle cell lymphoma, a rare but serious subtype of non-Hodgkin's lymphoma that causes malignant transformation of B-lymphocytes in lymph nodes. Bortezomib was developed by Millennium Pharmaceuticals and introduced in 2003 under the trade name Velcade.

Bortezomib acts by inhibiting the proteasome, a complex assembly of protein subunits present in both normal and cancer cells. Proteasomes provide the major pathway for degradation of damaged or superfluous proteins by chain cleavage at multiple points. They also participate in the regulated breakdown of proteins crucial in cell cycle progression, apoptosis, and generation of antigenic peptides of the immune system. Proteasome inhibition causes downregulation of transcription factors such as NF-κB, upregulation of the cell cycle regulatory proteins p27 and p53, and upregulation of apoptosis. Proteasomes are cylindrical structures that are composed of two major assemblies. The core 20S assembly, which is responsible for catalytic activity, contains 28 protein subunits organized in four stacked rings, each ring incorporating 7 proteins. This is capped by the 19S regulatory assembly, which recognizes proteins tagged for degradation by ubiquitin, a small protein, and

initiates degradation. The 20S substructure of a yeast proteasome is depicted below.[1] The unique boronic acid moiety of bortezomib (shown in red above) is critical for the high affinity of the drug at the active site of the enzyme.[2,3]

Surface and mesh representation of the 20S core of proteasome.

Thalidomide (Thalomid™, Celgene) is also efficacious against multiple myeloma. Originally it was introduced as a sedative/hypnotic agent in the 1960s, but had to be withdrawn, because it caused malformation of the fetus when taken by pregnant women. It is also a potent anti-inflammatory agent and is used for the treatment of leprosy. A derivative of thalidomide, lenalidomide (Revlimid, Celgene) was approved for multiple myeloma in 2004.

Since multiple myeloma can cause bone damage, zoledronic acid (Zometa™, Novartis), a bisphosphonate drug, is often prescribed to patients to inhibit bone resorption and prevent fractures.

**Thalidomide** (Thalomid™)   **Lenalidomide** (Revlimid™)   **Zoledronic acid** (Zometa™)

1. *Structure* **2006**, *14*, 451-456 (**2F16**); 2. *Nat. Rev. Cancer* **2004**, *4*, 349-360; 3. *Nat. Clin. Pract. Oncol.* **2006**, *3*, 374-387; **Refs. p. 202**

# ANCILLARY ANTICANCER AGENTS

Doxorubicin, the most widely used anthracycline anticancer antibiotic, is efficacious against a wide range of cancers, including several types of lymphoma and leukemia, as well as breast, ovary, bladder, stomach, and thyroid gland cancer.[1]

Doxorubicin, first isolated in 1967 from the fungus *Streptomyces peucetius* at Farmitalia in Milan, was introduced in the US in 1974 under the name Adriamycin by Upjohn. The discovery of doxorubicin was preceded by the isolation of its deoxygenated relative, daunorubicin (Cerubidine), in 1964.

**Doxorubicin (Adriamycin™)**

**Daunorubicin (Cerubidine™)**

Daunorubicin is mainly used for acute myeloid leukemia, the most common form of leukemia in adults. Several related anthracyclines have subsequently been developed, including idarubicin (Idamycin) for the treatment of acute myeloid leukemia, and epirubicin (Ellence) for the treatment of several types of solid tumors. Epirubicin differs from doxorubicin in the spatial arrangement of the hydroxyl group (shown in green); idarubicin lacks the methoxy group present in daunorubicin (shown in red).

Anthracyclines act by intercalating with DNA in a way that causes disruption of transcription and replication.[2] They also form a ternary complex with DNA and the enzyme topoisomerase II. Topoisomerase II is essential for DNA replication. It relieves torsional strain by cleaving and religating double stranded DNA so as to allow strand passing and uncoiling of supercoiled DNA. Complexation of anthracyclins with topoisomerase II and DNA hinders religation and leads to tumor cell apoptosis.

The most serious and dose-limiting side effect of anthracyclines is cardiotoxicity. Anthracyclines react with iron in cells to form complexes which promote the formation of oxygen and carbon centered radicals. These very reactive species cause tissue damage and are thought to be responsible for cardiotoxicity.

Doxorubicin is available in a liposome-encapsulated form (Doxil) for the treatment of Kaposi's sarcoma.[3] Doxorubicin is a component of the ABVD (adriamycin, bleomycin, vinblastine, dacarbazine), the CHOP (cyclophosphamide, adriamycin, vincristine, prednisone), and FAC (5-fluorouracil, adriamycin, cyclophosphamide) anticancer regimens.

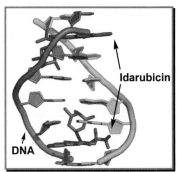

DNA stacking of idarubicin.

Tumors treated with anthracyclines can become multidrug resistant and even cross-resistant to several structurally unrelated drugs.[4,5] Multidrug resistance is a major impediment to successful chemotherapy. Multidrug resistance results mainly from the overexpression of ATP-dependant efflux pumps which favor the survival of resistant cell lines. These nonselective efflux pumps belong to the ATP-binding cassette (ABC) transporter family of glycoproteins.[6] Multidrug resistance also occurs with other agents (e.g., paclitaxel and vinblastine).

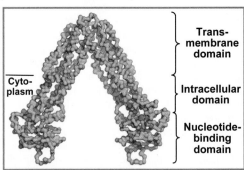

Structure of MsbA protein form *E. coli*, a homolog of the multidrug resistant ABC transporters.

Mitomycin C, an anticancer antibiotic commercialized by Bristol-Myers Squibb under the name Mutamycin, was isolated from a bacterial culture in the late 1950s at Kyowa, in Japan. It is used mainly in combination with other anticancer agents for the treatment of cervical, stomach, breast, bladder, head, neck and lung cancer.[7]

**Mitomycin C (Mutamycin™)**

Mitomycin C is an alkylating agent that, after activation in cells, forms crosslinks between DNA strands, thereby blocking DNA synthesis (*see* figure below).[8,9]

**Mitomycin C**

DNA alkylation

cross-linking

Alkylated DNA complex formed by covalent linkage of mitomycin to guanine of DNA.

Vorinostat, also known as suberoylanilide hydroxamic acid or SAHA, was introduced in 2006 by Merck under the name Zolinza as a second-line therapy for cutaneous T-cell lymphoma, a rare subtype of non-Hodgkin's lymphoma. In cutaneous T-cell lymphoma, the malignant T-cells migrate to the surface of the skin, leading to skin lesions.[10]

**Vorinostat (Zolinza™)**

Vorinostat is the first anticancer agent that acts by inhibiting the enzyme histone deacetylase (HDAC), which catalyzes the removal acetyl groups from the N-acetylated lysine residues of histones. Certain tumors overexpress histone deacetylases. Histones are the basic protein building blocks of chromatins, which contain DNA coiled around the histone units.[11] They compact DNA into a smaller volume, and also play a role in DNA replication, repair, mitosis, and gene expression. Chromatin structure and function depends on the acetylation and deacetylation of histones. Inhibition of histone deacetylases leads to accumulation of acetylated histones, which causes cell cycle arrest and apoptosis.[12]

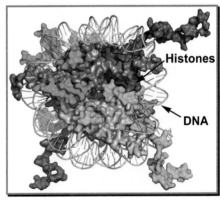

Histone-DNA complexation forming the nucleosome, the repeat element of chromatin.

Ixabepilone (Bristol-Myers Squibb), a semisynthetic derivative of the natural product epothilone A, is an anticancer agent currently in clinical development. Ixabepilone acts by stabilizing microtubule formation, as does paclitaxel. Ixabepilone was shown to be effective against tumor cell lines that are resistant to paclitaxel.[13]

**Ixabepilone**

1. *Expert Opinion on Drug Discovery* **2006**, *1*, 549-568; 2. *Nucleic Acids Res.* **1995**, *23*, 1710-1716 (**198D**); 3. *Nature (London)* **1996**, *380*, 561-562.

4. *Nat. Rev. Cancer* **2002**, *2*, 48-58; 5. *Nat. Rev. Drug Discov.* **2006**, *5*, 219-234; 6. *Science* **2001**, *293*, 1793-1800.

7. *Cancer Treat. Rev.* **2001**, *27*, 35-50; 8. *J. Mol. Biol.* **1995**, *247*, 338-359 (**199D**); 9. *Chem. Biol.* **1995**, *2*, 575-579.

10. *Oncogene* **2007**, *26*, 1351-1356; 11. *J. Mol. Biol.* **2002**, *319*, 1097-1113 (**1KX5**); 12. *Nat. Clin. Pract. Onc.* **2005**, *2*, 150-157.

13. *Expert Opin. Inv. Drug.* **2006**, *15*, 691-702.

Refs. p. 202

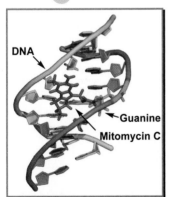

DNA

Guanine
Mitomycin C

# REFERENCES FOR PART V.

## An Overview of Cancer (page 184)[7]
### X-Ray Structures (1M14[3], 1NVU[5], 1E8X[1], 1O6L[4], 1I09[2], 1V18[6])

1. Walker, E. H. et al. Structural determinants of phosphoinositide 3-kinase inhibition by wortmannin, LY294002, quercetin, myricetin, and staurosporine. *Mol. Cell* **2000**, *6*, 909-919.

2. ter Haar, E. et al. Structure of GSK3beta reveals a primed phosphorylation mechanism. *Nat. Struct. Biol.* **2001**, *8*, 593-596.

3. Stamos, J., Sliwkowski, M. X. & Eigenbrot, C. Structure of the Epidermal Growth Factor Receptor Kinase Domain Alone and in Complex with a 4-Anilinoquinazoline Inhibitor. *J. Biol. Chem.* **2002**, *277*, 46265-46272.

4. Yang, J. et al. Crystal structure of an activated Akt/Protein Kinase B ternary complex with GSK3-peptide and AMP-PNP. *Nat. Struct. Biol.* **2002**, *9*, 940-944.

5. Margarit, S. M. et al. Structural evidence for feedback activation by Ras.GTP of the Ras-specific nucleotide exchange factor SOS. *Cell* **2003**, *112*, 685-695.

6. Ha, N.-C., Tonozuka, T., Stamos, J. L., Choi, H.-J. & Weis, W. I. Mechanism of phosphorylation-dependent binding of APC to β-catenin and its role in b-catenin degradation. *Mol. Cell* **2004**, *15*, 511-521.

7. Weinberg, R. A. The Biology of Cancer (Garland Science, **2007**).

## Capecitabine (Xeloda, page 187)[1-5]

1. Pentheroudakis, G.; Twelves, C. The rational development of capecitabine from the laboratory to the clinic *Anticancer Res.* **2002**, *22*, 3589-3596.

2. Walko, C. M.; Lindley, C. Capecitabine: A review *Clin. Ther.* **2005**, *27*, 23-44.

3. Ershler, W. B. Capecitabine monotherapy: safe and effective treatment for metastatic breast cancer *Oncologist* **2006**, *11*, 325-335.

4. Schmoll, H.-J.; Arnold, D. Update on capecitabine in colorectal cancer *Oncologist* **2006**, *11*, 1003-1009.

5. Schellens Jan, H. M. Capecitabine *Oncologist* **2007**, *12*, 152-155.

## Carboplatin (Paraplatin, page 188)[1,3-5]
### X-Ray Structure (1A84)[2]

1. Zee-Cheng, R. K. Y.; Cheng, C. C. Carboplatin - a second-generation platinum analog *Drugs of Today* **1987**, *23*, 563-573.

2. Gelasco, A.; Lippard, S. J. NMR Solution Structure of a DNA Dodecamer Duplex Containing a cis-Diammineplatinum(II) D(GpG) Intrastrand Cross-Link, the Major Adduct of the Anticancer Drug Cisplatin *Biochemistry (Mosc).* **1998**, *37*, 9230-9239.

3. Lokich, J.; Anderson, N. Carboplatin versus cisplatin in solid tumors: an analysis of the literature *Ann. Oncol.* **1998**, *9*, 13-21.

4. Lokich, J. What is the "best" platinum: cisplatin, carboplatin, or oxaliplatin? *Cancer Invest.* **2001**, *19*, 756-760.

5. Wang, D.; Lippard, S. J. Cellular processing of platinum anticancer drugs *Nat. Rev. Drug Discov.* **2005**, *4*, 307-320.

## Vinblastine (Velban, page 189)[1-5]

1. Noble, R. L. The discovery of the vinca alkaloids - chemotherapeutic agents against cancer *Biochem. Cell Biol.* **1990**, *68*, 1344-1351.

2. Duflos, A.; Kruczynski, A.; Barret, J.-M. Novel aspects of natural and modified Vinca alkaloids *Current Medicinal Chemistry: Anti-Cancer Agents* **2002**, *2*, 55-70.

3. Jordan, M. A.; Wilson, L. Microtubules as a target for anticancer drugs *Nat. Rev. Cancer* **2004**, *4*, 253-265.

4. Gueritte, F.; Fahy, J. The vinca alkaloids *Anticancer Agents from Natural Products* **2005**, 123-135.

5. Zhou, J.; Giannakakou, P. Targeting microtubules for cancer chemotherapy *Current Medicinal Chemistry: Anti-Cancer Agents* **2005**, *5*, 65-71.

## Paclitaxel (Taxol, page 190)[1,2,4-6]
### X-Ray Structure (1TUB)[3]

1. Crown, J.; O'Leary, M. The taxanes: an update *Lancet* **2000**, *355*, 1176-1178.

2. Wang, T. H.; Wang, H. S.; Soong, Y. K. Paclitaxel-induced cell death: where the cell cycle and apoptosis come together *Cancer* **2000**, *88*, 2619-2628.

3. Lowe, J.; Li, H.; Downing, K. H.; Nogales, E. Refined structure of alpha beta-tubulin at 3.5 A resolution *J. Mol. Biol.* **2001**, *313*, 1045-1057.

4. Oberlies, N. H.; Kroll, D. J. Camptothecin and Taxol: Historic Achievements in Natural Products Research *J. Nat. Prod.* **2004**, *67*, 129-135.

5. Fang, W. S.; Liang, X. T. Recent progress in structure activity relationship and mechanistic studies of taxol analogues *Mini-Reviews in Medicinal Chemistry* **2005**, *5*, 1-12.

6.  Kingston, D. G. I. Taxol and its analogs *Anticancer Agents from Natural Products* **2005**, 89-122.


## Cyclophophamide (Cytoxan, page 191)[1-5]

1.  Brock, N. The history of the oxazaphosphorine cytostatics *Cancer* **1996**, *78*, 542-547.

2.  Steinberg, A. D. Cyclophosphamide *Ther. Immunol.* **1996**, 23-43.

3.  Fleming, R. A. An overview of cyclophosphamide and ifosfamide pharmacology *Pharmacotherapy* **1997**, *17*, 146S-154S.

4.  Wright, J. E. Phosphoramide and oxazaphosphorine mustards *Cancer Therapeutics* **1997**, 23-79.

5.  Colvin, O. M. An overview of cyclophosphamide development and clinical applications *Curr. Pharm. Des.* **1999**, *5*, 555-560.


## Tamoxifen (Nolvadex, page 192)[1,2,4-6]
## X-Ray Structure (3ERT)[3]

1.  Grainger, D. J.; Metcalfe, J. C. Tamoxifen: teaching an old drug new tricks? *Nature Medicine (New York)* **1996**, *2*, 381-385.

2.  Osborne, C. K. Tamoxifen in the treatment of breast cancer *N. Engl. J. Med.* **1998**, *339*, 1609-1618.

3.  Shiau, A. K.; Barstad, D.; Loria, P. M.; Cheng, L.; Kushner, P. J.; Agard, D. A.; Greene, G. L. The structural basis of estrogen receptor/coactivator recognition and the antagonism of this interaction by tamoxifen *Cell* **1998**, *95*, 927-937.

4.  Jordan, V. C. Tamoxifen: A most unlikely pioneering medicine *Nat. Rev. Drug Discov.* **2003**, *2*, 205-213.

5.  Shang, Y. Molecular mechanisms of estrogen and SERMs in endometrial carcinogenesis *Nat. Rev. Cancer* **2006**, *6*, 360-368.

6.  Jordan, V. C.; Brodie, A. M. H. Development and evolution of therapies targeted to the estrogen receptor for the treatment and prevention of breast cancer *Steroids* **2007**, *72*, 7-25.


## Irinotecan (Camptosar, page 193)[1-3,5,6]
## X-Ray Structure (1T8I)[4]

1.  Pizzolato, J. F.; Saltz, L. B. The camptothecins *Lancet* **2003**, *361*, 2235-2242.

2.  Lorence, A.; Nessler, C. L. Camptothecin: Over four decades of surprising findings *Phytochemistry* **2004**, *65*, 2735-2749.

3.  Liu, L. F.; Desai, S. D. Mechanism of action of topoisomerase 1 poisons *Camptothecins in Cancer Therapy* **2005**, 3-21.

4.  Staker, B. L.; Feese, M. D.; Cushman, M.; Pommier, Y.; Zembower, D.; Stewart, L.; Burgin, A. B. Structures of Three Classes of Anticancer Agents Bound to the Human Topoisomerase I-DNA Covalent Complex *J. Med. Chem.* **2005**, *48*, 2336-2345.

5.  Li, Q.-Y.; Zu, Y.-G.; Shi, R.-Z.; Yao, L.-P. Review camptothecin: current perspectives *Curr. Med. Chem.* **2006**, *13*, 2021-2039.

6.  Pommier, Y. Topoisomerase I inhibitors: camptothecins and beyond *Nat. Rev. Cancer* **2006**, *6*, 789-802.


## Bleomycin (Blenoxane, page 194)[1-3,6,7]
## X-Ray Structure (1MXK, 1GJ2)[4,5]

1.  Boger, D. L.; Cai, H. Review of bleomycin: synthetic and mechanistic studies *Angew. Chem., Int. Ed. Engl.* **1999**, *38*, 448-476.

2.  Claussen, C. A.; Long, E. C. Nucleic acid recognition by metal complexes of bleomycin *Chem. Rev.* **1999**, *99*, 2797-2816.

3.  Lazo, J. S. Bleomycin *Cancer Chemother. Biol. Response. Modif.* **1999**, *18*, 39-45.

4.  Hoehn, S. T.; Junker, H.-D.; Bunt, R. C.; Turner, C. J.; Stubbe, J. Solution Structure of Co(III)-Bleomycin-OOH Bound to a Phosphoglycolate Lesion Containing Oligonucleotide: Implications for Bleomycin-Induced Double-Strand DNA Cleavage *Biochemistry (Mosc).* **2001**, *40*, 5894-5905.

5.  Zhao, C.; Xia, C.; Mao, Q.; Forsterling, H.; DeRose, E.; Antholine, W. E.; Subczynski, W. K.; Petering, D. H. Structures of $HO_2$-Co(III) bleomycin $A_2$ bound to $d(GAGCTC)_2$ and $d(GGAAGCTTCC)_2$: structure-reactivity relationships of Co and Fe bleomycins *J. Inorg. Biochem.* **2002**, *91*, 259-268.

6.  Chen, J.; Stubbe, J. Bleomycins: towards better therapeutics *Nat. Rev. Cancer* **2005**, *5*, 102-112.

7.  Hecht, S. M. Bleomycin group antitumor agents *Anticancer Agents from Natural Products* **2005**, 357-381.


## Imatinib (Gleevec, page 195)[1,2,4-6]
## X-Ray Structure (1IEP)[3]

1.  Capdeville, R.; Buchdunger, E.; Zimmermann, J.; Matter, A. Glivec (STI571, imatinib), a rationally developed, targeted anticancer drug *Nat. Rev. Drug Discov.* **2002**, *1*, 493-502.

2.  Cohen, P. Protein kinases-the major drug targets of the twenty-first century? *Nat. Rev. Drug. Discov.* **2002**, *1*, 309-315.

3.  Nagar, B.; Bornmann, W. G.; Pellicena, P.; Schindler, T.; Veach, D. R.; Miller, W. T.; Clarkson, B.; Kuriyan, J. Crystal structures of the kinase domain of c-Abl in complex with the small

molecule inhibitors PD173955 and imatinib (STI-571) *Cancer Res.* **2002**, *62*, 4236-4243.

4. Ren, R. Mechanisms of BCR-ABL in the pathogenesis of chronic myelogenous leukemia *Nat. Rev. Cancer* **2005**, *5*, 172-183.

5. Liu, Y.; Gray, N. S. Rational design of inhibitors that bind to inactive kinase conformations *Nature Chemical Biology* **2006**, *2*, 358-364.

6. Reiter, A.; Walz, C.; Cross, N. C. P. Tyrosine kinases as therapeutic targets in BCR-ABL negative chronic myeloproliferative disorders *Current Drug Targets* **2007**, *8*, 205-216.

## Sunitinib (Sutent, page 196)[1-4]

1. Cabebe, E.; Wakelee, H. Sunitib: a newly approved small-molecule inhibitor of angiogenesis *Drugs of Today* **2006**, *42*, 387-398.

2. Cooney, M. M.; van Heeckeren, W.; Bhakta, S.; Ortiz, J.; Remick, S. C. Drug insight: vascular disrupting agents and angiogenesis-novel approaches for drug delivery *Nat. Clin. Pract. Onc.* **2006**, *3*, 682-692.

3. Deeks, E. D.; Keating, G. M. Sunitinib *Drugs* **2006**, *66*, 2255-2266.

4. Rock, E. P.; Goodman, V.; Jiang, J. X.; Mahjoob, K.; Verbois, S. L.; Morse, D.; Dagher, R.; Justice, R.; Pazdur, R. Food and drug administration drug approval summary: sunitinib malate for the treatment of gastrointestinal stromal tumor and advanced renal cell carcinoma *Oncologist* **2007**, *12*, 107-113.

## Bortezomib (Velcade, page 197)[1-4]
## X-Ray Structure (2F16)[5]

1. Adams, J. The proteasome: a suitable antineoplastic target *Nat. Rev. Cancer* **2004**, *4*, 349-360.

2. Montagut, C.; Rovira, A.; Mellado, B.; Gascon, P.; Ross, J. S.; Albanell, J. Preclinical and clinical development of the proteasome inhibitor bortezomib in cancer treatment *Drugs of Today* **2005**, *41*, 299-315.

3. Papandreou, C. N. The proteasome as a target for cancer treatment: focus on bortezomib *American Journal of Cancer* **2005**, *4*, 359-372.

4. Caravita, T.; de Fabritiis, P.; Palumbo, A.; Amadori, S.; Boccadoro, M. Bortezomib: efficacy comparisons in solid tumors and hematologic malignancies *Nature Clinical Practice Oncology* **2006**, *3*, 374-387.

5. Groll, M.; Berkers, C. R.; Ploegh, H. L.; Ovaa, H. Crystal Structure of the Boronic Acid-Based Proteasome Inhibitor Bortezomib in Complex with the Yeast 20S Proteasome *Structure* **2006**, *14*, 451-456.

## Ancillary Anticancer Agents (page 198)[3,4,7-12]
## X-Ray Structure (198D, 199D, 1KX5)[1,2,5,6]

1. Dautant, A.; d'Estaintot, B. L.; Gallois, B.; Brown, T.; Hunter, W. N. A trigonal form of the idarubicin:d(CGATCG) complex; crystal and molecular structure at 2.0 .ANG. resolution *Nucleic Acids Res.* **1995**, *23*, 1710-1716.

2. Sastry, M.; Fiala, R.; Lipman, R.; Tomaxz, M.; Patel, D. J. Solution structure of the monoalkylated mitomycin C-DNA complex *J. Mol. Biol.* **1995**, *247*, 338-359.

3. Lasic, D. D. Doxorubicin in sterically stabilized liposomes *Nature (London)* **1996**, *380*, 561-562.

4. Bradner, W. T. Mitomycin C: A clinical update *Cancer Treat. Rev.* **2001**, *27*, 35-50.

5. Chang, G.; Roth, C. B. Structure of MsbA from E. coli: A homolog of the multidrug resistance ATP binding cassette (ABC) transporters *Science* **2001**, *293*, 1793-1800.

6. Davey, C. A.; Sargent, D. F.; Luger, K.; Maeder, A. W.; Richmond, T. J. Solvent Mediated Interactions in the Structure of the Nucleosome Core Particle at 1.9 Resolution *J. Mol. Biol.* **2002**, *319*, 1097-1113.

7. Gottesman, M. M.; Fojo, T.; Bates, S. E. Multidrug resistance in cancer: role of ATP-dependent transporters *Nat. Rev. Cancer* **2002**, *2*, 48-58.

8. Kelly, W. K.; Marks, P. A. Drug insight: Histone deacetylase inhibitors-development of the new targeted anticancer agent suberoylanilide hydroxamic acid *Nat. Clin. Pract. Onc.* **2005**, *2*, 150-157.

9. Larkin, J. M. G.; Kaye, S. B. Epothilones in the treatment of cancer *Expert Opin. Inv. Drug.* **2006**, *15*, 691-702.

10. Nadas, J.; Sun, D. Anthracyclines as effective anticancer drugs *Expert Opinion on Drug Discovery* **2006**, *1*, 549-568.

11. Szakacs, G.; Paterson, J. K.; Ludwig, J. A.; Booth-Genthe, C.; Gottesman, M. M. Targeting multidrug resistance in cancer *Nat. Rev. Drug Discov.* **2006**, *5*, 219-234.

12. Marks, P. A. Discovery and development of SAHA as an anticancer agent *Oncogene* **2007**, *26*, 1351-1356.

# PART VI.

*Mental disorders are among the most disabling of human illnesses. Unfortunately, they also are the least understood and least manageable of all bodily ailments.*

*Despite large research investment and the efforts of countless scientists, we are just at the start of the beginning of the quest for effective treatments of cognitive, behavioral, and mood disorders or neuro-degenerative illnesses.*

*The human and economic costs of mental disorders are incalculable.*

*It would be logical to increase greatly the levels of fundamental and applied research in this area because, in the words of Samuel Johnson (Rambler no. 178), "the future is purchased by the present".*

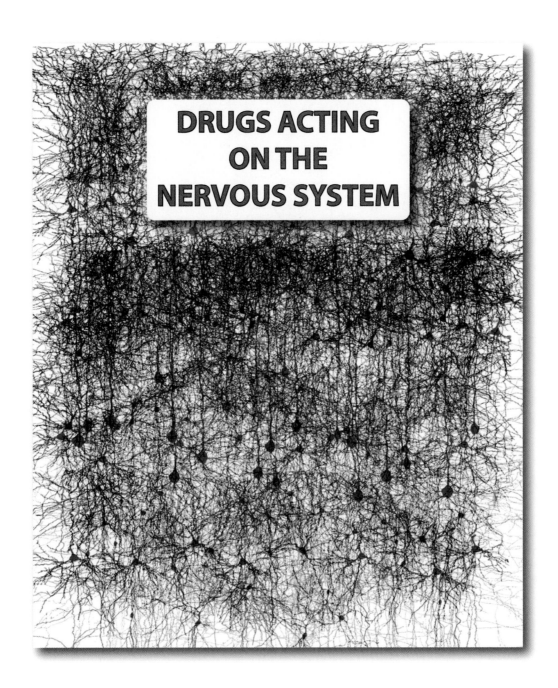

# DRUGS ACTING ON THE NERVOUS SYSTEM

Artwork is by IBM/EPFL Blue Brain Project: "**A Forest of Neurons**" - A dye is introduced into the neural soma and then developed in order to reveal the morphology. This image shows a minute fraction of the cells and connections within the microcircuitry of the neocortex. (Printed with permission from IBM.)

**Fentanyl** (Duragesic™) is an analgesic that is 100 times more potent than morphine.

**Pregabalin** (Lyrica™) is used for the treatment of neuropathic pain and chronic fatigue syndrome (fibromyalgia).

**Diazepam** (Valium™) is widely used as a sedative and muscle relaxant.

# PAIN AND ANALGESIA

# LIDOCAINE (XYLOCAINE™)

| Structural Formula | Ball-and-Stick Model | Space-filling Model |
|---|---|---|

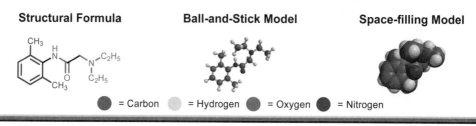

● = Carbon    ● = Hydrogen    ● = Oxygen    ● = Nitrogen

*Year of discovery:* 1943 (Niels Löfgren); *Year of introduction:* 1948 (Astra); *Drug category:* Local anesthetic/antiarrhythmic agent; *Main uses:* As a local anesthetic in dental practice; *Other brand names:* Xylocard, Lidamantle; *Related drugs:* Benzocaine, Prilocaine (Citanest), Bupivacaine (Marcaine), Tetracaine (Synera) and Procaine (Novocaine).

Human consciousness includes the constant awareness of an individual's state along a euphoria-dysphoria axis. At the extreme dysphoria end is pain, which in many respects is like a sixth sense, no less functional than the other five. Without a sense of pain, the warning sign of illness or injury to the body, human life would be far more tenuous. Vital though it is, pain must be controlled in many situations, for example during invasive medical or dental procedures. The sensing of pain at the local level causes the generation of molecules such as $PGE_2$ and other mediators that can activate ion channels of neuronal cells. That leads to voltage spikes that are transmitted by nerves from the neuron to the brain, where they are interpreted as pain.

Cocaine, a product of the coca shrub, is an anesthetic that was first used for ophthalmological surgery in 1884. However, its use was limited by toxicity and the possibility of addiction. An effort to find compounds with similar activity, but lower toxicity, was rewarded in 1905, when the German chemist Albert Einhorn prepared procaine, the first injectable synthetic aminoester anesthetic. It was widely used in surgery for 50 years.

**Procaine**

During the 1930s two Swedish scientists, Nils Löfgren and Bengt Lundquist, synthesized and tested a series of aminoamide analogs of procaine as local anesthetic agents. Their work led to the discovery of lidocaine which has a much faster onset of action (just 5-10 min) than procaine and a convenient duration of local anesthesia (ca. 2 h). In 1948, Astra, a Swedish pharmaceutical company, began large-scale production and marketing of lidocaine as Xylocaine.

Over the past half century, a detailed correlation of structure with local anesthetic activity has been made. Typically, active compounds have a hydrophobic (blue) and a hydrophilic (green) moiety connected by either an ester or an amide functional group (red).[1]

Aminoester    Aminoamide

**Prototypical Local Anesthetic Agents**

The balance between the overall hydrophobicity, hydrophilicity, and size of these compounds is critical. Molecules with large hydrophobic sections have increased potency and duration of action, but also increased toxicity. On the other hand, molecules that are more hydrophilic usually have a better therapeutic index.

Lidocaine blocks the generation and conduction of nerve impulses by the inhibition of sodium ion ($Na^+$) influx through voltage-gated sodium ion channels. As a result the transmission of a pain signal through the nerves is prevented.

The metabolism of lidocaine affords two metabolites, glycinexylidide (GX), and monoethylglycinexylidide (MEGX), that are both substantially less active than lidocaine itself. The efficacy of lidocaine is diminished when it is administered over an extended period of time because these metabolites compete for the receptors and block lidocaine action.

GX                    MEGX

1. *Handbook of Experimental Pharmacology* **2007**, *177*, 95-127; Refs. p. 233

# MORPHINE (AVINZA™)

| **Structural Formula** | **Ball-and-Stick Model** | **Space-filling Model** |
|---|---|---|

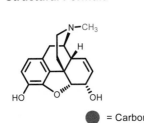  

● = Carbon    ◌ = Hydrogen    ● = Oxygen    ● = Nitrogen

*Year of discovery:* May 21st 1805 (Friedrich Wilhelm Adam Sertürner in Einbeck, Germany); *Year of introduction:* 1850s; Avinza was approved in 2002 in the US (Elan Corporation); *Drug category:* Opiate (narcotic) analgesic/phenantrene opioid receptor (μ-opioid receptor) agonist; *Main uses:* For the treatment of moderate to severe pain from malignant and non-malignant disease; *Other brand names:* Oramorph SR, MS Contin, Kadian, Roxanol; *Related drugs:* Oxycodone (OxyContin), Acetaminophen/Oxycodone (Percocet).

Opium, the sticky juice (latex) of the unripe seed pod of the poppy plant (*Papaver somniferum*), has been used for over a millennium for relief of pain. Opium contains morphine (5-20%) as the principal alkaloid, as well as lesser amounts of the related opioids codeine, thebaine, and papaverine.[1]

In 1805, F.W.A. Sertürner, a German pharmacy assistant, isolated morphine as a colorless crystalline powder from raw opium. The name derives from Morpheus, the Greek god of dreams. Sertürner described morphine to be over ten times as powerful as opium in promoting sleep. In the 1850s morphine became the standard medication for the management of pain during and after surgical procedures. However, by the late 1800s awareness by medical practitioners of morphine's addictive properties spurred a systematic search for a non-addictive analgesic. As a result, a large number of structurally related compounds were prepared from morphine and tested. All but two of these morphine analogs were addictive, the exceptions being nalorphine and naloxone, the structures of which are depicted below (changes from morphine shown in red).[2]

More than 230 tons of morphine are now used annually in medicine, mainly for relief of pain. Morphine can be administered orally or by injection. Most recently, the FDA approved Avinza, a slow release formulation of morphine sulfate that provides pain relief for 24 hours with a single oral dose.

The mechanism of action of morphine and related opioids is now known in considerable detail.[3] This advance was sparked by the

discovery in the 1970s of high-affinity binding sites for morphine in the brain. These are now known as μ, δ, and κ opioid receptors. Shortly thereafter, three classes of endogenous peptides were isolated from the brain that also bound strongly to these opioid receptors. These peptides, the *enkephalins*, *dynorphins* and *endorphins*, serve as endogenous opioids in the body's own pleasure/reward system.

At the customary doses, morphine is highly selective for the μ-receptors. However, at larger doses, μ-selective opioids also interact with the δ and κ receptors and this may result in a change in their pharmacological profiles. Agonists of the δ and κ receptors have limited analgesic effects and many side effects, thus they are seldom used in clinical practice. Morphine affects a wide range of physiological systems. It produces analgesia, sedation, changes in mood (e.g., euphoria and tranquility), and slower respiratory, cardiovascular, and gastrointestinal function. The effects of μ-receptor agonists can be prevented or reversed by the injection of a μ-receptor antagonist such as naloxone. Partial μ-receptor agonists (e.g., methadone) found use in the treatment of addiction to opioids such as heroin.

**Nalorphine**      **Naloxone**

1. *Natural Medicines* **1996**, *50*, 86-102; 2. *Handbook of Experimental Pharmacology* **2007**, *177*, 31-63; 3. *Rev. Neurosci.* **2006**, *17*, 393-402; **Refs. p. 233**

# ACETAMINOPHEN (TYLENOL™)

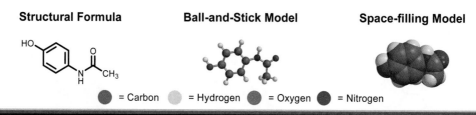

| **Structural Formula** | **Ball-and-Stick Model** | **Space-filling Model** |

● = Carbon ○ = Hydrogen ● = Oxygen ● = Nitrogen

*Year of discovery:* 1873; *Year of introduction:* 1955 (McNeil Laboratories). Now it is sold by Johnson & Johnson; *Other brand names:* Paracetamol, acetyl-*para*-aminophenol (APAP). *Drug category:* Analgesic/Antipyretic. *Main uses:* For relief of pain, alone or in combination with oxycodone. *Related drugs:* Acetaminophen/Propoxyphene: 325/50 (Darvocet), Acetaminophen/Oxycodone: 350/5 (Percocet).

During the late 1800s several simple synthetic derivatives of aniline were tested for the relief of pain and fever. Two of these substances, acetanilide and phenacetin, were found to be effective and gained wide acceptance as analgesics; s*ee* figure below (aniline portion highlighted in red).[1]

**Acetanilide**    **Phenacetin**

Prolonged use of these analgesic substances at high doses is dangerous since they reduce the oxygen-carrying capacity of blood to a degree that can be fatal. Studies during the 1950s showed that both acetanilide and phenacetin are metabolized to a common substance, acetaminophen, which is actually responsible for the analgesic effect. It is also significantly less toxic. In 1955, McNeil Laboratories commercialized acetaminophen as Tylenol. Over the past 50 years Tylenol has become the most widely used analgesic and antipyretic agent and is available globally. It is a component of more than 100 medications. The remarkable popularity of acetaminophen is partly due to the lack of gastric irritation, in contrast to aspirin. Acetaminophen does not inhibit the enzymes COX-1 and COX-2, whereas aspirin does. The exact mechanism of action of acetaminophen has been controversial and is still unclear.[2]

There is a dark side to acetaminophen, however, because the toxic dose is not far above the doses frequently used. This is especially problematic because of its widespread use, and because the danger of toxicity is greatly increased when taken with alcohol. When used in excess or with alcohol, acetaminophen can cause liver failure and even death. This toxicity is attributed to the production of a highly reactive metabolite, NAPQI, formed by enzymatic oxidation in the liver. The oxidizing enzyme is a member of the cytochrome P450 class that normally serves as a detoxification system. When a therapeutic dose of acetaminophen is administered, NAPQI is efficiently eliminated as the corresponding glutathione derivative (*see* scheme below). However, in the case of higher doses, the body's glutathione reserves are quickly depleted and the reactive NAPQI covalently binds to important proteins of liver cells, causing cell death and eventual liver failure.[3]

Glutathione-SH →

**NAPQI**        **Glutathione Derivative**

The combination of acetaminophen (350 mg) with oxycodone (5 mg) is an important pain-relieving medicine. It is available only by prescription and is commonly used for short-term relief of moderate to severe pain. The combination, first marketed as the brand Percocet, is now an inexpensive generic product. It is considerably more effective than the separate ingredients because of synergism arising from differing mechanisms of analgesia. The synergistic mixture, which is frequently abbreviated as oxycod + APAP, provides rapid relief for musculoskeletal and neuralgic pain over about a 6-hour period.

≡    **Oxycodone**

1. *Acute Pain* **1997**, *1*, 33-40; 2. *CNS Drug Rev.* **2006**, *12*, 250-275; 3. *Drug Metab. Rev.* **2005**, *37*, 581-594; Refs. p. 233

# FENTANYL (DURAGESIC™)

| **Structural Formula** | **Ball-and-Stick Model** | **Space-filling Model** |

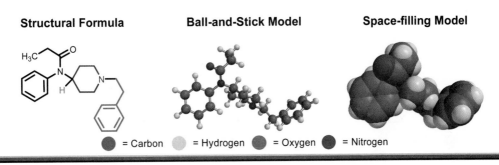

= Carbon ⚪ = Hydrogen ⚫ = Oxygen ⚫ = Nitrogen

*Year of discovery:* Late 1950s (Janssen Pharmaceutica); *Year of introduction:* 1960s (as Sublimaze); *Drug category:* Phenylpiperidine/opioid μ-receptor agonist; *Main uses:* As an analgesic for the treatment of persistent severe and chronic pain and also for spinal or epidural anesthesia; *Other brand names:* Actiq (Fentanyl citrate by Cephalon), Fentora (Cephalon); *Related drugs:* Alfentanil (Alfenta), Sulfentanil (Sulfenta), Remifentanil (Ultiva), Carfentanil (Wildnil).

The goal of developing analgesic compounds that are potent but not addicting has motivated the synthesis and evaluation of thousands of new molecules. In the late 1950s, a simple phenylpiperidine derivative, fentanyl, was synthesized by Paul Janssen, founder of Janssen Pharmaceutica in Belgium. Fentanyl was found to be 100 times more potent than morphine, and to bind to the same μ-opioid receptors in the central nervous system as morphine. A number of more potent and shorter acting analogs of fentanyl were also developed at Janssen including alfentanil, sulfentanil, remifentanil and carfentanil (*see* below). The common structural feature of morphine and the active fentanyl analogs is the substituted piperidine ring (shown in blue).[1]

**Remifentanil (Ultiva™)**

**Morphine**

**Sulfentanil (Sulfenta™)**

**Carfentanil (Wildnil™)**

Remifentanil is the shortest acting opioid (5-10 min) for human use.[2] Sulfentanil is 1000 times as potent as morphine and is used in heart surgery, which involves considerable damage to highly innervated tissues in the chest. Carfentanil is only used for the sedation of large animals since it is 10,000 times more potent than morphine and ultra fast acting (1-2 minutes). These huge differences in potency are achieved with very minor changes of the groups on the piperidine ring (shown as red and green).[3]

Fentanyl and its analogs are superior to morphine as analgesics for several reasons: (1) respiratory depression is less likely to occur with injection of the drug; (2) fentanyl does not give rise to bronchoconstriction and vasodilation, because it does not cause histamine release, in contrast to morphine; (3) it takes only a few minutes to reach peak analgesic effect after intravenous administration (15 minutes with morphine) and (4) recovery from the analgesia occurs more rapidly.

Since its introduction in the 1960s, fentanyl has been widely used as an analgesic adjuvant in surgical procedures with intravenous or epidural administration. Slow-release (48h) transdermal patches of fentanyl, such as Duragesic, are now widely used since fentanyl is readily absorbed through the skin due to its high lipophilicity. By 2003, more than 5.7 million prescriptions for fentanyl were written annually. Access to fentanyl is restricted because of its potency and abuse potential.

1. *J. Pain Symptom Manage.* **2005**, *29*, S67-S71; 2. *Drugs* **2006**, *66*, 365-385; 3. *Seminars in Anesthesia, Perioperative Medicine and Pain* **2005**, *24*, 108-119; Refs. p. 233

# SODIUM THIOPENTAL (SODIUM PENTOTHAL™)

| Structural Formula | Ball-and-Stick Model | Space-filling Model |
|---|---|---|

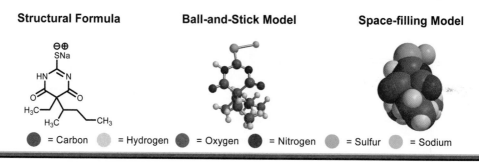

● = Carbon    ○ = Hydrogen    ● = Oxygen    ● = Nitrogen    ● = Sulfur    ○ = Sodium

*Year of discovery:* 1933 (Ernest H. Volwiler & Donalee L. Tabern at Abbott Laboratories); *Year of introduction:* 1935; *Drug category:* Barbiturate/anesthetic; *Main uses:* For inducing pre-surgical anesthesia in combination with sedatives; *Related drugs:* Pentobarbital (Nembutal), Methohexital (Brevital), Butalbital (Fiorinal).

Barbiturates are derivatives of barbituric acid, which can be made simply by heating diethyl malonate (blue) and urea (red).

The first medically useful barbiturate, the sedative barbital, was prepared by the famous German organic chemist Emil Fischer at the end of the 19th century. It was marketed in 1904 as a sleeping aid under the trade name Veronal by Bayer.[1]

**Barbituric acid**     **Barbital**

During the first half of the 20th century over 2500 barbituric acid derivatives were prepared and some fifty were actually marketed. Today only a few are in medical use. Barbiturates can be taken orally or intravenously and have a wide range of physiological effects. Most importantly, they depress the central nervous system and can produce mild sedation, sleep, or even coma. Barbiturates can also serve as anesthetics and anticonvulsants. At present, the barbiturate class is typically used for deep sedation and consists of ultra-short, short, intermediate, and long-acting agents.[2]

**Pentobarbital (Nembutal™)**     **Thiopental**

Abbott Laboratories prepared pentobarbital, a long-acting oral barbiturate that was sold

as Nembutal. Later, by replacing an oxygen atom of pentobarbital with sulfur, an ultra-short-acting intravenous barbiturate, thiopental, was developed. The sodium salt of thiopental is widely used in pre-surgical anesthesia and marketed as Sodium Pentothal

A standard intravenous dose of sodium thiopental will render a person unconscious within 1 minute. The duration of anesthesia is usually no longer than 10 minutes. For this reason thiopental is a preferred anesthetic during minor surgical procedures. If prolonged anesthesia is needed, inhaled anesthetics (e.g., halothane, sevoflurane) are administered after the initial injection of thiopental.[3]

Thiopental and other barbiturates work by binding to the gamma-aminobutyric acid-A (GABA$_A$) receptor. GABA$_A$ is a ligand-gated chloride ion channel and it is responsive to binding of GABA, a major endogenous inhibitory neurotransmitter in the brain (below). When GABA binds to GABA$_A$, the conformation of the receptor changes, causing the chloride ion channel to open and leading to inhibition of neuron-firing. Barbiturates are potent GABA$_A$ agonists that mimic its action.

Barbiturates have been, and still are, often abused. An overdose of these drugs may result in death. In fact, a high dose of thiopental is part of the drug cocktail used in death sentences by lethal injection in the US.

**GABA**

1. *Neuropsychiatr. Dis. Treatm.* **2005**, *1*, 329-343; 2. *Analytical Toxicology for Clinical, Forensic and Pharmaceutical Chemists* **1997**, 347-368; 3. *Clin. Pharmacokinet.* **1998**, *35*, 95-134; Refs. p. 233

# GABAPENTIN (NEURONTIN™)

| Structural Formula | Ball-and-Stick Model | Space-filling Model |
|---|---|---|

● = Carbon ○ = Hydrogen ● = Oxygen ● = Nitrogen

*Year of discovery:* 1983 (Parke-Davis, now Pfizer); *Year of introduction:* 1994 (For the control of partial seizures); 2002 (for neuropathic pain); *Drug category:* Anticonvulsant/Analgesic; *Main uses:* For the treatment of postherpetic neuralgia, pain after shingles, postoperative chronic pain, migraine, bipolar disorder, and epilepsy; *Related drugs:* Pregabalin (Lyrica).

Gamma-aminobutyric acid (GABA, *see* page 212) is a natural inhibitor of neurotransmission in mammalian brain. The synthesis and evaluation of gabapentin resulted from a search to find a treatment for epileptic seizures caused by impairment of GABA function. Orally administered gabapentin was found to enter the brain efficiently and to relieve seizures. Gabapentin was approved in 1994 for the treatment of partial seizures in epileptic patients and marketed as Neurontin.[1] Later, it was also approved as an effective treatment of neuropathic pain and for the prevention of frequent migraine headaches. Neuropathic pain usually does not respond to treatment with non-steroidal anti-inflammatory drugs (NSAIDs) or opioids.

Gabapentin belongs to a class of drugs now known as gabapentinoids. These compounds are the mono- or dialkyl 3-substituted derivatives of GABA (*see* below). The latest member of this class, pregabalin, is approximately three times as potent as gabapentin by all routes of administration. Pregabalin is indicated for the management of both diabetic peripheral neuropathy (DPN) and postherpetic neuralgia (PHN). It gained approval in 2004 and is now marketed as Lyrica by Pfizer.[2]

R[1] and R[2] = H or alkyl

**Gabapentinoid**     **Pregabalin (Lyrica™)**

Gabapentin and pregabalin are unusually safe and well-tolerated by patients. They are also very soluble in water and efficiently absorbed. Gabapentin is not metabolized in humans and is excreted unchanged in the urine with a half-life of 4-6 hours.

Although gabapentin was designed as a lipophilic derivative of GABA, it does not interact with GABA receptors, GABA transporters, or GABA transaminase. The molecular target of gabapentin remained obscure for over a decade. However, it is now known that gabapentin binds to the $\alpha_2\delta$ subunit of a multiprotein complex that regulates a specific voltage-dependent calcium ion ($Ca^{2+}$) channel. This ion channel plays a role in neurotransmitter release at presynaptic nerve endings. Voltage-dependent $Ca^{2+}$ channels involve several protein subunits ($\alpha 1$, $\alpha 2$, $\beta$, $\gamma$ and $\delta$) which collectively determine $Ca^{2+}$ channel conduction. The $\alpha 2$ and $\delta$ subunits are spatially next to each other and are referred to as the $\alpha_2\delta$ region. Together they enhance calcium ion flux through the ion channel. Both gabapentin and pregabalin bind to the $\alpha_2\delta$ region with nanomolar (nM) affinity and inhibit the influx of $Ca^{2+}$ ions. As a result, the overall excitability of the neuron is decreased and the transmission of the pain signal is reduced.[3]

Recently, gapabentin was shown to be a substrate of the system L amino acid transporter that carries large neutral amino acids (e.g., leucine, isoleucine, and valine) from the circulation into the brain. This finding explains the efficient uptake of gabapentin into the brain. This transport process can be enhanced even further by administering gabapentin enacarbil, a gabapentin prodrug.

**Gabapentin Enacarbil**

1. *Handbook of Experimental Pharmacology* **2007**, *177*, 145-177; 2. *Clin. Ther.* **2007**, *29*, 26-48; 3. *Curr. Opin. Pharmacol.* **2006**, *6*, 108-113; Refs. p. 233

# DIAZEPAM (VALIUM™)

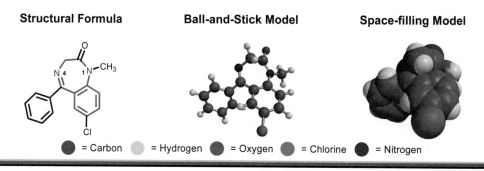

**Structural Formula**    **Ball-and-Stick Model**    **Space-filling Model**

● = Carbon    ○ = Hydrogen    ● = Oxygen    ● = Chlorine    ● = Nitrogen

*Year of discovery:* 1959 (by Leo Sternbach at Roche); *Year of introduction:* 1963 (Hoffman-La Roche); *Drug category:* Benzodiazepine/anticonvulsant/central nervous system depressant; *Main uses:* For the treatment of anxiety, seizures, muscle spasms, insomnia, and symptoms of opiate or alcohol withdrawal; *Other brand names:* Stesolid, Diazemuls, Seduxen, Diapam, Antenex, and Bosaurin; *Related drugs:* Oxazepam (Alepam), Nitrazepam (Alodorm), Alprazolam (Xanax).

During the first half of the 20th century, barbiturates (e.g., pentobarbital, pheno-barbital, and sodium thiopental; *see* page 212) were the standard medicines for sedation and for the treatment of insomnia. However, the therapeutic and toxic doses for barbiturates are so close that there have been many deaths caused by an overdose. In addition, barbiturates are addictive and sudden withdrawal can be fatal. Eventually, a new class of sedatives, the benzodiazepines (BZD) was discovered by Leo Sternbach of Hoffman-La Roche.[1] The core unit of the benzodiazepines is a bicyclic structure in which a benzene ring and a seven-membered ring containing two nitrogens share a common edge (The core ring-system, 1,4-benzodiaze-pine, is shown in red above.) The first useful BZD, chlordiazepoxide, was an antianxiety drug marketed as Librium. Unfortunately, chlordiazepoxide causes ataxia (loss of balance). It was soon followed by a safer drug, diazepam, which was marketed under the trade name Valium in 1963.

**Chlordiazepoxide (Librium™)**    **Alprazolam (Xanax™)**

Diazepam soon became a major drug and by the mid-1970s more than 8000 tons of BZDs were made annually. In the late 1970s alprazolam, a BZD that is 1000 times as potent as diazepam, was developed by Upjohn and marketed as Xanax. It has been used principally as a mild sedative, sleeping aid, and muscle relaxant. The use of benzodiazepines is limited by the fact that they suffer from side effects, such as amnesia and dependence.[2] Long-term use is contra-indicated.

Benzodiazepines bind to $GABA_A$ receptors in the central nervous system at what is called the benzodiazepine site.[3] When benzodiaze-pines bind to this receptor site, a confor-mational change occurs in the receptor that enhances the binding of GABA. Although barbiturates bind to the same receptor complex, they do so at a different subunit with different consequences. Benzodiazepines and barbiturates both enhance the efficiency of the binding of GABA by increasing its affinity for the receptor. However, barbiturates also open chloride ion channels, even in the absence of GABA, whereas BZDs do not. This difference in activity may explain why BZDs are less dangerous, less toxic, and less likely to cause respiratory depression than barbiturates.

The abrupt withdrawal of BZDs after long-term use causes a significant drop in the efficiency of the $GABA_A$ receptors that, in turn, can lead to withdrawal symptoms such as anxiety, insomnia, and seizures.

1. Baenninger, A. & et al. Good Chemistry: The Life and Legacy of Valium Inventor Leo Sternbach (McGraw-Hill, New York, **2003**); 2. *J. Clin. Psychiatry* **2005**, *66*, 28-33; 3. *Med. Chem. Rev.-Online* **2005**, *2*, 251-256; Refs. p. 234

# SUMATRIPTAN (IMITREX™)

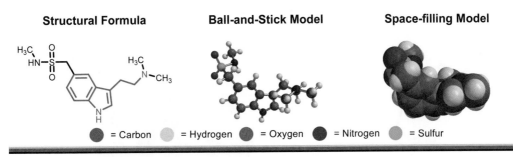

| **Structural Formula** | **Ball-and-Stick Model** | **Space-filling Model** |

● = Carbon  ○ = Hydrogen  ● = Oxygen  ● = Nitrogen  ○ = Sulfur

*Year of discovery:* 1984 (GSK); *Year of introduction:* 1993; *Drug category:* Triptan/antimigraine drug/highly selective 5-HT$_{1D}$ receptor agonist; *Main uses:* For the treatment of severe migraine headaches; *Related drugs:* Zolmitriptan (Zomig), Naratriptan (Amerge), Almotriptan (Axert), Rizatriptan (Maxalt).

Sumatriptan is used for the treatment of migraine headache either alone (Imitrex) or in combination with naproxen (Trexima).[1]

Migraine, an intense, disabling episodic headache on one or both sides of the head, affects an estimated 15% of the world's population. It is more severe than common headache, which accounts for 75% of all headache. The onset (triggering event), the intensity and the duration of migraine vary widely from one person to another. Migraine can last anywhere from a few hours to days and the accompanying pain and feeling of nausea ranges from moderate to severely debilitating. Although the etiology of migraine headache is unclear, one hypothesis is that it is the result of cranial nerve root irritation by inflammatory mediators. Migraine has also been linked to low levels of the neurotransmitter serotonin (5-hydroxytryptamine or 5-HT), which is involved in neuronal cell formation and maintenance, sleep, cognition, appetite, and mood. Although there appears to be a genetic predisposition to migraine, the relevant genes have not been identified. Serotonin is formed in the body in two steps from the amino acid tryptophan (*see* below).

they do not alleviate severe migraine. To date the most successful drug treatments of severe migraine are based on serotonin agonists that activate the 5-HT$_{1D}$ receptor.

Currently seven 5-HT receptor families and fourteen receptor subtypes have been identified. Although the various 5-HT receptor subtypes have very similar amino acid sequences, they induce very different pharmacological responses.

Selective agonists of the 5-HT$_{1D}$ and 5-HT$_{1B}$ receptors, which constrict the blood vessels in the skull, were found to alleviate migraine. Sumatriptan, the first 5-HT$_{1D}$-selective agonist, emerged in the 1980s from a systematic search for novel therapeutic agents for migraine.[2] In this work, a large number of serotonin analogs (triptans) were synthesized and evaluated for efficacy in animal models. It was found later that triptans which show selectivity for either 5-HT$_{1D}$ or 5-HT$_{1B}$ receptor subtypes and have minimal affinity for the other 5-HT receptor subtypes can ameliorate migraine headaches.

Several second-generation triptan drugs (e.g., naratriptan, zolmitriptan, and rizatriptan) that are more potent than sumatriptan have been approved since 1993 (*see* below).

**Tryptophan** → 1. hydroxylation 2. decarboxylation - CO$_2$ → **Serotonin**

There are three approaches to the management of migraine: (1) avoidance of triggering events; (2) control of symptoms; and (3) the use of preventive drugs. Although analgesic drugs such as acetaminophen (Tylenol), aspirin, or ibuprofen are helpful,

**Naratriptan (Amerge™)**  **Rizatriptan (Maxalt™)**  **Zolmitriptan (Zomig™)**

1. *Neuropsychiatr. Dis.Treatm.* **2006**, *2*, 293-297; 2. *Exp. Opin. Pharmacother.* **2006**, *7*, 1503-1514; **Refs. p. 234**

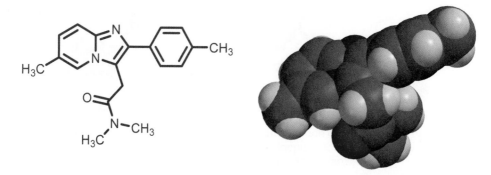

**Zolpidem** (Ambien™) is a widely used agent for the treatment of insomnia and time zone change related sleep disturbances.

## *One in every eight people suffers from insomnia.*

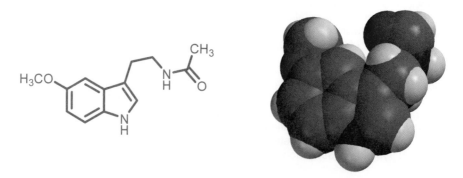

**Melatonin** is produced in the brain when the eye is not receiving light, because its biosynthesis is inhibited by light. It is a natural bioregulator that induces and maintains sleep.

# HYPNOTICS (INSOMNIA)
# AND
# ANTISMOKING

# ZOLPIDEM (AMBIEN™)

| Structural Formula | Ball-and-Stick Model | Space-filling Model |
|---|---|---|

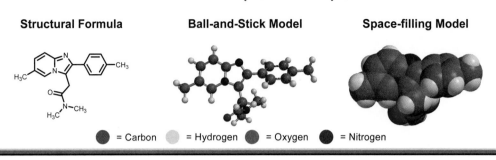

● = Carbon    ● = Hydrogen    ● = Oxygen    ● = Nitrogen

*Year of discovery:* 1982 (Synthélabo); *Year of introduction:* 1992; *Drug category:* Non-benzodiazepine sedative/hypnotic agent; *Main uses:* Short-term treatment of insomnia; *Related drugs:* Eszopiclone (Lunesta), Zaleplon (Sonata).

Insomnia, the inability to fall into and maintain the state of sleep, affects most people at one time or another, and is a chronic problem for many, especially travelers, shift workers, and the elderly. It is more prevalent in women and patients with medical or psychiatric conditions. Zolpidem, developed by the French company, Synthélabo (later Sanofi Aventis) in the early 1980s, is used for the short-term treatment of insomnia. It was introduced in the US in 1992 under the name Ambien. An extended release formulation of zolpidem, marketed under the name Ambien CR was approved in 2005.

Zolpidem has a rapid onset of action and a half-life of ca. 1 hour. It initiates sleep in a short period of time (usually 10-30 minutes) and prolongs total sleep time. Extended release zolpidem provides plasma concentrations beyond three hours of administration, thereby improving the quality of sleep and sustaining deep sleep. Zolpidem has little effect on the stages of sleep.

Zolpidem has a greater safety margin than barbiturates (*see* page 212) and benzodiazepines (*see* page 214). Anecdotal incidents of "sleep driving" and eating have been reported.[1]

The hypnotic effect of Zolpidem is due to modulation of GABA ($\gamma$-aminobutyric acid) receptors (*see* page 212), as is in the case of barbiturates and benzodiazepines. GABA is an important inhibitory neurotransmitter affecting ca. 20-50% of all synapses of the central nervous system. There are three types of membrane-bound GABA receptors. $GABA_A$ and $GABA_C$ receptors are ligand-gated ion channels, and $GABA_B$ receptors are G protein-coupled receptors that regulate other ion channels. The $GABA_A$ receptor is a pentameric protein complex, most commonly containing $\alpha_n$, $\beta_m$ and $\gamma_o$ subunits (where n+m+o = 5). GABA binds to the $\beta$ subunit of the receptor, leading to the opening of a chloride ion channel, hyperpolarization of the membrane, and reduction of neuronal activity. Zolpidem binds to the $\alpha$-receptor subunits (as do benzodiazepines), and facilitates the opening of chloride channels in response to GABA. However, unlike benzodiazepines that activate all $\alpha$-receptor subtypes non-selectively, zolpidem binds preferentially to the $\alpha_1$ subtype, which is thought to explain its reduced muscle relaxant and antiepileptic effects.[2]

There are two other approved hypnotic agents that also bind the $\alpha$-subunit of GABA receptors. Zaleplon was introduced in 1999 as Sonata by King Pharmaceuticals and eszopiclone was marketed as Lunesta by Sepracor since 2004. Although eszopiclone is less effective than benzodiazepines, it can be used for long-term treatment of insomnia without serious withdrawal symptoms upon discontinuation.

**Zaleplon (Sonata™)**

**Eszopiclone (Lunesta™)**

1. *CNS Drugs* **2005**, *19*, 65-89; 2. *Angew. Chem., Int. Ed. Engl.* **1999**, *38*, 2853-2864; **Refs. p. 234**

# RAMELTEON (ROZEREM™)

| Structural Formula | Ball-and-Stick Model | Space-filling Model |
|---|---|---|

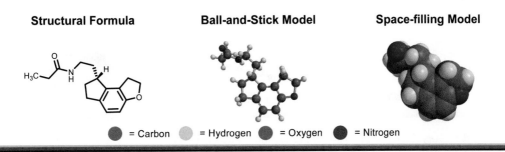

● = Carbon   ● = Hydrogen   ● = Oxygen   ● = Nitrogen

*Year of discovery:* Late 1990s; *Year of introduction:* 2005 (Takeda); *Drug category:* Hypnotic, selective melatonin receptor agonist; *Main uses:* Treatment of insomnia.

Ramelteon, the first hypnotic agent that acts by selectively activating melatonin receptors, was developed in the 1990s at Takeda Pharmaceuticals in Japan. It was introduced for the treatment of short and long-term insomnia in the US in 2004.

Ramelteon was shown to significantly reduce the time to fall asleep and increase total sleep time in people with transient or chronic insomnia, without causing troublesome residual effects. In contrast to traditional hypnotics that target GABA receptors (e.g., barbiturates, benzodiazepines, zolpidem, etc), ramelteon does not influence behavior, impair cognition, memory, alertness, or the ability to concentrate. Furthermore, the use of ramelteon does not appear to cause dependence, abuse, or withdrawal symptoms.[1]

Melatonin is a mammalian hormone that mediates circadian rhythm, the 24-hour physiological sleep-wake cycle. It is formed via a series of biological processes in the brain from the amino acid tryptophan by the way of the neurotransmitter serotonin.

The synthesis of melatonin is inhibited by light, which signals from the retina to the pineal gland via the suprachiasmatic nucleus of the hypothalamus. The synthesis commences at darkness, peaks in the middle of the night and drops thereafter. Melatonin assists in inducing and maintaining sleep, but, because of its relatively short half-life (ca. 40 min) its use in the treatment of insomnia is limited to initiating sleep and easing jetlag and time zone adjustments.

Melatonin and ramelteon exert their effect by binding to melatonin receptors. Three subtypes of melatonin receptors have been identified, of which two are involved in the regulation of sleep. MT$_1$ appears to regulate sleepiness, and MT$_2$ is involved in circadian rhythm. Activation of these receptors by melatonin or ramelteon causes inhibition of the enzyme adenylate cyclase and reduction of the levels of intracellular cyclic adenosine monophosphate (cAMP). Ramelteon exhibits a 3 to 6-fold higher affinity toward MT$_1$ and MT$_2$ receptors than melatonin and has a longer half-life (1-2.6 h). Ramelteon does not bind to other sites that normally promote sleep (e.g., benzodiazepine, dopamine, and opioid receptors). The role of ramelteon in the management of insomnia has yet to be established.

Melatonin occurs naturally in many plants and animals. Since it is a powerful antioxidant, it may function to protect the host from reactive chemical fragments formed as byproducts of metabolism.

1. *Drugs of Today* **2006**, *42*, 255-263; **Refs. p. 234**

**Tryptophan**          **Serotonin**

**Melatonin**

# VARENICLINE (CHANTIX™)

| **Structural Formula** | **Ball-and-Stick Model** | **Space-filling Model** |
|---|---|---|

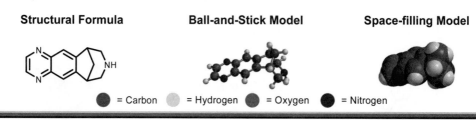

● = Carbon    ○ = Hydrogen    ● = Oxygen    ● = Nitrogen

*Year of discovery:* Late 1990s; *Year of introduction:* 2006 (Pfizer); *Drug category:* Partial agonist on the α4β2 nicotinic acetylcholine receptor; *Main uses:* Aid to smoking cessation; *Related drugs:* Bupropion (Zyban).

Varenicline, a modulator of the α4β2 subtype of nicotinic acetylcholine receptors, is a non-nicotinic aid to smoking cessation introduced by Pfizer as Chantix in 2006.

The smoking of tobacco is a major health risk that can lead to cancer, heart disease, and chronic obstructive pulmonary disease. In the US and Europe, approximately 20-25% of the population are smokers. Although smoking is the leading cause of preventable disease and death in the developed world, the problem has persisted because it is very difficult for smokers to quit.

The alkaloid, nicotine (structure below) is primarily responsible for the physiological effects of smoking, including alertness, improved concentration, memory, and enhanced pleasure. Nicotine is present in the leaves of tobacco in amounts ranging from 0.3 to 5%. Nicotine is efficiently absorbed from the lungs and is rapidly transported via blood into the brain where it causes the release of chemical messengers such as norepinephrine, acetylcholine, and dopamine (*see* pages 221-222). Nicotine acts by binding to nicotinic acetylcholine receptors (nAChR), and opening ligand-gated ion channels that allow calcium, sodium, and potassium to pass through the cell membrane.[1]

The strongly addictive properties of nicotine stem from its ability to cause the release of dopamine in the mesolimbic pathway, the reward circuit of the brain responsible for positive reinforcement and feelings of pleasure. Although nicotine leads to the desensitization of nicotinic acetylcholine receptors, the brain compensates for this effect by upregulating the number of these receptors. The net result is an increase in reward pathway sensitivity. Nicotine addiction is physiological, persistent, and psychological. Relapse is common even after months of cessation.

Varenicline acts on the α4β2 nicotinic acetylcholine receptor, one of the major subtypes in the central nervous system that is important in the dependence-forming effect of nicotine. Binding of varenicline (12 hour half-life) to this receptor leads to a sustained increase in dopamine levels in the mesolimbic system, which is less than that induced by nicotine (1-2 hour half-life) but high enough to alleviate nicotine craving and withdrawal symptoms. It also blocks the activation of these receptors by nicotine, because it binds to the receptor more strongly.[2]

Other therapies to aid smoking cessation are primarily based on prevention of nicotine withdrawal symptoms and include various low dose nicotine formulations such as in a patch, nasal spray, gum, or sublingual tablet.

A non-nicotine based aid, bupropion, originally developed as an antidepressant by GlaxoSmithKline, was approved for smoking cessation in 1997 (marketed as Zyban).

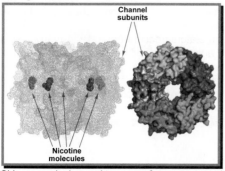

Side on mesh view and top on surface representation of nicotine binding to nAChR.

**Nicotine**     **Bupropion (Zyban™)**

1. *Neuron* **2004**, *41*, 907-914 (1UW6); 2. *Nat. Rev. Drug Discov.* **2006**, *5*, 537-538; **Refs. p. 234**

# THE BRAIN, NEUROTRANSMISSION AND MOLECULAR NEUROTRANSMITTERS

The least developed and most challenging area of molecular medicine is that encompassing disorders of the brain. This fact is hardly surprising considering the complexity of the brain and its remarkably powerful capabilities.[1]

The human brain contains about $10^{11}$ neurons and about $10^{15}$ synapses, the connections between them. In addition, there are about $10^{12}$–$10^{13}$ glial cells, which support the functions of neuronal cells. Each neuron has a cell body (soma with a nucleus, protein synthesizing and processing machinery, and mitochondria that provide energy as ATP. Also, each neuron is associated with:

(1) dendrites – long, thin, highly-branched, stringy structures that allow reception of signals from (but generally not to) other neurons.

(2) an axon – a long fiber (up to 1 m) which carries signals away from the soma to axonal termini (synaptic knobs). Axons are usually extensively branched allowing signals to be sent to many other neurons via dendrites.

Axons are covered by sheaths of a fatty substance, myelin, which provides insulation and allows the passage of a low current through the axon at speeds of up to 120 m/sec (or 270 mph). Each neuron can make a large number (>25,000) of connections with other neurons. The combined length of nerve fibers in each human brain has been estimated as about 4 million miles.

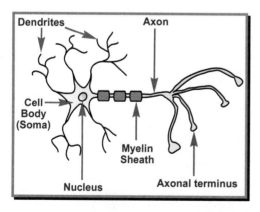

Signaling between neurons occurs by chemical neurotransmission from an axonal terminus to a dendrite across a synaptic gap of about $2 \times 10^{-5}$ mm (20 nm) that is mediated by about 50 different chemical neurotransmitters. Several steps are involved in this neurotransmission. The neurotransmitter molecule which is synthesized in the soma and stored near the axonal terminus (presynaptic) is released when the neuron attains its "firing" potential. The neurotransmitter diffuses across the synaptic gap and binds to a cognate receptor (postsynaptic) which becomes activated, changes shape and triggers the opening of an ion channel on the postsynaptic dendrite. In this process the original signal is amplified and transmitted via the dendrite to the receiving neuron. The greater the amount of neurotransmitter released at the synapse, the stronger will be the connection and signal.

In summary, neurons transmit signals via axons and receive signals from other neurons via dendrite fibers. The flow of information occurs across the synapse and is carried by neurotransmitter molecules that activate specific receptors across the synaptic gap.

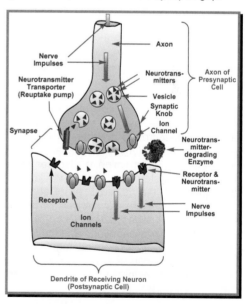

There are several types of receptors for neurotransmitters including ligand-gated ion channels and G protein-coupled receptors (see page 78). Once a neurotransmitter has signaled, it may either be inactivated biochemically or recycled by a transporter protein – a reuptake pump – that carries it back to the axonal terminus for reuse.

The fifty or so neurotransmitters fall into several structurally distinct classes. Some examples are listed below.

## (1) Acetylcholine;

**Acetylcholine**

Acetylcholine, the first neurotransmitter to be identified, was possibly also the first to arise during evolution. It occurs in both the central and peripheral nervous system and plays a key role in countless body functions; it might be even called the dominant neurotransmitter. It is involved in several disorders (e.g., myasthenia gravis and Alzheimer's disease). In the case of Alzheimer's dementia, the levels of acetyl-choline are subnormal and possibly respon-sible in part for the characteristic deficits in cognition and memory. Inhibition of enzymes that break down acetylcholine and increase the levels of this neurotransmitter is a general approach for the palliative treatment of the condition (*see* page 225).

## (2) Monoamines (dopamine, norepinephrine, serotonin, histamine);

**Dopamine**    **Norepinephrine**    **Serotonin**

In the brain, dopamine is crucial for many processes including the control of movement. In Parkinson's disease for example, the loss of dopamine-secreting neurons in the brain leads to slowness of movement and rigidity (*see* page 224). Dopamine is also central to the reward system of the brain, and is associated with the development of addiction in case of certain substances including nicotine (*see* page 220). Dopamine is important for concentration, memory, and regulation of mood. Dopamine regulates the secretion of prolactin and also is a precursor to other neurotransmitters such as epineph-rine and norepinephrine.

Serotonin has many functions in the central nervous system, including regulation of aggression, mood, sleep, body temperature etc. A deficit in serotonin may contribute to migraine, depression, bipolar disorder, and anxiety.

## (3) Amino acids (glycine, GABA, glutamic acid);

**Glycine**        **GABA**        **Glutamic acid**

GABA is an inhibitory neurotransmitter that acts by binding to transmembrane proteins that causes the opening of ion channels. Drugs that act as agonists of GABA are useful sleep aids and sedatives.

## (4) Peptides (neuropeptide Y, vasopressin, endorphin, glucagon, somatostatin);

**Vasopressin**

Substance P, a peptide neurotransmitter, is involved in the sensing of pain while endorphins act as natural opiates.

## (5) Purines (adenosine, ATP, GTP);

**Adenosine**            **Guanosine triphosphate**

Small-molecule (non-peptidic) neurotrans-mitters are biosynthesized close to the synaptic knobs whereas peptidic neurotrans-mitters are usually made in the soma and transported to the synaptic knobs. The neurotransmitters are stored in granules.

Each neurotransmitter has its own set of functions and binds to a cognate receptor with either an excitatory or an inhibitory effect on the postsynaptic cell.

The balance of the various neurotrans-mitters in the brain is critical to the preservation of mental and physical health. Mental disease often can be linked to an imbalance of neurotransmitters or deficits in the downstream signaling.

The brain requires oxygen, glucose and other molecules for proper function. Other molecules which might be harmful are not allowed to cross the blood-brain barrier.[2]

1. Siegel, G. J., Albers, R. W., Brady, S. & Price, D. L. Basic Neurochemistry (Elsevier/Academic Press, **2005**). 2. *Chem. Eng. News* **2007**, *85*, 33-36; Refs. p. 234

# NEURODEGENERATIVE
# AND
# PSYCHIATRIC
# DISEASES

# LEVODOPA (LARODOPA™)

| Structural Formula | Ball-and-Stick Model | Space-filling Model |
|---|---|---|

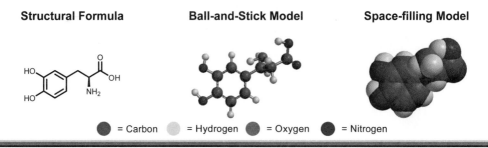

= Carbon   = Hydrogen   = Oxygen   = Nitrogen

*Year of discovery:* Early 1960s (University of Vienna); *Year of introduction:* 1970 (Roche); *Drug category:* Precursor to the neurotransmitter dopamine; *Main uses:* Treatment of Parkinson's disease; *Other names:* Dopar, L-DOPA, 3,4-Dihydroxy-L-phenylalanine; *Related drugs:* Bromocriptine (Parlodel), Pergolide (Permax), Ropinirole (Requip), Pramipexole (Mirapex).

Levodopa (L-DOPA), the metabolic precursor of the mammalian neurotransmitter dopamine, is used for the management of Parkinson's disease, a neurodegenerative disorder. This disease affects about 1% of the population over the age of 65. The earliest symptoms are difficulty in movement, muscular rigidity, tremor, and impairment of postural balance. Certain genetic mutations have been associated with its development. The disease may be multifactorial since Parkinson-like symptoms can be also induced by environmental toxins, certain drugs, inflammation, stroke, or head trauma.[1]

The primary defect in Parkinson's disease is the loss of pigmented dopamine-secreting neurons in the *substantia nigra pars compacta* region of the brain that project to the striatum. These neurons participate in mediating the activity of neural circuits of the basal ganglia that regulate the flow of information from the cerebral cortex to the motor neurons of the spinal cord and control movement. Loss of these neurons leads to reduced excitation of the motor cortex, causing slowness of movement and rigidity. A characteristic feature of Parkinson's disease is the abnormal accumulation of ubiquinated proteins in the affected areas of the brain that form inclusions termed as Lewy bodies, which are implicated in neuronal damage. Neurodegeneration can be the result of mitochondrial dysfunction, oxidative stress, excitotoxicity or inflammation.[2]

Dopamine in the brain is produced from tyrosine through the intermediacy of levodopa. Levodopa was introduced for the treatment of Parkinson's disease in the late 1960s in Europe, soon after it was recognized that persons suffering from the condition have reduced levels of dopamine in the brain.

Tyrosine → (Tyrosine-3-mono-oxygenase) → Levodopa

L-Amino acid decarboxylase ↓

Dopamine

Levodopa, unlike dopamine, can cross the blood brain barrier, where it is decarboxylated to form dopamine and $CO_2$. Decarboxylation can also occur in the periphery. To prevent this process, levodopa is generally administered with carbidopa, an inhibitor of the enzyme L-amino acid decarboxylase.

Carbidopa

Currently levodopa is the most widely used and effective agent for the treatment of Parkinson's disease. At early stages of the disease, the symptoms can be almost completely suppressed by levodopa. However, the disease is progressive and the beneficial effects of the drug eventually decrease. Deep-brain electrical stimulation via implanted electrodes is the treatment of last resort.

1. *Lancet* **2004**, *363*, 1783-1793; 2. *Science* **2003**, *302*, 819-822; **Refs. p. 234**

# DONEPEZIL (ARICEPT™)

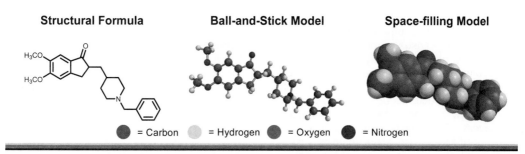

**Structural Formula**  **Ball-and-Stick Model**  **Space-filling Model**

● = Carbon  ○ = Hydrogen  ● = Oxygen  ● = Nitrogen

*Year of discovery:* Late 1980s; *Year of introduction:* 1996 (Eisai); *Drug category:* Acetylcholine esterase inhibitor; *Main uses:* Treatment of Alzheimer's disease; *Related drugs:* Tacrine (Cognex), Rivastigmine (Exelon), Galantamine (Razadyne).

Donepezil, an acetylcholine esterase inhibitor, is used for the management of Alzheimer's disease, globally the most common neurodegenerative disorder. In the US alone it affects some 5 million people, mainly those over 70. Although the exact causes of the disease are unclear, one variant, early onset Alzheimer's has been linked to genetic abnormalities.

Alzheimer's is a progressive disease that results in the irreversible loss of neurons from the cortex and hippocampus and leads to impairment of memory, judgment, decision making, spatial orientation, and language. A characteristic feature of the disease is the formation of extracellular plaques and intracellular neurofibrillary tangles in the brain that consist of insoluble aggregates of β-amyloid peptides. In addition, a protein (called the tau protein) that is involved in neuronal cell structure is dysfunctional in Alzheimer's disease. β-Amyloid peptides, which usually contain 42 amino acids, have a high propensity to aggregate. They are formed by the abnormal cleavage of the transmembrane protein, β-amyloid peptide precursor (APP), by proteolytic enzymes called secretases.[1,2]

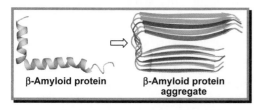

β-Amyloid protein       β-Amyloid protein aggregate

The exact mechanism by which these proteins damage neurons is uncertain. It has been speculated that the aggregates might (1) paralyze neuronal transport and thus provoke an inflammatory response followed by the release of neurotoxic cytokines, or (2) overactivate N-methyl-D-aspartate receptors to a degree that results in the generation of neurotoxic sideproducts.[3]

The loss of brain neurons reduces the levels of several neurotransmitters, the most critical of which is acetylcholine. Donepezil, developed by Eisai and marketed as Aricept in the US since 1996, inhibits the enzyme acetylcholine esterase, which catalyzes the breakdown of acetylcholine. Increased acetylcholine levels in the brain lead to a modest improvement of cognition, but this therapy does not halt the progressive loss of neurons. Several other acetylcholine esterase inhibitors are available, including galantamine (Razadyne™, Janssen Pharma) and rivastigmine (Exelon™, Novartis).

Memantine (Namenda™, Forest Labs), a novel agent for Alzheimer's disease, acts by blocking the N-methyl-D-aspartate receptors.

**Galantamine (Razadyne™)**  **Rivastigmine (Exelon™)**  **Memantine (Namenda™)**

These drugs lead to modest symptomatic improvement, but do not cure the disease. Current research strategies are focused on the inhibition of secretases to impede the formation of β-amyloid aggregates, activating the immune system to clear β-amyloid peptides or restoring normal tau function.[4]

1. *Eur. J. Biochem.* **2002**, *269*, 5642-5648 (**1IYT**); 2. *Proc. Natl. Acad. Sci. U. S. A.* **2005**, *102*, 17342-17347 (**2BEG**); 3. *Nat. Med.* **2004**, 10 Suppl, S2-9; 4. *Nat. Rev. Drug Discov.* **2002**, *1*, 859-866; **Refs. p. 235**

# ANTIEPILEPTIC AGENTS

Epilepsy is a recurring periodic seizure disorder that arises from disregulated electrical and neuronal activity within the brain and results in the loss of mental awareness or consciousness, involuntary movement, or convulsion. It affects about 1% of the human population with varying frequency, duration, and severity. The condition can be caused by a variety of factors, including genetics, injury to the head or stroke, infections of the central nervous system, stress, use of alcohol or certain drugs, or autoimmune cerebral vasculitis. Mild (partial) seizures involve a limited part of the brain and last up to 1-2 minutes. Generalized seizures involve the whole brain, loss of consciousness, and convulsion.

Although there is currently no cure for epilepsy, with appropriate medication, seizures can be suppressed, and about 70% of patients can lead normal lives. Antiepileptic agents are available that selectively downregulate the abnormally high firing rates of neurons that cause seizures without disrupting the normal activity of the nervous system. Treatment often involves the use of more than one agent and must be tailored to the individual patient. [1-3]

The first synthetic antiepileptic agent, phenobarbital, a barbiturate derivative (see page 212), was synthesized in the early 1900s and introduced in 1912 as Luminal by Bayer in Germany. Although phenobarbital has hypnotic and sedative properties, it is also a potent antiseizure agent, and is still used for this purpose. The phenyl group (shown in red) is essential for antiseizure activity.

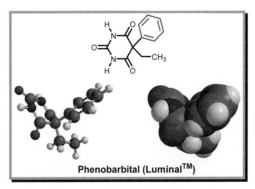

**Phenobarbital (Luminal™)**

Phenobarbital is efficacious for the prevention of most types of epileptic seizures but not for for absence seizures, which are characterized by a short period of impaired consciousness

associated with staring. Phenobarbital acts by activating neuroinhibitory GABA receptors.

Another early anticonvulsant, phenytoin was discovered in 1936 and marketed since 1953 in the US as Dilantin by Parke-Davis. Phenytoin is used for the prevention of both partial and generalized seizures (tonic-clonic type). The latter involves loss of consciousness and alternating periods of sustained muscle contraction and relaxation. The phenyl groups in phenytoin (shown in red) are required for potency.

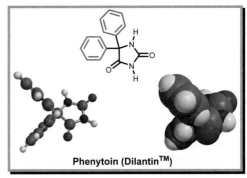

**Phenytoin (Dilantin™)**

During partial and generalized tonic-clonic seizures, neurons fire and depolarize at abnormally high frequency, disrupting brain function. Phenytoin binds to sodium channels in the depolarized state of the neuronal membrane and delays the influx of sodium, which is essential to attain the potential required for firing. The delay is sufficient to prevent the very fast firing that gives rises to seizures, but small enough to allow normal firing rates.

Ion channels are transmembrane protein assemblies that regulate the flow of ions into and out of cells. They are essential for signaling in the nervous system and modulate the transmission of nerve impulses generally, for example to control sensation and movement. The three main types of ion channels are: (1) *voltage-gated ion channels*; (2) *ligand-gated ion channels*, which respond to chemical transmitters; and (3) *mechanically gated channels,* which are triggered by mechanical stimuli. Ion channels are highly selective for specific ions (e.g., $Na^+$, $K^+$, $Ca^{2+}$). The figure below depicts a potassium ion channel. The Nobel Prize in Chemistry for 2003 was awarded to Roderick MacKinnon "for structural and mechanistic studies on ion channels".[4] Ion channels are targets for a number of important medicines, including amlodipine (*see* page 72), gabapentin (*see* page 213) and lidocaine (*see* page 208)

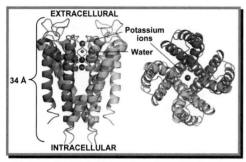

Top-on view and side-on view of a potassium ion channel.

Carbamazepine (Tegretol), another antiepileptic agent used for the prevention of partial and generalized tonic-clonic seizures, was synthesized in 1953 at Geigy laboratories in Basel, and introduced in the US in 1968.

Carbamazepine contains the ring system of the tricyclic antidepressants (*see* page 230) and also a carbamyl group (shown in red), which is essential for antiseizure activity. As with phenytoin, carbamazepine acts by downregulating sodium ion channels so as to inhibit high-frequency firing of neurons. Carbamazepine is also used for the treatment of schizophrenia and trigeminal neuralgia, a condition that results from nerve compression and causes intense facial pain.

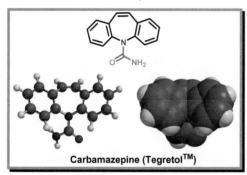

**Carbamazepine (Tegretol™)**

Valproic acid and its sodium salt (Depakene) are also useful antiepileptic medicines. The antiseizure action of valproic acid was serendipitously discovered in the 1960s at the University of Lyon. It has been available in the US from Abbott Laboratories since 1967.

The efficacy of valproic acid is due to a combination of three pharmacological effects. As with phenytoin and carbamazepine, valproic acid prolongs the recovery of voltage-activated sodium ion channels. It also mediates voltage-gated calcium ion channels, causing a small reduction in the low threshold

calcium current. In addition, it leads to an increase in the amount of GABA that can be recovered from the brain after the drug is administered.

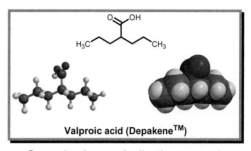

**Valproic acid (Depakene™)**

Several other antiepileptic agents have been developed that can be used as monotherapy or in combination with the other drugs. Gabapentin, which acts by regulating voltage-gated calcium channels, (*see* page 213) is effective against partial seizures. Ethosuximide, developed by Parke-Davis in the 1960s and sold under the name Zarontin, is the primary agent for the prevention of absence seizures. It acts by reducing low threshold calcium ion currents in the hypothalamus. Lamotrigine, developed by GlaxoSmithKline and marketed under the name Lamictal since 1994, is useful against a range of seizures. Lamotrigine acts by mediating voltage-gated sodium ion channels. Tiagabine, a selective GABA uptake inhibitor, was introduced in 1997 under the name Gabitril by Cephalon. It is used for the treatment of partial epileptic seizures. Topiramate, developed by Ortho-McNeil (part of Johnson & Johnson), was introduced in 1996 as Topamax for the prevention of several types of epileptic seizures, but not for absence seizures. It is also used for the treatment of migraine.

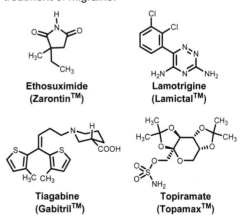

| **Ethosuximide** (Zarontin™) | **Lamotrigine** (Lamictal™) |
| **Tiagabine** (Gabitril™) | **Topiramate** (Topamax™) |

1. *N. Engl. J. Med.* **1996**, *334*, 168-175; 2. *Lancet Neurol.* **2004**, *3*, 729-735; 3. *Nat. Rev. Neurosci.* **2004**, *5*, 553-564; 4. *Science* **1998**, *280*, 69-77 (**1BL8**); **Refs. p 235**

# ANTIANXIETY AGENTS

Anxiety disorders affect nearly 40 million people in the US alone. Brief episodes of anxiety, due to a presentation, a performance, or an examination, are considered a normal aspect of the "fight or flight" response that prepares the body for action. Feelings of intense irrational fear or panic, coupled with dizziness, trembling, nausea, sweat, and rapid heartbeat are symptomatic of anxiety disorder, of which there are five clinically distinguishable variants:

(1)   panic disorder;

(2)   obsessive-compulsive disorder (OCD);

(3)   post-traumatic stress disorder (PTSD);

(4)   phobias (specific and social phobia);

(5)   generalized anxiety disorder (GAD).

These conditions require treatment since they may progress to depression or other mental problems that can negatively impact the individual's life. Traumatic events such as death of a loved one, divorce, or life-threatening experiences (e.g., combat) can trigger anxiety disorders. Other factors that may contribute are genetic predisposition, abnormalities in personality or brain chemistry, and stress.

Once diagnosed, the symptoms of anxiety disorder are often effectively treated with medications, cognitive-behavioral therapy, or relaxation techniques.[1-3]

The introduction of diazepam (Valium™, *see* page 214) was the first major advance in the treatment of anxiety. Other benzodiaze-pines such as clonazepam (Klonopin™, Roche) and alprazolam (Xanax™, Upjohn) are now used specifically to treat panic attacks (*see* structures below). However, the potential for abuse, drug dependence or even addiction limits their use to the most serious illness.

**Clonazepam (Klonopin™)**          **Alprazolam (Xanax™)**

In 1986, buspirone (BuSpar™, BMS) was approved for the treatment of GAD. Buspirone

is not a benzodiazepine and is less potent than benzodiazepines in the treatment of anxiety disorders, but it shows no tendency for addiction or abuse and it does not possess anticonvulsant or sedative properties. However, it suffers from a long onset of action (1-3 weeks). Adverse effects (dizziness and headache) are minor. One treatment strategy is to initiate therapy with a combination of buspirone and a benzodiazepine, and to reduce gradually the benzodiazepine doses. This approach minimizes the occurrence of withdrawal symptoms.

Buspirone is an agonist of presynaptic and a partial agonist of postsynaptic 5-HT$_{1A}$ receptors and, thus, modulates the levels and function of the neurotransmitter serotonin. Buspirone has moderate affinity for dopamine type 2 receptors (DA$_2$) at which it acts as a mixed agonist/antagonist. It also increases the metabolism of norepinephrine, decreases the concentration of acetylcholine and indirectly interferes with GABA transmission.

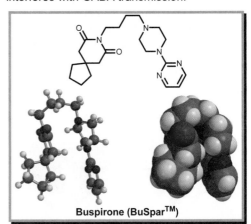

**Buspirone (BuSpar™)**

When depression and anxiety occur simultaneously, a selective serotonin reuptake inhibitor (SSRI) or a serotonin-norepinephrine reuptake inhibitor (SNRI) can be used effectively (*see* also page 230). Other drugs have been used to treat anxiety disorders, including β-blockers (page 68), tricyclic antidepressants (TCAs, page 230) and monoamine oxidase inhibitors (MAOIs, *see* page 229). Currently the first-line treatment option is SSRIs or SNRIs; benzodiazepines are considered only as a second-line therapy.

1. *Human Psychopharmacology* **2002**, *17*, 383-400; 2. *Biol. Psychiatry* **2002**, *51*, 109-120; 3. *Eur. Neuropsycho-pharmacol.* **2006**, *16*, S119-S127; Refs. p. 236

# ANTIDEPRESSANTS

## Introduction

Depression is a serious mental health problem that, at any particular time, affects over a 120 million people according to the World Health Organization. It is estimated that in the US about 15% of the adult population will suffer from depression at least once in their lives. It can be severely debilitating and impact every aspect of an individual's life and economic situation. The symptoms of depression include profound sadness, loss of energy, appetite, motivation and ability to concentrate, insomnia, withdrawal, hopelessness, and low self-esteem. In extreme cases depression can lead to suicide, causing an estimated 850,000 deaths each year. Prolonged depression is a spectrum of disorders at the ends of which are major depression (major depressive disorder) and bipolar disorder (manic depressive disorder).

Depression is usually diagnosed by a primary care physician or a psychiatric specialist. The treatment for depression varies greatly from one patient to another and can involve drug therapy, psychotherapy, or a combination of the two.[1,2] The four main classes of antidepressant agents, as discussed in this section, are:

(1)     Monoamine Oxidase Inhibitors (MAOIs);

(2)     Tricyclic Antidepressants (TCA);

(3)     Selective Serotonin Reuptake Inhibitors (SSRIs)

(4)     Serotonin-Norepinephrine Reuptake Inhibitors (SNRIs).

Finally, lithium salts, generally administered in the form of lithium carbonate ($Li_2CO_3$), have the remarkable property of diminishing abnormal mood changes. Lithium carbonate is frequently used in combination with an antipsychotic or an antidepressant for the management of bipolar disorder. The dose must be adjusted according to blood levels, which should be monitored carefully (every 2 months).[3]

## Monoamine Oxidase Inhibitors (MAOIs)

For over four centuries *laudanum*, an elixir prepared from poppy, was used to treat pain and sorrow. It was later recognized that opium

(*see* page 209) was responsible for the analgesic and antidepressant effects of laudanum. Another old natural remedy for melancholia (depressed mood) was the extract of the snakeroot plant (*Rauwolfia serpentina*) that contains the alkaloid reserpine (shown below).[4]

**Reserpine**                     **Iproniazid**

In the 1950s it was discovered by N.S. Kline that treatment with reserpine decreased the level of monoamine neurotransmitters norepinephrine and serotonin (*see* page 222) and actually *induced* depression in normal individuals. This led to the hypothesis that monoamine neurotransmitters play a critical role in the development of mood disorders. Coincidentally, Kline also discovered the antidepressant activity of iproniazid (Marsilid™, Hoffmann-La Roche), the first entirely synthetic antidepressant agent. Iproniazid (structure shown above), was originally developed as an antitubercular drug along with isoniazid (*see* page 140). The research team led by Kline demonstrated that iproniazid significantly improved the mood of tubercular patients suffering from depression. Although Marsilid enjoyed success for a few years, it had to be withdrawn from use because of liver toxicity.

In 1951, it was shown that iproniazid inhibits the enzyme monoamine oxidase (MAO) that is responsible for the oxidative deactivation of monoamine neurotransmitters such as serotonin (*see* below). When MAO is inhibited, the amount of serotonin available for neurotransmission is elevated, along with the mood of the patient.

**Serotonin**                     **5-Hydroxyindole acetic acid**

There are two isoforms of MAO, MAO-A and MAO-B. The enzyme MAO-A metabolizes both dietary amines (e.g., tyramine in cheese, beer, and wine) and neurotransmitters. However, MAO-B metabolizes only the latter.

Older MAO inhibitors such as tranyl-cypromine (Parnate™) inhibit both enzymes and can cause elevation of tyramine levels and dangerous hypertension unless the user abstains from foods that provide tyramine.

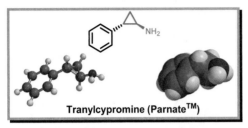

**Tranylcypromine (Parnate™)**

Tranylcypromine, a cyclopropylamine (shown in red) derivative, is indicated for the treatment of major depression that does not respond to other antidepressants. It is also used for post-traumatic stress disorder.

Until recently, MAOIs were used sparingly due to the risks mentioned above. In 2006, selegiline (Emsam™, BMS), a selective MAO-B inhibitor, was approved for the treatment of major depression in the form of a transdermal patch. When administered in small doses, selegiline selectively inhibits MAO-B with the result that dietary restrictions are unnecessary. However, at higher doses (all oral formulations) the inhibition of MAO-A becomes significant. Because MAOIs have a strong tendency to affect the metabolism of other drugs, they should not be used in combination with other antidepressants.

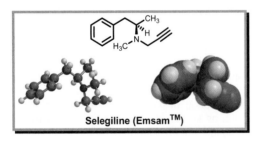

**Selegiline (Emsam™)**

**Tricyclic Antidepressants (TCAs)**

The discovery of tricyclic antidepressants was serendipitous. Chlorpromazine, a tricyclic benzothiazine derivative, was observed to have antischizophrenic activity in the early 1950s. Its success stimulated the search for other antischizophrenic compounds. Roland Kuhn, a Swiss psychiatrist working at Ciba-Geigy, tested a large number of tricyclic compounds that were structurally analogous to chlorpromazine, and found that the amine side chain (shown in blue) is essential for

pharmacological activity. One of these compounds, imipramine, had a beneficial effect on depressed patients. Imipramine was then marketed as Tofranil, and for several years was the only available antidepressant because of the withdrawal of iproniazid, the first MAO inhibitor.

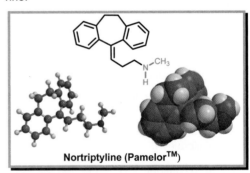

**Chlorpromazine**          **Imipramine (Tofranil™)**

Tricylic antidepressants are inhibitors of neurotransmitter reuptake, including the reuptake of serotonin, norepinephrine, and dopamine. Imipramine causes numerous side effects and also has a narrow therapeutic window, which can lead to an overdose, or even death. During the 1960s and 1970s, over a dozen other TCAs were approved including nortriptyline (Pamelor™), a second-generation drug. The first-generation TCA drugs all contain a tertiary amino group and inhibit the reuptake of both serotonin and norepinephrine. However, nortriptyline, a secondary amine (functional group shown in green), selectively inhibits the reuptake of norepinephrine.

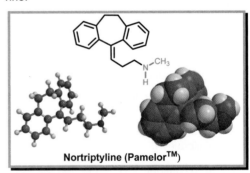

**Nortriptyline (Pamelor™)**

Most TCAs have been shown to inhibit ion channels ($Na^+$, $K^+$ and $Ca^{2+}$) and to cause dangerous (sometimes life-threatening) arrhythmia or sudden cardiac death at high doses. For this reason TCAs have been largely replaced by selective serotonin reuptake inhibitors (SSRIs).

**Selective Serotonin Reuptake Inhibitors (SSRIs)**

The first widely used selective serotonin reuptake inhibitor was discovered by Bryan B. Malloy and coworkers at Eli Lilly Co. in 1970. The starting points for their studies were the findings that:

(1)  MAOs are effective as antidepressants because they elevate concentrations of amine neurotransmitters;

(2)  the antihistamine diphenhydramine is a weak antidepressant;

(3)  diphenhydramine blocks the transport (reuptake) of norepinephrine and serotonin (discovered by Dr. Arvid Carlsson in Sweden in the 1960s).

By synthesizing and testing numerous analogs of diphenhydramine, the Lilly group discovered highly active compounds of the phenoxypropylamine class.

**Diphenhydramine (Benadryl)**

**Phenoxypropylamines** ($R^1$, $R^2$, $R^3$ are various substituents)

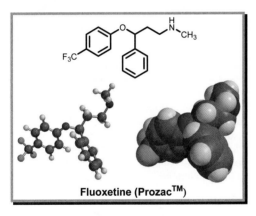

**Fluoxetine (Prozac™)**

Fluoxetine, the compound with the best overall profile, was identified as a highly selective serotonin reuptake inhibitor and introduced into clinical practice in 1988 as Prozac. Owing to its superior safety profile and effectiveness as an antidepressant, Prozac soon became a historic drug with annual sales of over $3 billion. A number of second-generation SSRIs were developed and approved during the 1990s including sertraline (Zoloft™, Pfizer), paroxetine (Paxil™, GSK) and escitalopram (Lexapro™, Lundbeck), each of which attained annual sales in the billions.

Although SSRIs are safer than TCAs, they are not devoid of side effects. Most common are sexual side effects (e.g., anorgasmia, erectile dysfunction and decreased libido), sleep disorders (e.g., insomnia or excessive sleep) and gastrointestinal discomfort. Abrupt discontinuation of treatment with SSRIs can lead to withdrawal symptoms (e.g., vertigo and electric shock sensations).

**Sertraline (Zoloft™)**   **Paroxetine (Paxil™)**

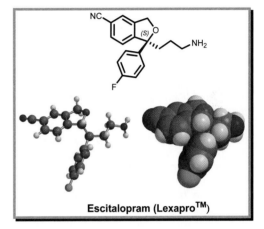

**Escitalopram (Lexapro™)**

### Serotonin-Norepinephrine Reuptake Inhibitors (SNRIs)

Venlafaxine (Effexor™, Wyeth) and duloxetine (Cymbalta™, Eli Lilly) belong to a class of antidepressants that simultaneously inhibit the reuptake of both serotonin and norepinephrine. These dual reuptake inhibitors can be effective for patients who do not respond to other drugs.

Bupropion is a widely used dual reuptake inhibitor of dopamine and norepinephrine, which functions both as an antidepressant and an aid to smoking cessation (*see* page 220).

**Venlafaxine (Effexor™)**   **Duloxetine (Cymbalta™)**

1. *NeuroRx* **2006**, *3*, 42-56; 2. *Pharmacol. Ther.* **2007**, *113*, 134-153; 3. *Am. J. Med.* **2006**, *119*, 478-481; 4. *J. Neural Transm.* **2001**, *108*, 707-716; Refs. p. 235

# ANTIPSYCHOTICS

There is a wide range of disorders of the brain that are disabling, chronic and medically important. These are generally grouped into broad classes that depend on symptoms and medical observations because the detailed biochemical, physiological, genetic, or situational causes are unknown. Most of these involve some degree of depression admixed with varying proportions of other abnormalities including anxiety, misperception of reality, inability to reason, antisocial behavior or withdrawal, low self-esteem, delusions, hallucinations, paranoia, mania, or suicidality.

Simple depression is a treatable mental condition (see page 229) that is very wide-spread and affects one person in five during a lifetime. It is also serious, since it can impair a person's ability to function, and can even lead to suicide. Chronic dysthymic disorder is a more severe form of depression that involves profound mood disturbance, frequent recurrence, and resistance to treatment. Admixed with mania or mood disorders, it is known as bipolar disorder, an oversimplification of what is a multifactorial condition. More than half of all suicides are linked to mood disorders at some level. Mood swings are somewhat ameliorated by the use of lithium, valproate, or carbamazepine (see page 227). The incidence of the various mood disorders has been estimated at about five percent of the total population.

The mental illness known as schizophrenia affects about one percent of the population globally. It is severely disabling because it can involve impairment of cognition, reasoning, language ability, and memory as well as delusional, antisocial, and hallucinatory behavior. Schizophrenia may be associated with abnormalities of the brain that can be detected by imaging studies and also defects in neurotransmission, especially that involving dopamine, GABA, and acetylcholine. Most schizophrenics self-medicate with nicotine by smoking heavily.

There are innumerable factors that have been implicated in the pathogenesis of depression, mood, and cognitive disorders. These include stress, insomnia, low levels of neurotransmitters, and defects in G protein receptors and calcium signaling.

The earliest antipsychotic agents, chlorpromazine and haloperidol showed efficacy but caused a number of side effects. They were replaced by second-generation antipsychotics that are now used almost exclusively in clinical practice. These drugs, the atypical antipsychotics, show somewhat greater efficacy and a diminished side effect profile as compared to the first-generation medicines.[1, 2]

The first atypical antipsychotic, clozapine, a tricyclic compound that was developed by Sandoz in the early 1960s (see structure below), has been used widely for the treatment of schizophrenia. Unfortunately, the use of clozapine can cause a fatal blood disorder, agranulocytosis, in about 1% of those treated.

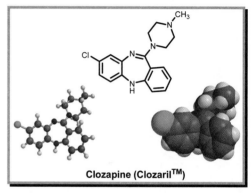

**Clozapine (Clozaril™)**

Other atypical antipsychotics in wide clinical use for both schizophrenia and bipolar disorder are quetiapine (Seroquel™, AstraZeneca) and ziprasidone (Geodon™, Pfizer). Although these do not appear to have serious side effects at normal doses, their efficacy is limited.

**Quetiapine (Seroquel™)**     **Ziprasidone (Geodon™)**

**Olanzapine (Zyprexa™)**     **Aripiprazole (Abilify™)**

In addition, the atypical antipsychotics olanzapine (Zyprexa™, Eli Lilly) and aripiprazole (Abilify™, Otsuka Pharmaceuticals) have been shown to be somewhat helpful but to carry an increased risk for weight gain and other metabolic disorders.

1. Curr. Drug Ther. 2006, 1, 1-7; 2. CNS Drugs 2006, 20, 389-409; Refs. p. 236

# REFERENCES FOR PART VI.

## Lidocaine (Xylocaine, page 208)[1-5]

1. Budenz Alan, W. Local anesthetics in dentistry: then and now. *J. Calif. Dent. Assoc.* **2003**, *31*, 388-396.

2. Sabrine, A. & Lyons, G. New local anesthetic analgesics. *Pain* **2003**, 795-801.

3. Fozzard, H. A., Lee, P. J. & Lipkind, G. M. Mechanism of local anesthetic drug action on voltage-gated sodium channels. *Curr. Pharm. Des.* **2005**, *11*, 2671-2686.

4. White, J. L. & Durieux, M. E. Clinical pharmacology of local anesthetics. *Anesthesiology Clinics of North America* **2005**, *23*, 73-84.

5. Yanagidate, F. & Strichartz, G. R. Local anesthetics. *Handbook of Experimental Pharmacology* **2007**, *177*, 95-127.

## Morphine (Avinza, page 209)[1-5]

1. Zenk, M. H. & Tabata, M. Opium. Its history, merits and demerits. *Natural Medicines* **1996**, *50*, 86-102.

2. Bodnar, R. J. & Klein, G. E. Endogenous opiates and behavior: 2005. *Peptides* **2006**, *27*, 3391-3478.

3. Marie, N., Aguila, B. & Allouche, S. Tracking the opioid receptors on the way of desensitization. *Cell. Signal.* **2006**, *18*, 1815-1833.

4. McClung, C. A. The molecular mechanisms of morphine addiction. *Rev. Neurosci.* **2006**, *17*, 393-402.

5. Zoellner, C. & Stein, C. Opioids. *Handbook of Experimental Pharmacology* **2007**, *177*, 31-63.

## Acetaminophen (Tylenol, page 210)[1-5]

1. Brune, K. The early history of non-opioid analgesics. *Acute Pain* **1997**, *1*, 33-40.

2. Brown, T. M., Dronsfield, A. T. & Ellis, P. M. Pain relief: from coal tar to paracetamol. *Educ. Chem.* **2005**, *42*, 102-105.

3. Josephy, P. D. The Molecular Toxicology of Acetaminophen. *Drug Metab. Rev.* **2005**, *37*, 581-594.

4. Kalso, E. Oxycodone. *J. Pain Symptom Manage.* **2005**, *29*, S47-S56.

5. Bertolini, A. et al. Paracetamol: new vistas of an old drug. *CNS Drug Rev.* **2006**, *12*, 250-275.

## Fentanyl (Duragesic, page 211)[1-5]

1. Van Gestel, S. & Schuermans, V. Thirty-three years of drug discovery and research with Dr. Paul Janssen. *Drug Dev. Res.* **1986**, *8*, 1-13.

2. Berger, J. M. Opioids in anesthesia. *Seminars in Anesthesia, Perioperative Medicine and Pain* **2005**, *24*, 108-119.

3. Stanley, T. H. Fentanyl. *J. Pain Symptom Manage.* **2005**, *29*, S67-S71.

4. Battershill, A. J. & Keating, G. M. Remifentanil: a review of its analgesic and sedative use in the intensive care unit. *Drugs* **2006**, *66*, 365-385.

5. Mystakidou, K., Katsouda, E., Parpa, E., Vlahos, L. & Tsiatas, M. Oral Transmucosal Fentanyl Citrate: Overview of Pharmacological and Clinical Characteristics. *Drug Delivery* **2006**, *13*, 269-276.

## Sodium Thiopental (Sodium Pentothal, page 212)[1-5]

1. McLeish, M. J. Thiopental sodium. *Analytical Profiles of Drug Substances and Excipients* **1992**, *21*, 535-572.

2. Brandenberger, H. Barbiturates. *Analytical Toxicology for Clinical, Forensic and Pharmaceutical Chemists* **1997**, 347-368.

3. Russo, H. & Bressoll, F. Pharmacodynamics and pharmacokinetics of thiopental. *Clin. Pharmacokinet.* **1998**, *35*, 95-134.

4. Lopez-Munoz, F., Ucha-Udabe, R. & Alamo, C. The history of barbiturates a century after their clinical introduction. *Neuropsychiatr. Dis. Treatm.* **2005**, *1*, 329-343.

5. White, P. F. Intravenous (non-opioid) anesthesia. *Seminars in Anesthesia, Perioperative Medicine and Pain* **2005**, *24*, 101-107.

## Gabapentin (Neurontin, page 213)[1-5]

1. Revill, P., Bolos, J., Serradell, N. & Bayes, M. Gabapentin enacarbil. *Drugs of the Future* **2006**, *31*, 771-777.

2. Sills, G. J. The mechanisms of action of Gabapentin and Pregabalin. *Curr. Opin. Pharmacol.* **2006**, *6*, 108-113.

3. Dickenson, A. H. & Ghandehari, J. Anti-convulsants and anti-depressants. *Handbook of Experimental Pharmacology* **2007**, *177*, 145-177.

4. Dooley, D. J., Taylor, C. P., Donevan, S. & Feltner, D. Ca2+ channel $\alpha2\delta$ ligands: novel modulators of neurotransmission. *Trends Pharmacol. Sci.* **2007**, *28*, 75-82.

5. Tassone, D. M., Boyce, E., Guyer, J. & Nuzum, D. Pregabalin: a novel γ-aminobutyric acid analogue in the treatment of neuropathic pain, partial-onset seizures, and anxiety disorders. *Clin. Ther.* **2007**, *29*, 26-48.

## Diazepam (Valium, page 214)[1-5]

1. Baenninger, A. & et al. Good Chemistry: The Life and Legacy of Valium Inventor Leo Sternbach (McGraw-Hill, New York, **2003**).

2. O'Brien, C. P. Benzodiazepine use, abuse, and dependence. *J. Clin. Psychiatry* **2005**, *66*, 28-33.

3. Sigel, E. The benzodiazepine recognition site on GABA$_A$ receptors. *Med. Chem. Rev.--Online* **2005**, *2*, 251-256.

4. Beracochea, D. Anterograde and retrograde effects of benzodiazepines on memory. *TheScientificWorld* **2006**, *6*, 1460-1465.

5. Sieghart, W. GABAA receptors as targets for different classes of drugs. *Drugs of the Future* **2006**, *31*, 685-694.

## Sumatriptan (Imitrex, page 215)[1-5]

1. Lanfumey, L. & Hamon, M. 5-HT1 receptors. *Current Drug Targets: CNS & Neurological Disorders* **2004**, *3*, 1-10.

2. Cady, R. & Schreiber, C. Sumatriptan: update and review. *Exp. Opin. Pharmacother.* **2006**, *7*, 1503-1514.

3. Krymchantowski, A. V. The use of combination therapies in the acute management of migraine. *Neuropsychiatr. Dis. Treatm.* **2006**, *2*, 293-297.

4. Ramadan, N. M. & Buchanan, T. M. New and future migraine therapy. *Pharmacol. Ther.* **2006**, *112*, 199-212.

5. Goadsby, P. J. Serotonin receptor ligands: treatments of acute migraine and cluster headache. *Handbook of Experimental Pharmacology* **2007**, *177*, 129-143.

## Zolpidem (Ambien, page 218)[1-5]

1. Kupfer, D. J.; Reynolds, C. F., III Management of insomnia *N. Engl. J. Med.* **1997**, *336*, 341-346.

2. Sanger, D. J.; Depoortere, H. The pharmacology and mechanism of action of zolpidem *CNS Drug Reviews* **1998**, *4*, 323-340.

3. Lancel, M.; Steiger, A. Sleep and its modulation by drugs that affect GABAA receptor function *Angew. Chem., Int. Ed. Engl.* **1999**, *38*, 2853-2864.

4. Sateia, M. J.; Nowell, P. D. Insomnia *Lancet* **2004**, *364*, 1959-1973.

5. Harrison, T. S.; Keating, G. M. Zolpidem: A review of its use in the management of insomnia *CNS Drugs* **2005**, *19*, 65-89.

## Ramelteon (Rozerem, page 219)[1-5]

1. Hastings, M. H.; Reddy, A. B.; Maywood, E. S. A clockwork web: circadian timing in brain and periphery, in health and disease *Nat. Rev. Neurosci.* **2003**, *4*, 649-661.

2. Turek Fred, W.; Gillette Martha, U. Melatonin, sleep, and circadian rhythms: rationale for development of specific melatonin agonists *Sleep Med.* **2004**, *5*, 523-532.

3. Nguyen, N. N.; Yu, S. S.; Song, J. C. Ramelteon: a novel melatonin receptor agonist for the treatment of insomnia *Formulary* **2005**, *40*, 146-150, 152-155.

4. Saper, C. B.; Scammell, T. E.; Lu, J. Hypothalamic regulation of sleep and circadian rhythms *Nature* **2005**, *437*, 1257-1263.

5. Owen, R. T. Ramelteon: profile of a new sleep-promoting medication *Drugs of Today* **2006**, *42*, 255-263.

## Varenicline (Chantix, page 220)[2-6]
## X-Ray Structure (1UW6)[1]

1. Celie, P. H. N.; Van Rossum-Fikkert, S. E.; Van Dijk, W. J.; Brejc, K.; Smit, A. B.; Sixma, T. K. Nicotine and carbamylcholine binding to nicotinic acetylcholine receptors as studied in AChBP crystal structures *Neuron* **2004**, *41*, 907-914.

2. Laviolette Steven, R.; van der Kooy, D. The neurobiology of nicotine addiction: bridging the gap from molecules to behaviour *Nat. Rev. Neurosci.* **2004**, *5*, 55-65.

3. Foulds, J. The neurobiological basis for partial agonist treatment of nicotine dependence: varenicline *Int. J. Clin. Pract.* **2006**, *60*, 571-576.

4. Niaura, R.; Jones, C.; Kirkpatrick, P. Varenicline *Nat. Rev. Drug Discov.* **2006**, *5*, 537-538.

5. Sorbera, L. A.; Castaner, J. Varenicline tartrate: aid to smoking cessation nicotinic a4b2 partial agonist *Drugs of the Future* **2006**, *31*, 117-122.

6. Tonstad, S. Varenicline for smoking cessation *Expert. Rev. Neurother.* **2007**, *7*, 121-127.

## The Brain, Neurotransmission and Molecular Neurotransmitters (page 221)[1, 2]

1. Siegel, G. J.; Albers, R. W., Brady, S. & Price, D. L. Basic Neurochemistry (Elsevier/Academic Press, **2005**).

2. Everts, S. Brain Barricade. *Chem. Eng. News* **2007**, *85*, 33-36.

## Levodopa (Larodopa, page 224)[1-5]

1. Yudofsky, S. C. Parkinson's disease, depression, and electrical stimulation of the brain *N. Engl. J. Med.* **1999**, *340*, 1500-1502.

2. Dawson, T. M.; Dawson, V. L. Molecular pathways of neurodegeneration in Parkinson's disease *Science* **2003**, *302*, 819-822.

3. Nussbaum, R. L.; Ellis, C. E. Alzheimer's disease and Parkinson's disease *N. Engl. J. Med.* **2003**, *348*, 1356-1364.

4. Samii, A.; Nutt, J. G.; Ransom, B. R. Parkinson's disease *Lancet* **2004**, *363*, 1783-1793.

5. Schapira, A. H. V.; Bezard, E.; Brotchie, J.; Calon, F.; Collingridge, G. L.; Ferger, B.; Hengerer, B.; Hirsch, E.; Jenner, P.; Le Novere, N.; Obeso, J. A.; Schwarzschild, M. A.; Spampinato, U.; Davidai, G. Novel pharmacological targets for the treatment of Parkinson's disease *Nat. Rev. Drug Discov.* **2006**, *5*, 845-854.

## Donepezil (Aricept, page 225)[2-4,6,7]
## X-Ray Structure (1IYT, 2BEG)[1,5]

1. Crescenzi, O.; Tomaselli, S.; Guerrini, R.; Salvadori, S.; D'Ursi, A. M.; Temussi, P. A.; Picone, D. Solution structure of the Alzheimer amyloid b-peptide (1-42) in an apolar microenvironment. Similarity with a virus fusion domain *Eur. J. Biochem.* **2002**, *269*, 5642-5648.

2. Wolfe, M. S. Therapeutic strategies for Alzheimer's disease *Nat. Rev. Drug Discov.* **2002**, *1*, 859-866.

3. Bossy-Wetzel, E.; Schwarzenbacher, R.; Lipton Stuart, A. Molecular pathways to neurodegeneration *Nat. Med.* **2004**, *10 Suppl*, S2-9.

4. Ross, C. A.; Poirier, M. A. Protein aggregation and neurodegenerative disease *Nat. Med.* **2004**, S10-S17.

5. Luhrs, T.; Ritter, C.; Adrian, M.; Riek-Loher, D.; Bohrmann, B.; Dobeli, H.; Schubert, D.; Riek, R. 3D structure of Alzheimer's amyloid-b(1-42) fibrils *Proc. Natl. Acad. Sci. U. S. A.* **2005**, *102*, 17342-17347.

6. Blennow, K.; de Leon, M. J.; Zetterberg, H. Alzheimer's disease *Lancet* **2006**, *368*, 387-403.

7. Robertson, E. D.; Mucke, L. 100 Years and Counting: Prospects for Defeating Alzheimer's Disease *Science* **2006**, *314*, 781-784.

## Antiepileptic Agents (page 226)[1,3-8]
## X-Ray Structure (1BL8)[2]

1. Brodie, M. J.; Dichter, M. A. Drug therapy: antiepileptic drugs *N. Engl. J. Med.* **1996**, *334*, 168-175.

2. Doyle, D. A.; Cabral, J. M.; Pfuetzner, R. A.; Kuo, A.; Gulbis, J. M.; Cohen, S. L.; Chait, B. T.; MacKinnon, R. The structure of the potassium channel: molecular basis of K+ conduction and selectivity *Science* **1998**, *280*, 69-77.

3. Browne, T. R.; Holmes, G. L. Epilepsy *N. Engl. J. Med.* **2001**, *344*, 1145-1151.

4. Avanzini, G.; Franceschetti, S. Cellular biology of epileptogenesis *Lancet Neurol.* **2003**, *2*, 33-42.

5. McCorry, D.; Chadwick, D.; Marson, A. Current drug treatment of epilepsy in adults *Lancet Neurol.* **2004**, *3*, 729-735.

6. Rogawski Michael, A.; Loscher, W. The neurobiology of antiepileptic drugs *Nat. Rev. Neurosci.* **2004**, *5*, 553-564.

7. Schachter Steven, C. Epilepsy: major advances in treatment *Lancet. Neurol.* **2004**, *3*, 11.

8. Pollard, J. R.; French, J. Antiepileptic drugs in development *Lancet Neurol.* **2006**, *5*, 1064-1067.

## Antianxiety Agents (page 228)[1-5]

1. Blanco, C., Antia, S. X. & Liebowitz, M. R. Pharmacotherapy of social anxiety disorder. *Biol. Psychiatry* **2002**, *51*, 109-120.

2. Bourin, M. & Lambert, O. Pharmacotherapy of anxious disorders. *Human Psychopharmacology* **2002**, *17*, 383-400.

3. Goodman, W. K. Selecting pharmacotherapy for generalized anxiety disorder. *J. Clin. Psychiatry* **2004**, *65*, 8-13.

4. Baldwin, D. S. Serotonin noradrenaline reuptake inhibitors: a new generation of treatment for anxiety disorders. *International Journal of Psychiatry in Clinical Practice* **2006**, *10*, 12-15.

5. Dinan, T. Therapeutic options: Addressing the current dilemma. *Eur. Neuropsychopharmacol.* **2006**, *16*, S119-S127.

## Antidepressants (page 229)[1-5]

1. Ban, T. A. Pharmacotherapy of depression: a historical analysis. *J. Neural Transm.* **2001**, *108*, 707-716.

2. Babu, R. P. K. & Maiti, S. N. Norepinephrine reuptake inhibitors for depression, ADHD and other neuropsychiatric disorders. *Heterocycles* **2006**, *69*, 539-567.

3. Freeman, M. P. & Freeman, S. A. Lithium: Clinical Considerations in Internal Medicine. *Am. J. Med.* **2006**, *119*, 478-481.

4. Holtzheimer, P. E., III & Nemeroff, C. B. Advances in the treatment of depression. *NeuroRx* **2006**, *3*, 42-56.

5. Rosenzweig-Lipson, S. et al. Differentiating antidepressants of the future: Efficacy and safety. *Pharmacol. Ther.* **2007**, *113*, 134-153.

MOLECULES AND MEDICINE

## Antipsychotics (page 232)[1-5]

1. Akhondzadeh, S. Pharmacotherapy of schizophrenia: the past, present and future. *Curr. Drug Ther.* **2006**, *1*, 1-7.

2. Fineberg, N. A., Gale, T. M. & Sivakumaran, T. A review of antipsychotics in the treatment of obsessive compulsive disorder. *J. Psychopharmacol.* **2006**, *20*, 97-103.

3. Horacek, J. et al. Mechanism of action of atypical antipsychotic drugs and the neurobiology and schizophrenia. *CNS Drugs* **2006**, *20*, 389-409.

4. Stroup, T. S., Alves, W. M., Hamer, R. M. & Lieberman, J. A. Clinical trials for antipsychotic drugs: design conventions, dilemmas and innovations. *Nat. Rev. Drug. Disc.* **2006**, *5*, 133-146.

5. Seitz, D. P., Gill, S. S. & van Zyl, L. T. Antipsychotics in the treatment of delirium: a systematic review. *J. Clin. Psychiatry* **2007**, *68*, 11-21.

# GLOSSARY

## A

**Å**
Ångström: $10^{-10}$ meter or one ten-billionth of a meter or 0.1 nanometer (nm).

**ACE**
*see* angiotensin-converting enzyme.

**acetylase**
an enzyme that catalyzes the attachment of an acetyl group to an alcohol to form an acetate ester.

**acetylcholine**
a neurotransmitter that is involved in the propagation of nerve impulses. *See* a detailed description in *"The Brain, Neurotransmission and Molecular Neurotransmitters"*, page 221.

**action potential**
a change in electric potential across the membrane of a nerve cell sufficient to produce an electrical signal in the nerve fiber (neuron-firing).

**active site (of an enzyme)**
a region of an enzyme which binds one or more substrates and accelerates the formation of products.

**Addison's disease**
(chronic adrenal insufficiency) a condition in which the adrenal gland produces insufficient amounts of glucocorticoid and mineralocorticoid hormones.

**adenine**
one of the bases present in RNA and DNA; structure:

**adenosine**
a nucleoside in which adenine (blue) is attached to ribose (red) *via* a carbon-nitrogen bond.

**adiponectin**
a regulatory protein that increases insulin sensitivity and lowers blood glucose levels.

**adrenal gland**
orange-colored, triangle-shaped gland located on the top of each kidney that regulates the stress response. The central part, the medulla, produces epinephrine and norepinephrine (adrenaline and noradrenaline). The cortex, which surrounds the medulla, produces corticosteroids, e.g., cortisol.

**adrenoceptor**
(also known as adrenergic receptor) G protein-coupled receptors on the surface of many cells that are activated by the hormones adrenaline and noradrenaline. See *"Atenolol"*, page 68.

**adrenocorticoid steroids**
steroid hormones (e.g., cortisol, cortisone, aldosterone) produced by the cortex of the adrenal glands.

**Agar gel**
a medium for growing cultures of bacteria and fungi that is prepared from red algae or seaweed. It consists of chains of the sugar galactose.

**agonist**
a substance that binds to and activates a receptor, including mimics of the endogenous signaling substance.

**agranulocytosis**
a severe reduction in the number of white blood cells in which the count of white blood cells (normally about 3,000-10,000 per $mm^3$ of blood) drops below $500/mm^3$.

**AIDS**
Acquired Immunodeficiency Syndrome. The illness caused by infection by the human immunodeficiency virus (HIV).

**Ala**
abbreviation for the amino acid alanine; structure:

**alkaloid**
basic nitrogen-containing natural substances produced by plants that have a physiological effect on humans.

**alkylating agent**
a substance that attaches an alkyl group to a nitrogen atom of RNA or DNA.

**allergen**
a substance that causes an allergic reaction.

**allosteric**
binding site of an enzyme that is different from its active site, which when occupied, changes the shape and activity of an enzyme.

**amphoteric**
a compound that can behave either as an acid or a base.

**anandamide**
(arachidonylethanolamide) a derivative of arachidonic acid that acts on cannabinoid receptors in the brain; structure:

**anaphylaxis**
dangerously extreme allergic reaction to an antigen subsequent to a sensitizing earlier exposure.

**androgens**
male steroid hormones (e.g., testosterone) that play a role in the development and maintenance of masculine characteristics.

**anemia**
abnormally low red cell levels in blood.

**angio-**
a combining form pertaining to blood vessels.

MOLECULES AND MEDICINE

**angiogenesis**
generation of new capillaries from pre-existing blood vessels.

**angiotensin-converting enzyme (ACE)**
enzyme catalyzing the conversion of the protein angiotensin I into angiotensin II, a peptide that elevates blood pressure.

**anoxia**
(hypoxia, ischemia) inadequate level of oxygenation.

**antagonist**
a substance that blocks the binding of an endogenous activator to its receptor and prevents the biological response.

**anthranilic acid**
an aromatic amino acid that forms physiologically as a metabolite of the amino acid tryptophan; structure:

**anthrax**
a potentially lethal infectious disease caused by the Gram-positive bacterium *Bacillus anthracis*.

**antibody**
a large Y-shaped protein (immunoglobulin) in the immune system that recognizes specific targets (antigens) such as bacteria and viruses. *See* detailed description in "*A Brief Overview of the Immune System*", page 112.

**anticonvulsant**
a class of drugs used to prevent seizures (convulsions) especially those that occur in epilepsy.

**antigen**
a substance that triggers the production of antibodies (immunoglobulins), which bind to it. *See* detailed description in "*A Brief Overview of the Immune System*", page 112.

**antipyretic**
a drug that reduces fever.

**APC**
antigen-presenting cell.

**apoptosis**
biochemically programmed or regulated cell death.

**arachidonylglycerol**
(2-AG) an endogenous regulator in the human brain that acts on the cannabinoid (marijuana) receptors; structure:

**Arg**
abbreviation for the amino acid arginine; structure:

**arrhythmia**
(also known as cardiac arrhythmia) a condition in which the rhythm of the heart is irregular, abnormally slow or fast.

**Asp**
abbreviation for the amino acid aspartic acid; structure:

**aspartic protease**
protease enzyme that usually has two aspartic acid residues in the catalytically active site (e.g., renin, pepsin and HIV protease).

**ATP**
adenosine triphosphate; a multifunctional nucleotide which provides chemical activation within the cell and powers many physical and biochemical processes; structure:

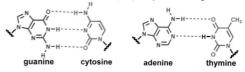

**ATPase**
a class of enzymes that catalyze the cleavage of adenosine triphosphate (ATP) to adenosine diphosphate (ADP) and phosphate ion.

**atrial fibrillation**
uncontrolled and irregular heartbeat.

**base pairs**
two structural units (two bases) on opposite strands of DNA or RNA connected by hydrogen bonding.

guanine    cytosine        adenine        thymine

**bile acids**
detergent-like steroid carboxylic acids that are biosynthesized in the liver from cholesterol (e.g., cholic acid).

**bioavailability**
the ratio of the amount of drug administered to the amount of drug absorbed by and distributed to the body.

**biomarker (molecular)**
a measurable indicator of a biological response or a disease state.

**blastocyst**
mammalian embryo at the earliest stage of development.

**blood-brain barrier**
cellular lining separating the brain from the bloodstream; in essence a filtration system that only allows the passage of certain molecules from the blood into the brain and serves to protect the central nervous system from harmful substances.

**body mass index (BMI)**
the relationship between body weight (in kg) and height (in meters) that is an indicator of increased health risk for cardiovascular disease. A BMI value between 25 and 30 indicates moderately excessive weight and a BMI of 30 or over indicates obesity.

$$BMI = \frac{weight\ (kg)}{height\ (m)^2}$$

**bone resorption**
the biological process by which bone is removed by

osteoclasts or by the action of acids and enzymes.

**bradykinin**
a nonapeptide hormone with multiple biological activities.

**bronchodilator**
a medication that relaxes the smooth muscles surrounding the airways and improves airflow to the lungs.

**bronchospasm**
constriction of the smooth muscles surrounding the air passages of the lung producing narrowing of the airways and causing difficulty in breathing.

**brush border**
cylindrical protrusions on the surface of intestinal cells that greatly increase the surface area of these cells and promote the efficient absorption of nutrients.

**buccal**
pertaining to the cheek or mouth cavity.

**capsid**
the protective protein shell that surrounds a viral particle. See "On Viruses and Viral Diseases", page 147.

**carbohydrate**
the collective name of sugars and larger molecules containing them (e.g., starches and cellulose) derived from the general formula $C_n(H_2O)_n$.

**carcinoma**
a malignant growth formed by mutated epithelial cells.

**cardio-**
a combining form pertaining to heart.

**catabolic**
pertaining to the breakdown of complex substances to simple molecules in living organisms.

**catalase-peroxidase enzyme (KatG)**
an enzyme that can catalyze the decomposition of hydrogen-peroxide ($H_2O_2$) to water ($H_2O$) and oxygen gas ($O_2$), or decomposition of toxic organic peroxides (ROOR).

**cation**
a positively charged ion.

**CCR5 receptor**
(chemokine receptor 5) a receptor expressed on the surface of T-cells, macrophages and dendritic cells that plays a key role as a co-receptor in HIV infections. See "Maraviroc", page 156.

**cDNA**
complementary DNA - synthesized from a messenger RNA template (mRNA) and subsequently converted to double-stranded DNA.

**cerebrum**
an anterior part of the brain that integrates sensory and neuronal function and coordinates voluntary activity.

**cervix**
lower, narrow portion of the uterus.

**cestode**
parasitic flatworm (also referred to as tapeworm) that can infest the intestines of humans and other vertebrates.

**chemokine receptors**
transmembrane proteins found mainly on the surface of certain immune cells (e.g., leukocytes) that interact with signaling proteins of the chemokine class.

**chemotaxis**
the characteristic movement of a cell (e.g., immune cell) in response to a chemical stimulus.

**chronic myeloid leukemia**
a form of cancer involving proliferation of blood cells.

**collagen**
a fibrous structural protein found in the connective tissue of mammals including skin, cartilage, ligaments, tendon, bone and teeth.

**conformation**
a particular spatial (three-dimensional) arrangement of a flexible molecule. See "Understanding Structural Diagrams of Organic Molecules", page 4.

**congestive heart failure**
loss of pumping power of the heart due to muscular degeneration and weakness.

**co-receptor**
a receptor that is required in addition to a primary receptor for a biochemical event to occur (e.g., entry of HIV into a T-cell using the CCR5 co-receptor).

**coronary**
pertaining to the heart.

**corpus luteum**
hormone-releasing structure in the ovary.

**cortex**
outer layer of the cerebrum of the brain of vertebrates that is critical to many brain functions.

**corticosteroids**
class of steroid hormones (e.g., cortisone) that are secreted by the adrenal gland and are involved in a wide range of physiological processes.

**cortisol**
a steroid hormone that is involved in stress response, inflammation and immune regulation.

**Crohn's disease**
a severe inflammation of the intestinal tract.

**Cushing's disease**
a disorder characterized by extreme levels of cortisol in the blood due to overactivity of the pituitary.

**cytidine**
a nucleoside in which cytosine (blue) is attached to ribose (red); structure:

**cytochrome P450**
a family of iron-containing enzymes which catalyze the addition of oxygen to organic substances in living systems.

**cytokine**
protein secreted by immune cells that regulates the immune response and also can play a role in inflammation and cell proliferation.

**cytotoxic**
substances toxic to cells that either kill cells or prevent their reproduction.

**decidua**
lining of the uterine during pregnancy.

**dehydrogenase**
an enzyme that causes oxidation by removing hydrogen from a molecule (usually with formation of $H_2O$).

**deoxyribose**
the 5-carbon sugar that serves as a building block for DNA.

**diabetic peripheral neuropathy**
a painful nerve disorder caused by diabetes which most often affects the extremities.

**dihydrofolate reductase (DHFR)**
an enzyme that reduces folic acid to tetrahydrofolic acid in two stages with the intermediacy of dihydrofolic acid (see more in "Methotrexate", page 46).

**disease-modifying agent**
a substance capable of slowing the progression of a particular disease (e.g., arthritis).

**dissociation energy**
the energy required to break a chemical bond in a molecule with formation of two fragments.

**diuretic**
a substance that stimulates the urinary secretion of sodium chloride from blood circulating through the kidneys.

**DNA**
deoxyribonucleic acid; (polynucleotide) the macromolecule that carries genetic information in living organisms.

**d-orbitals**
a set of five orbitals of elements of row three or higher in the periodic table that can hybridize with s and p orbitals.

**dyscrasia**
an abnormal or pathological condition.

**ectoparasite**
a parasite that can survive outside the host (e.g., ticks, lice or fleas).

**eczema**
inflammation of the outer layer of skin characterized by redness, itching, dryness or blistering.

**edema**
a swelling of the body, often as a result of excessive fluid retention.

**endometrium**
inner lining of the uterus.

**endoplasmatic reticulum**
an extensive network of functional membranes within eukaryotic cells involved in chemical modification or transport of molecules.

**enteric coating**
a coating for medications that permits passage through the stomach and into the small intestine where the active drug is released.

**enzyme**
substances (usually proteins) that accelerate and control biochemical transformations in living organisms.

**epidermal growth factor**
a protein that plays an important role in cell growth and proliferation.

**estrogen**
a female steroid hormone which is important for reproduction and health.

**etiology**
the study of the cause of a disease or medical condition.

**eukaryotic cell**
cell with a membrane-bound nucleus containing the genetic material (DNA) and discrete organelles.

**fatty acid**
a carboxylic acid with a long unbranched chain that can be either saturated or unsaturated. Since natural fatty acids are derived from the joining of acetic acid units, they contain an even number of carbon atoms (8 or more).

**fibrinolysis**
a process in which a fibrin clot (blood clot) is broken down.

**follicle stimulating hormone**
hormone produced in the pituitary gland that stimulates the maturation of the Graafian follicles.

**gene expression**
the translation of the information encoded in the gene into various types of RNA and subsequently to a variety of proteins.

**genitalia**
reproductive organs.

**genome**
the hereditary material (DNA) contained in the nucleus of a cell.

**glaucoma**
a condition of increased pressure inside the eye that damages the optic nerve.

**Glu**
abbreviation for the amino acid glutamic acid; structure:

**glucagon**
a hormone produced by the pancreas which raises blood glucose levels.

**Gly**
abbreviation for the amino acid glycine; structure:

**glycoprotein**
a protein that has carbohydrate (sugar) molecules covalently attached.

**glycosides**
molecules in which a sugar group (red) is bound to a non-sugar group through a bond to the carbon next to the ring oxygen.

**Golgi apparatus**
a membrane-bound organelle in eukaryotic cells, located close to the nucleus where proteins are converted to glyco- and lipoproteins.

**gonadotropin**
protein hormones secreted by the pituitary (e.g., luteinizing hormone and follicle stimulating hormone).

**GPCR**
G protein-coupled receptors. *See "Information Flow into the Cell by Chemical Signaling"*, page 78.

**Graafian follicle**
a fluid filled structure in the ovary in which eggs develop.

**graft**
living tissue transplanted surgically.

**guanine**
one of the four bases present in RNA and DNA, possessing the following structure:

**guanosine**
a nucleoside in which guanine (blue) is attached to ribose (red) *via* a linkage (glycosidic bond) that connects N7 of guanine and C1 of ribose.

**half-life**
the time required for one half of a given substrate (e.g., drug molecule) to undergo chemical transformation (e.g., metabolism).

**HDL**
high-density lipoprotein, a transporter of lipids (e.g., cholesterol) between body tissues and the liver through the blood.

**heme-iron complex**
a non-protein component of several important biomolecules (e.g., cytochrome P450, hemoglobin) which consists of an $Fe^{2+}$ ion (orange) surrounded by a porphyrin molecule (green).

**hemolytic anemia**
a low red cell count in blood due to the destruction of red blood cells by toxic substances (e.g., certain drugs) or infectious agents.

**hemopoietic**
pertaining to the formation of blood or blood cells.

**hemorrhagic stroke**
condition caused by the rupture of a blood vessel in the brain.

**hepato-**
a combining form pertaining to the liver.

**herpes viruses**
a family of viruses that contain double-stranded DNA and cause common infections in humans such as genital herpes, chickenpox, shingles, mononucleosis and measles.

**heterozygous**
possessing two different forms (alleles) of a specific gene.

**His**
abbreviation for the amino acid histidine; structure:

**histamine**
a regulator of various body functions including immunity. It is released from mast cells in response to an allergen (allergic reaction). It is produced from the amino acid histidine by decarboxylation.

**HIV**
human immunodeficiency virus.

**HLA**
human leukocyte antigen molecule. *See* detailed description in *"A Brief Overview of the Immune System"*, page 112.

**HMG-CoA reductase**
an enzyme that catalyzes the conversion of HMG-CoA (3-hydroxy-3-methylglutaryl coezyme A) to mevalonic acid, an intermediate in the pathway of cholesterol biosynthesis.

**Hodgkin's lymphoma**
a cancer of the lymphatic tissue causing enlargement of the lymph nodes, liver, and spleen.

**homozygous**
possessing two identical forms (alleles) of a specific gene. Individuals who are homozygous for a specific trait are called homozygotes.

**hormone**
a chemical messenger (e.g., insulin) produced in one part of the body that is carried by the blood to another part of the body where it regulates a specific physiological function.

**hydrolysis**
a chemical reaction in which water cleaves certain covalent bonds (e.g., esters) in organic compounds.

**hydroxyl group**
O-H group (red).

**hyper-**
combining form for abnormally high.

**hyperglycemia**
condition involving elevated blood glucose levels.

**hyperlipidemia**
elevated levels of lipids and lipoproteins in the blood.

**hypertension**
repeatedly elevated blood pressure exceeding 130/90 mmHg that is commonly referred to as high blood pressure.

**hypnotic**
an agent that induces sleep.

**hypo-**
combining term for abnormally low.

**hypoglycemia**
condition involving abnormally low blood glucose levels.

**hypogonadism**
defect in the function of the gonads (testes or ovaries).

**hypothalamic gonadotropin releasing hormone**
peptide hormone produced in the hypothalamus that stimulates the release of follicle stimulating hormone and luteinizing hormone from the pituitary.

**hypothalamus**
a gland located above the brain stem in the brain that connects the nervous system with the endocrine (hormonal) system.

**hypoxia**
deficiency of oxygen in the body.

**Ig**
(immunoglobulin) *See* detailed description in *"A Brief Overview of the Immune System"*, page 112.

**Ile**
abbreviation for the amino acid isoleucine; structure:

**immunoglobulin**
(Ig) a class of Y-shaped proteins used by the immune system to recognize and bind to foreign substances (also known as an antibody).

**inflammation**
a condition characterized by redness, heat, swelling, pain or irritation.

**initiation complex (in protein synthesis)**
a complex formed at the beginning of protein synthesis (translation) in the ribosome which consists of the 30S ribosomal subunit, mRNA, N-formyl-methionine tRNA and three initiation factors.

**insulin**
a polypeptide hormone composed of 51 amino acid residues which is produced by the pancreas and regulates carbohydrate metabolism and various other processes.

**insulin resistance**
a condition in which normal levels of insulin fail to produce the normal response in insulin-sensitive tissues (e.g., fat, muscle, and liver).

**integrase**
enzyme in a retrovirus (e.g., HIV) that enables the viral genome to be integrated into the DNA of the infected cell.

**intercalating agent**
a substance that inserts between the stacked bases of the DNA helix and impairs replication.

**interferon**
a group of regulatory proteins produced by immune cells.

**interleukin**
a family of protein regulators that serve as

messengers between the various cells of the immune system.

**ischemia**
deficiency in the oxygen supply of a tissue due to the obstruction of blood vessels (i.e., inadequate blood flow).

**ischemic stroke**
lack of oxygen in the brain due to blockage of an artery (e.g., by a blood clot).

**Kaposi's sarcoma**
a form of cancer caused by a herpes virus that usually produces lesions on the skin and various mucous membranes.

**kcal**
kilocalories (1000 calories): the energy (heat) required to raise the temperature of 1 kg of water from 20 °C to 21 °C. 1 kcal equals one dietary calorie (1 Cal).

**kinase**
an enzyme that catalyses the transfer of a phosphate group from a tri- or diphosphate donor (ATP or ADP) to an acceptor molecule (e.g., another enzyme).

**lactone**
cyclic ester.

**LDL**
low-density lipoprotein, a transporter of lipids (e.g., cholesterol) in the blood.

**Leu**
abbreviation for the amino acid leucine; structure:

**leukemia**
a cancer of the bone marrow and other blood forming organs characterized by the production of increased numbers of immature or abnormal blood cells, usually white blood cells.

**leukocyte**
white blood cell.

**leukotriene**
twenty-carbon molecules that mediate inflammation and asthmatic or allergic reactions.

**ligand-binding pocket**
an open space in an enzyme that accomodates a specific small molecule (ligand).

**lipid bilayer**
a structure composed of bipolar molecules (e.g., phospholipids) that have inner and outer hydrophilic (water-loving) surfaces and a hydrophobic (water-repelling) inner section.

**lipoproteins**
lipids (fats) that are covalently bound to proteins.

**liposome**
a spherically jacketed assembly consisting of phospholipids and cholesterol that can be used for transporting a drug to a target tissue.

**lipoxygenase**
an enzyme that catalyzes the dioxygenation of polyunsatuarted fatty acids. *See "Other Eicosanoids*

*in Inflammation*", page 41.

**lupus erythematosus**
a chronic autoimmune disorder that causes the immune system to attack the body's own tissues including the heart, joints, lungs, blood vessels, and skin.

**lupus nephropathy**
inflammation of the kidney caused by systemic lupus erythematosus.

**luteinizing hormone**
hormone secreted in the pituitary gland that stimulates ovulation in females and production of androgen in males.

**lymphocyte**
a type of white blood cells of the immune system that includes B- and T-cells, and natural killer cells.

**lymphoma**
a cancer that develops in the lymphatic system.

**macrocyclic ester**
also known as macrocyclic lactone; a cyclic ester with a large ring containing a minimum of 12 atoms.

**macrolide**
an antibiotic molecule containing a large lactone ring (12-16 membered) with one or more attached sugars.

**macromolecule**
large molecules, typically consisting of thousands of atoms (e.g., synthetic plastic, rubber, polypeptides, proteins, DNA).

**macular degeneration**
a condition in which the inner lining of the eye (macula) deteriorates causing loss of central vision.

**malignancy**
a cancer that can invade nearby tissue and spread to other parts of the body.

**MAO**
(monoamine oxidase) an enzyme that catalyzes the oxidative deamination of monoamines (e.g., tyramine as well as neurotransmitters). *See* MAO inhibitors on page 229.

**meningitis**
an often life-threatening condition that involves the inflammation of the protective membranes which cover the central nervous system (e.g., brain, spinal cord); most often caused by viruses or bacteria.

**metabolic syndrome**
a cluster of diseases arising from deficits of energy production and metabolism. See "*An Overview of Metabolic Syndrome, A Precursor of Diabetes, Heart Disease and Stroke*", page 56.

**metaphase**
a stage of the cell cycle during which the chromosomes align in the center of the cell before being separated.

**metastable state**
a short-lived non-equilibrium state.

**metastatic**
spreading from a primary site to other sites in the body.

**microtubule**
hollow cylindrical protein structures that participate in many cellular processes including transport and cell division.

**mitosis**
duplication of the cell's genetic information to form two identical daughter cells.

**mitotic spindle**
a cell apparatus that segregates chromosomes during cell division.

**molecular modeling**
the prediction of molecular shape (conformation) and behavior using computer software.

**monoclonal antibody**
a purified antibody that is derived from only one clone of cells (e.g., B-cells) and recognizes only one antigen.

**motor cortex**
region of the cerebral cortex involved in the regulation of voluntary movement.

**mRNA**
a single-stranded molecule of ribonucleic acid which serves as the template of protein synthesis. It is generated by the transcription of DNA.

**mucosa**
the moist lining inside the hollow organs in the body (e.g., gastrointestinal and respiratory tracts).

**multiple sclerosis**
a progressive inflammatory disease that affects the myelin sheath of the nerve cells in the brain and spinal cord.

**muscarinic receptor**
acetylcholine receptor that is more sensitive to the chemical substance muscarine than nicotine.

**mutation**
change in the structure of an organisms' DNA (or RNA in the case of viruses), usually due to transcription error or chemical damage.

**myelin**
an insulating sheath composed of phospholipids and proteins that surrounds the axons of many neurons.

**myeloid leukemia**
cancer characterized by the abnormal proliferation of blood cells originating in the bone marrow.

**myositis**
an autoimmune disorder that leads to inflammation and degeneration of muscle tissue.

**NADPH**
nicotinamide adenine dinucleotide phosphate reduced form, a reducing agent required in anabolic processes such as fatty acid or cholesterol biosynthesis. The hydrogen that is transferred in reductions is highlighted in red.

**nematode**
(parasitic roundworm) a worm having an unsegmented cylindrical body with pointed ends (e.g., hookworm, pinworm).

**neovascularization**
formation of new blood vessels.

**nephro-**
a combining term for kidney.

**neuralgic pain**
neuralgia – a painful disorder of the nerves which involves short episodes of excruciating pain.

**nM**
nanomolar, concentration of one billionth of a mole per liter of solution.

**nm**
nanometer: $10^{-9}$ meter or one billionth of a meter.

**NNRTI**
non-nucleoside reverse transcriptase inhibitor.

**non-immunogenic**
substance that does not induce a specific immune response.

**nuclear factor**
a protein that regulates gene expression by controlling the binding of RNA polymerase to DNA.

**nuclear receptor**
ligand activated proteins that migrate into the nucleus and promote or suppress gene expression.

**nucleoside**
a compound consisting of a heterocyclic base and a sugar linked covalently.

**nucleotide**
a compound consisting of a heterocyclic base, a sugar, and phosphate group. Nucleotides are the building blocks of nucleic acids.

**occlusive stroke**
brain damage caused by blockage of a cerebral blood vessel and oxygen deficiency.

**orbital**
a region around the nucleus of an atom in which an electron is distributed.

**organelle**
any structure within a cell that has a specialized function (e.g., mitochondria, Golgi apparatus, nucleus).

**osmotic pressure**
fluid pressure produced by two solutions (containing different concentrations of solute) that are separated in space by a semipermeable membrane which allows the passage of solvent molecules but not solute.

**osteoarthritis**
degenerative arthritis, a condition in which inflammation of the joints results in damage to the cartilage and tissues in joints.

**osteoblast**
a type of bone cell that is responsible for bone deposition.

**osteoclast**
a type of bone cell that removes bone.

**ovary**
female reproductive organ in which the eggs are produced.

**ovulation**
discharge of the egg from the ovary.

**palliative**
alleviating symptoms but not curative.

**parathyroid gland**
a small endocrine gland located in the neck that produces parathyroid hormone.

**parathyroid hormone**
a peptide hormone that regulates the level of calcium in the blood.

**parenteral**
administration of a drug by injection or infusion.

**parietal cell**
cell in the stomach that secretes gastric hydrochloric acid.

**pathogen**
infectious agent, a microorganism that causes illness to a host.

**pathological**
resulting from disease.

**pepsin**
digestive protease enzyme that in combination with gastric acid is responsible for the breakdown of dietary proteins.

**peptides**
molecules consisting of multiple $\alpha$-amino acids joined by amide bonds (-OC-NH-). *See "Proteins and Three-Dimensional Protein Structure"*, page 112.

**peptidoglycan**
a polymer consisting of peptides and sugars that forms a protective homogeneous layer outside the plasma membrane of bacteria. Gram-positive bacteria usually have a thicker peptidoglycan layer than Gram-negative bacteria.

**peptidomimetic**
a protein-like molecule that is designed to mimic certain properties of natural peptides.

**peripheral nervous system**
system nerves and neurons outside of the spinal cord and brain (central nervous system).

**Phe**
abbreviation for the amino acid phenylalanine; structure:

**phosphate**
salts or ester derivatives of phosphoric acid (see structure) composed of phosphorus and oxygen.

**phosphorylase**
an enzyme that catalyzes the addition of an inorganic phosphate to an acceptor molecule (e.g., glycogen); see kinase.

**phylogenetic**
pertaining to the evolutionary history of a particular group of organisms.

**pituitary gland**
a pea-sized endocrine gland in the brain controlling growth, development, and the function of other endocrine glands.

**placenta**
a temporary organ in the uterus of pregnant placental vertebrates that maintains the fetus through the

umbilical cord.

**plasma membrane**
the membrane that serves as a protecting boundary of a living cell. It consists of a lipid bilayer and embedded proteins.

**plasmid**
A circular double-stranded DNA that replicates independently of the chromosomal DNA. Plasmids commonly occur in bacteria. Such plasmids are used in recombinant DNA techniques in which genes are inserted and expressed in microbial cultures.

**platelet-derived growth factor**
a protein responsible for the regulation of cell growth which is especially important in new blood vessel formation.

**platinum coordination complex**
a neutral substance consisting of a central platinum cation with four attached ligands, all in one plane.

**polyene**
polyunsaturated organic compounds that have alternating single and double bonds. They are distinguished by the number of double bonds [e.g., diene (2), triene (3) or heptaene (7)].

diene        triene

heptaene

**polymerase**
an enzyme that catalyzes the synthesis of polymers (e.g., the synthesis of double-stranded DNA from monomers on a single-stranded template).

**polyol**
a compound that contains more than one hydroxyl groups. A triol has three whereas a hexaol has six hydroxyl groups.

triol        hexaol

**postherpetic neuralgia**
a condition that results from nerve damage caused by the herpes zoster virus (shingles).

**postsynaptic**
part of the nerve cell that receives an impulse from the synapse. See "*The Brain, Neurotransmission and Molecular Neurotransmitters*", page 221.

**PPARs**
(peroxisome proliferator-activated receptors) a group of intracellular transcription factors that up- or downregulate gene expression. They play a role in metabolism as well as cell differentiation.

**pressor**
substance producing an increase in blood pressure (e.g., by vascular constriction).

**presynaptic**
part of the nerve cell that sends an impulse across the synapse. See "*The Brain, Neurotransmission and Molecular Neurotransmitters*", page 221.

**prodrug**
an inactive (or minimally active) drug precursor that undergoes chemical modification in the body to form the active drug.

**progesterone**
reproductive steroid hormone.

**prophylaxis**
prevention of a particular disease.

**prostaglandins**
a group of lipid compounds containing 20 carbons that mediate many physiological processes. See "*How Do Anti-Inflammatory Drugs Work?*", page 40.

**protease enzymes**
also known as proteinases. Enzymes that cleave the peptide linkage in proteins to generate smaller proteins or polypeptides. They are involved in a large number of physiological processes (e.g., in the digestion of food or the clotting of blood). Currently, there are three major classes of proteases that utilize amino acid residues in the catalytically active sites: (1) serine, (2) cysteine, and (3) aspartic acid type. Metalloproteases utilize a catalytic metal ion in the active site.

**protectin**
a family of polyunsaturated compounds, generated from docosahexaenoic acid, which have strong anti-inflammatory properties. See "*Other Eicosanoids in Inflammation*", page 41.

**proteins**
large polypeptides essential for the structure, function and regulation of the body's cells, tissues and organs.

**proteolytic enzyme**
also known as protease or peptidase; enzyme that catalyzes the breakdown (hydrolysis) of large peptides into smaller peptides or into the individual amino acid components.

**proton pump**
a membrane-bound protein that is capable of moving protons ($H_3O^+$) across the membrane of a cell and creating a gradient of both pH ($H_3O^+$ concentration) and electric charge.

**psoriasis**
an inflammatory skin disease characterized by red, itchy scaly, patches.

**pulmonary embolism**
blockage of the artery carrying blood from the heart to the lungs by a blood clot, tumor cells or fat.

**purines**
heterocyclic organic compounds derived from the parent purine structure (blue). Adenine and guanine make up the building blocks of nucleic acids.

purine        adenine

caffeine        uric acid

**pyrimidines**
heterocyclic organic compounds derived from the parent pyrimidine structure (blue). Cytosine, thymine and uracil make up the building blocks of nucleic acids.

pyrimidine

cytosine

thymine

uracil

**radical**
a highly reactive molecule having one valence electron in an unfilled orbital (e.g., hydroxyl radical HO·).

**receptor**
a protein molecule that binds to and responds to a hormone, neurotransmitter or molecular regulator.

**receptor tyrosine kinase**
membrane receptor that acts by transferring a phosphate from ATP to a tyrosine residue of a target protein.

**regulatory protein**
a protein that regulates gene expression, cell cycle progression, the activity of other proteins, or other cellular processes.

**renin**
enzyme secreted in the kidney that converts angiotensinogen to angiotensin I.

**renin-angiotensin system (RAS)**
hormone system involved in regulating blood pressure.

**replication fork**
a region of DNA in which the enzymes copying DNA are bound to untwisted, single-stranded DNA.

**resolvin**
endogenous polyunsaturated lipid mediator with strong anti-inflammatory properties. Resolvins are derived from ω-3-eicosapentaenoic acid.

**respiratory syncytial virus (RSV)**
an RNA virus that causes mild cold and cough in adults, or more severe respiratory problems in children.

**retina**
a layer of neural cells in the back of the eyeball that is sensitive to light and generates nerve impulses directed to the brain via the optic nerve.

**reverse transcriptase**
retroviral enzyme that copies viral RNA into complementary DNA (cDNA) and later to double-stranded DNA.

**rheumatoid arthritis**
a chronic, progressive inflammatory autoimmune disorder that causes the immune system to attack and inflame the joints.

**rhinitis**
inflammation of the nasal passages caused by viruses, bacteria, and irritants.

**RNA**
ribonucleic acid; macromolecule essential for the translation of DNA into protein products; in some viruses it is the carrier of genetic information.

**sarcoma**
a cancer of the connective tissue such as muscle, fat, blood vessel, bone, and cartilage.

**sedative**
a substance that acts on the central nervous system to relieve anxiety and induce calmness and/or sleep.

**seizure**
a sudden, involuntary movement of muscles due to the uncontrolled discharge of neurons in the brain.

**semisynthetic**
compounds that are made by chemical synthesis using starting materials isolated from natural sources.

**Ser**
abbreviation for the amino acid serine; structure:

**shingles**
a viral infection of sensory nerves by a human herpesvirus that causes itching, intense pain and blisters on the skin.

**signal transduction**
the conversion of a signal from outside of the cell into a signal inside the cell that leads to a functional change. See "*Information Flow into the Cell by Chemical Signaling*", page 221.

**signaling**
see detailed description in "*Information Flow into the Cell by Chemical Signaling*".

**SNRI**
serotonin-norepinephrine reuptake inhibitor (*see* page 230).

**somatic cells**
cells forming the body excluding the germline cells.

**spin (of an electron)**
a discrete quantum property of electrons that distinguishes two axes of rotation (or spin) generally described by the quantum numbers: +1/2 and -1/2. Any given atomic or molecular orbital can be occupied by only two electrons of opposite spin. A shorthand notation of electron spin is a half-headed arrow. For example, the spins of four electrons in three $2p$ orbitals can be depicted as follows.

$2p_x$    $2p_y$    $2p_z$

**SSRI**
selective serotonin reuptake inhibitor (*see* page 230).

**steady state**
a state in which an equilibrium has been achieved.

**steatohepatitis**
non-alcoholic fatty liver disease or "silent" liver disease characterized by increased fat content.

**stem cell**
an undifferentiated cell which can divide countless times and develop into many other kinds of cells of the body.

**stem cell factor**
growth factor that regulates the differentiation and proliferation of stem cells.

**steroid hormone**
steroid compounds produced in the body that function as hormones.

**stromal cancer**
cancer forming in the connective, non-functional tissue framework of an organ, gland, and other structure.

**subcutaneous**
applied under the skin.

**substantia nigra pars compacta**
region of the midbrain responsible for dopamine production.

**suprachiasmatic nucleus**
region of the brain located in the hypothalamus that regulates the circadian rhythm.

**sympathetic nervous system**
part of the autonomous nervous system (involuntary nervous system) that is responsible for mobilizing the body's energy in response to stress and physical activity.

**synergistic effect (synergism)**
greater than additive activity when two or more drugs are combined.

**systemic agent (drug)**
substance that affects the entire body by traveling via the bloodstream to all parts of the body.

**systemic infection**
infection affecting the entire body.

**tau protein**
proteins present in neurons in the central nervous system that play a role in microtubule formation and maintenance, and Alzheimer's disease.

**TCA**
tricyclic antidepressants (*see* page 230).

**therapeutic dose**
the amount of the drug that is required to achieve the desired medical result

**therapeutic index**
also known as the margin of safety; the ratio of the therapeutic dose and toxic dose. Officially it is the lethal dose for 50% of the population ($LD_{50}$) divided by the minimum effective dose ($ED_{50}$) for 50% of the population.

$$\text{therapeutic index} = \frac{LD_{50}}{ED_{50}}$$

**Thr**
abbreviation for the amino acid threonine; structure:

**thymidine**
a nucleoside in which thymine (blue) is attached to deoxyribose (red).

**thyroid hormones**
small iodine-containing organic molecules produced by the thyroid gland. These compounds are essential in fat, carbohydrate and protein metabolism as well in cell development and differentiation.

**TNF-α**
tumor necrosis factor alpha; a cytokine produced by T-cells and macrophages which plays an important role in the immune response, in systemic inflammation and in cell death (apoptosis).

**topical**
applied directly to the part of the body that is affected.

**topoisomerase**
enzyme that catalyzes the breakage, passage, and religation of one or both DNA strands and thereby relieves strain.

**transcription**
conversion of the genetic information stored in DNA to a messenger RNA (mRNA).

**transdermal**
crossing of a drug through the skin (e.g., from a patch).

**transition state**
the arrangement of atoms during a chemical reaction that has the highest energy on the path from the reactants to the products.

**transpeptidase**
a bacterial enzyme that cross-links peptidoglycan chains to form rigid cell walls.

**transpeptidation**
cross-linking of peptidoglycan chains by the transpeptidase enzyme during bacterial cell wall synthesis.

**trematode**
parasitic flatworms (flukes) that have external suckers to attach to the host.

**triglycerides**
esters of glycerol with three fatty acids that are carried through the bloodstream to the various tissues.

**tRNA**
transfer RNA – the RNA molecule that transports the individual amino acid building blocks to the growing polypeptide chain in the ribosome where protein synthesis takes place.

**Trp**
abbreviation for the amino acid tryptophan; structure:

**tubulin**
the basic protein building block of microtubules.

**typhoid fever**
a disease caused by the bacterium *Salmonella typhi*. Most common symptoms are high fever, diarrhea and headache.

**Tyr**
abbreviation for the amino acid tyrosine; structure:

**tyramine**
a naturally-occurring amine derived from the amino acid tyrosine. Certain foods are rich in tyramine (e.g., cheese and wine).

**ubiquitin**
a small protein, which when attached to other proteins via covalent bonding, marks them for degradation.

**ulcer**
a non-healing wound.

**upregulation**
a process in which a cell increases the production of a certain protein (e.g., receptor) in response to a stimulus.

**Val**
abbreviation for the amino acid valine; structure:

**vascular endothelial growth factor**
signaling protein involved in the regulation of the formation of new blood vessels.

**vasculitis**
inflammation of the wall of blood vessels.

**vasodilation**
the relaxation of smooth muscles in the wall of blood vessels that result in the widening of the vessel and the reduction of blood pressure.

**ventricle (heart)**
a chamber in the heart which is filled with blood and from which the blood is pumped out into circulation. The human heart has a ventricle on the right and on the left sides.

**vertigo**
a balance disorder characterized by dizziness.

**voltage-gated ion channel**
transmembrane ion channels activated by electrical potential difference.

**wild-type**
the most common form of an organism that is found in nature.

**X-ray crystallography**
a technique that uses the diffraction of X-rays through the regular lattice of molecules in a crystal. The data are analyzed by computer to generate an electron density map from which the structure of the molecule is obtained.

**yeast**
single-celled eukaryotic fungi that reproduce by budding.

# INDEX